AF744204

University of Southampton
WITHDRAWN

# WORLD INDEX OF STRATEGIC MINERALS

# World Index of Strategic Minerals

## Production, Exploitation and Risk

D. Hargreaves
and
S. Fromson

Gower

© Gower Publishing and David Hargreaves, 1983

All rights reserved. No part of this publication may be reproduced or utilized in any form or by any means, electronic or mechanical, including photocopying, recording, or by any information storage and retrieval systems, without permission in writing from the Publisher.

Published in the United Kingdom in 1983 by Gower Publishing Company Limited, Croft Road, Aldershot, Hampshire GU11 3HR, England.

Published in the United States of America in 1983 by Facts on File, Inc., 460 Park Avenue South, New York, N.Y. 10016.

**British Library Cataloguing in Publication Data**

Hargreaves, David.
World Index of Strategic Minerals.
1. Mines and mineral resources
2. Strategic minerals
I. Title
553 HD9506.A2

ISBN 0-566-00489-5

Editorial Services by
Cambridge Information and Research Services Ltd
PO Box 147, Cambridge CB1 1NY

SOUTHAMPTON
LIBRARY
UNIVERSITY

Printed in Great Britain

# CONTENTS



## PART ONE: STRATEGIC RISK ASSESSMENT



## PART TWO: THE MINERALS

## PART THREE: THE COUNTRIES



## PART FOUR: THE COMPANIES

# ILLUSTRATIONS



*Each mineral review in Section Two includes tables, pie charts and graphs on the components of risk assessment, resources/reserves, production, consumption and prices. Each country review in Section Three includes tables on the components of risk assessment and the output and ratings of the principal minerals. Each company review in Section Four includes a table on mineral output, share of the western and total world output and company rank numbers.*

# INTRODUCTION

This book provides a comprehensive guide to the strategic minerals of the world. The term 'strategic' has a number of constituents, which are addressed both singly, and as a composite, as the work proceeds. For example, a mineral may be of strategic importance because of the critical nature of its end use e.g. defence: it may fall in this class because sources of supply are very limited, and politically unstable: because the routes by which the mineral is moved to key refining and processing points are at risk; or because the financial structure of the relevant mining enterprises is vulnerable, or subject to disruptive outside pressure. Again, a mineral will be more 'strategic', in the sense of affecting its users' vital interests, if it is difficult to substitute and if the lead-times for its production are long. The same factors can make for extreme price volatility. Considerations such as these have led most major governments, notably the US but on a smaller scale the UK, Germany and France among others, to hold stockpiles of selected strategic minerals. Consumers also attempt to form long-term arrangements with suppliers, buttressed if necessary by diplomatic, financial and even overt military support.

One object of this book is to analyse the individual factors which make certain minerals, and some countries and companies, more strategic than others. By assigning values, it attempts to quantify the various constituents of a mineral's strategic role. Inevitably judgement is called for here, and judgement can be fallible or change over time. But to the authors the justification of their classification is as clear as the problems it poses, in a matter as vital to the livelihood and security of many businesses and governments. Ranking of minerals by their risk factor will be conducted anyway—better then to establish an objective and data-sensitive basis. Thus, by placing chromite strategically above antimony and copper, the book highlights the technological and defence role of the former metal, the geographical concentration of reserves and production, the vulnerability to political manipulation and the dearth of indigenous supplies in the consumer countries. In assessing and cross-referencing over 30 countries, 37 minerals and 60-plus producing companies in this way it covers essentially the whole world minerals industry in terms of its vital components: location, labour, politics, finance and economics, the energy factor and resource dependence.

The second object of the book, and as important as the first, is to present a wide range of data on strategic minerals and the processes by which they are mined, refined and brought to the market. Some of this data is relatively accessible, at least by those like ourselves professionally involved in commodity analysis: some, like the breakdown of mineral output by company and country, has meant researching into a number of different sources, some of which are difficult to find and others which—not necessarily by chance—are difficult to interpret. Above all however, we feel that the merit of our presentation is in the breadth of information about minerals that has been brought together in one publication: and in the consistent standards that we have applied to make the information meaningful for readers concerned with the many sensitive issues on which strategic minerals touch.

## USING THE HANDBOOK

Access to the contents is by reference to the four broad sections:

1. Assessing the strategic risks of the minerals and countries.

2. The minerals in detail: including a strategic factor, occurrence and production details, applications and trade in the mineral.

3. The countries in detail: including a country risk factor for each, output and ratings of principal minerals and details of investment and development.

4. The companies in detail: including major products, subsidiary companies and countries of operation.

Let us assume a major funding institution has been approached to refinance a new platinum project in Canada, to be developed by Cominco Ltd. The book presents a profile on platinum, on Canada and on Cominco, each interrelated in terms of investment potential by reference to investment risk. Access to the information is as follows:

1. A study of Section 1 to appreciate the relevance of the individual factors which make up the strategic ratings and which indicate the comparative values.

2. Reference to platinum in Section 2, The Minerals, page 115. In addition to the basic information on the metal's occurrence, reserves, production, consumption, transportation, end uses, trade and prices—it gives the critical decision making data. It derives a high total strategic factor based on limited geographical distribution, the politically-biased producer influence of South Africa and the USSR, critical end uses and a dearth of US productive capacity.

3. Reference to Canada in Section 3, page 182. This shows a country rating of 4.4, comfortably below the average in spite of labour, political and financial difficulties which are fully explained. It shows the full range of mineral output, the structure of the mining industry and current investment and development levels. It indicates a country with a secure background of mining development.

4. Reference to Cominco Ltd in Section 4, The Companies, page 269. This identifies a large, Canadian-based group with a long history of mining development; its shares, its volume and range of products, its labour force, countries of operation, capital employed and ranking in the world minerals industry.

This approach enables a detailed appraisal to be made of a mineral in a world context and in the narrower aspect of a particular company within a chosen country.

An important feature of the book is its cross-referencing capabilities. For example information on gold can be drawn from the relevant minerals section and it can then be followed to the individual producing countries and companies—and through the Strategic Risk Assessments (Section 1) to show the relevance of each risk factor to the metal. A table listing the companies and their representation in the principal minerals is set out on pages 252 to 257.

An appendix gives details of sources for further information on mining and individual metals.

David Hargreaves and Sarah Fromson

# ABBREVIATIONS AND DEFINITIONS

| | |
|---|---|
| ABMS | American Bureau of Metal Statistics |
| AIEC | Asociation of Iron Exporting Countries |
| AISI-SAE steels | Standardised identification code for plain-carbon and low alloy steels (USA) |
| Alkali | Group of six extremely reactive metals occuring in a vertical line on the left of the periodic table |
| b/d | Barrel per day |
| Calcination | Heating to dry off water of hydration, generally on ore |
| CIF | Cost, insurance and freight (included in price) |
| CIPEC | Council of Copper Exporting Countries |
| C&F | Cost and freight |
| COMECON | Council for Mutual Economic Assistance (Soviet Bloc countries) |
| COMEX | Commodity Exchange New York |
| Dielectric | An insulating material, which is often used to store electric charge between two electrodes without conduction |
| Eastern bloc | The Comecon bloc, China |
| FOB | Free on board |
| GATT | General agreement on tariffs and trade |
| GDP | Gross domestic product |
| GNP | Gross national product |
| GSA | General Services Administration (USA organisation dealing with minerals stockpiling policies amongst other areas) |
| GWh | Giga watt hours (1,000,000,000 watt hours) |
| IBA | International Bauxite Association |
| ITA | International Tin Agreement |
| ITC | International Tin Council |
| Known resources | Reserves plus not yet economically exploitable deposits |
| kWh | Kilo watt hours (1,000 watt hours) |
| LDC | Less developed country |
| LME | London Metal Exchange |
| $m^3$ (cu m) | Cubic metre |
| MWh | Mega watt hours (1,000,000 watt hours) |
| NATO | North Atlantic Treaty Organisation |
| Newton (N) | Metric unit of force |
| NYMEX | New York Merchantile Exchange |
| OAU | Organisation of African Unity |
| OECD | Organisation for European Cooperation & Development |
| ohm | Unit of electrical resistance |
| Periodic Table | List of the chemical elements in order of their atomic numbers so arranged that elements with similar properties are adjacent |
| PGM | Platinum group metals (platinum, palladium, rhodium, osmium, iridium, ruthenium) |
| Piezoelectric | A dielectric material with an asymmetric internal structure such that the centres of positive and negative charge are not coincident under certain conditions |
| ppm | Parts per million |
| Rare-earths, lanthanides actinides | Specific sub-sets of the transition metal group, grouped by position on the periodic table |
| Reserves | Ore deposits considered economically exploitable with present techniques |
| SALT | Strategic Arms Limitation Talks |
| SEATO | South East Asia Treaty Organisation |

| | |
|---|---|
| Superalloys | Complex solid solution hardened alloy, further strengthened by a precipitation hardening mechanism and generally used at high temperatures where strength and oxidation resistance are important |
| Tempering | A toughening heat treatment process applied to certain types of alloys, particularly steels |
| Third World | The LDC's of Africa, Asia, S America |
| Total resources | Known resources plus not yet discovered resources |
| Toughness | A measure of the energy required for mechanical failure |
| Transition metals | Group of metals with many similar physical and chemical properties, which occur in a horizontal band across the centre of the periodic table |
| USBM | United States Bureau of Mines |
| USSR | Union of Socialist Soviet Republics |
| WBMS | World Bureau of Metal Statistics |
| Western Bloc | North America, Europe and Japan, Australia |

# PHYSICAL UNITS

| | | |
|---|---|---|
| 1 barrel (bbl) (oil) | = | 0.159 cubic metres, 35 imperial gallons |
| 1 carat (diamond) | = | 200 milligrams |
| 1 carat (gold) | = | one twenty-fourth part |
| 1 centimetre (cm), millimetre (mm) micrometre (um) | = | $\frac{1}{100}$ metre, $\frac{1}{1000}$ metre, $\frac{1}{1000,000}$ metre |
| 1 flask (fl) (mercury) | = | 76 pounds, avoirdupois |
| 1 horsepower | = | 736 watt |
| 1 joule (J) (energy) | = | 1 watt-second (Ws) |
| 1 kilogram (kg) | = | 2.2046 pounds, avoirdupois |
| 1 kWh (unit of heat) | = | 3,412 British thermal units |
| 1 long ton | = | 2,240 pounds, avoirdupois |
| 1 long ton unit (1tu) | = | 1% of a long ton, or 22.4 pounds |
| mcf | = | 1,000 cubic feet |
| 1 metre | = | 3.28 feet (1.093 yards) |
| 1 metric ton (mt) or tonne | = | 2,204.6 pounds, avoirdupois or 1,000 kilograms |
| 1 metric ton unit (mtu) | = | 1% of a metric ton, or 22.0 pounds |
| 1 $MN/m^2$ (stress) | = | 160 pounds per square inch (psi) |
| 1 pound (lb) | = | 453.6 grams |
| 1 short ton (st) | = | 2,000 pounds, avoirdupois |
| 1 short ton unit (stu) | = | 1% of one short ton or 20 pounds |
| 1 troy ounce | = | 1.09714 avoirdupois ounces |

# SECTION ONE:

# STRATEGIC RISK ASSESSMENT

# THE ASSESSMENT BASIS

As a political lever, the strategic minerals concept has been a major factor in every serious conflict since World War Two. It has gained particular momentum from the sharp increase in demand for high technology materials resulting from the growth of the electronics industry since 1960. It has taken on a further dimension following the political changes in several of the supply countries and it has been thrown into a state of confusion by the rapidly changing balance of the world economy. The international market for industrial raw materials is now of major significance for both commercial and military reasons.

All categories of consumer reliant on mineral imports—from individuals and companies to countries and political unions—need to establish their views on security of supply and the costs to them of any disruption. To arrive at an adequate assessment a series of questions should be addressed including:

—where are the major resources of these minerals located?

—where are the major production sources?

—how politically safe are these sources?

—how prone to labour disruption are they?

—are they threatened externally?

—are they economically sound?

—are they vulnerable to energy supply disruptions?

—are they vulnerable to the outside supply of other natural resources?

—are the producer countries vulnerable within the currently volatile framework of international finance?

Other considerations, extending beyond the boundaries of producer countries, need also to be taken on board. These include risks associated with transporting raw or semi processed materials and their liability to manipulation through collusive action on prices, embargoes etc. Finally the consumer must understand how the demand pattern for his particular mineral is changing, how it is likely to do so in the future and the impact any shortage would make on his industry and the economy generally. The availability of substitutes and the longer term likelihood of substitution are important elements in this assessment.

These questions are addressed for each of the 37 minerals in Section Two and for all but the three Communist states in the 34 country profiles in Section Three.

This section draws together the principal findings of the work setting out the comparative levels of risk attaching firstly to the minerals and secondly to the countries. Comparisons are made both in terms of an overall assessment and in terms of the components which go to make up this assessment and which are discussed individually in Sections Two and Three.

All ratings of this nature are essentially time dependent and the assessments and discussions which follow are based on a timescale to 1986. The assessments are made from the standpoint of consumers in the industrialised West namely the USA, Western Europe and Japan.

# MINERAL RISK ASSESSMENT

All widely known and used metals are covered, including iron, the backbone of heavy industry due to its ready availability and low cost. Metals excluded are those of only limited industrial use, such as arsenic and thallium, or which rarely occur, such as hafnium, which is extracted solely from nuclear-grade zirconium.

Of the semi-metals, only diamond, germanium, indium, selenium and silicon have been considered. Germanium, indium and selenium are all by-products of base metals and have established a secure position in several industrial end-uses, particularly in electronics. Diamond has been included because of the importance of industrial stones, natural or synthetic, in the machine tool and heavy equipment areas plus the dependence of certain African mineral producer countries on the revenues derived from diamond sales. An understanding of silicon and its properties is necessary to a full appreciation of the iron, steel, aluminium and magnesium sectors since silicon is an important alloying addition to these metals.

It was decided at the outset to exclude the non metallic minerals from the survey. The energy related minerals, such as oil and coal, have been excluded because of the special factors relating to the international energy market which, as events over the last decade have shown, operates quite independently from the international metal markets. Other minerals—sand, gravel, mica, clay, feldspar, etc—are normally in good local supply and do not warrant this type of analysis.

The assessment for each mineral is arrived at in the following way and as itemised in Table 1. Firstly the risk is broken down into four primary components associated with the principal stages of:

**TABLE 1: COMPONENTS AND SUB COMPONENTS OF THE MINERAL RISK ASSESSMENTS**

| *Components* | *Sub Components* | |
|---|---|---|
| | *Likelihood of Supply Disruption* | *Cost of Such Disruption* |
| 1. Production Risks | (a) Existing Capacity<br>(b) Labour Disputes<br>(c) Violent Conflict | (d) Range of Primary Supply Sources<br>(e) Time Lags for New Supplies |
| 2. Transportation Risks | (a) Primary (raw material) links<br>(b) Secondary (semi/finished material) links | |
| 3. Application/Use Risks | | (a) Total Economic Impact<br>(b) Effect on Key Industries<br>(c) Availability of Substitutes<br>(d) Longer Term substitutability |
| 4. Trade Risks | (a) Collusive Price Agreements<br>(b) Embargoes | |

—production

—transportation

—consumption,

—trade.

These components are then analysed in their turn under the two headings of the likelihood of supply disruption and the economic cost resulting from that disruption. Table 1 lists the 13 sub components for which individual risk assessments are made and how they relate within the assessment matrix.

Ratings from 1 to 10 have been employed to quantify the assessment for each of the 13 sub-components. A low rating implies that the risks or costs of supply disruption are considered low.

How each mineral compares within these sub-components can be seen in Table 3. As discussed later in this section these subcomponent ratings form the basis of the total strategic rating. They are now commented on individually in terms of the questions which have been addressed in arriving at their rating.

## 1.1 PRODUCTION RISKS

Table 1 identifies five sub-components which fall within this category. Three come under the heading of the likelihood of supply disruption and two under that for the cost of such disruption.

### 1.1(a) Existing Capacity

Is existing capacity sufficient to meet current demand? Total capacity includes both primary production plus that from scrap. Recycled material is of particular importance for lead, where approximately 40 per cent of consumption is met by scrap recovery, and of growing importance for aluminium, where recycling requires only five per cent of the energy needed to produce primary metal from ore.

Is existing capacity sufficient to meet current demand? Surpluses arose in copper and aluminium in 1982 when the USA and Japan made notable production cutbacks in the face of weakening, recession-hit demand. Molybdenum provides another instance of sudden and severe production cutbacks occurring shortly after major expansion in the late 1970's. In many of the high-volume base metals large stocks have accumulated and these can be considered as 'above-ground' mines, which are available to be utilised in the event of a sudden increase in requirements.

### 1.1(b) Labour Disputes

What is the historical record of labour relations in the major production areas of a metal? Do the disputes arise during the mining or the refining links of the production chain?

Copper is a prime example of a metal whose supplies have suffered as a result of sporadic strikes in Southern Africa and in North and South America and have been interrupted at all stages along the extraction route.

Generally by-product metals suffer less from labour problems since they are extracted in the later refining stages of a major metal and a smoothing effect often operates in the sense that flue dusts, slags etc may be accumulated for up to six months before the low-volume by-products

**TABLE 2: MINERAL PRODUCTION RISKS: EXISTING CAPACITY**

| | | | | | |
|---|---|---|---|---|---|
| Aluminium | 1 | Gold | 1 | Rhenium | 1 |
| Antimony | 1 | Indium | 3 | Selenium | 1 |
| Beryllium | 5 | Iron | 1 | Silicon | 1 |
| Bismuth | 2 | Lead | 1 | Silver | 1 |
| Cadmium | 2 | Lithium | 1 | Tantalum | 1 |
| Chromium | 3 | Magnesium | 1 | Tellurium | 1 |
| Cobalt | 1 | Manganese | 2 | Tin | 1 |
| Columbium | 2 | Mercury | 1 | Titanium | 1 |
| Copper | 2 | Molybdenum | 1 | Tungsten | 1 |
| Diamond | 1 | Nickel | 1 | Uranium | 1 |
| Gallium | 1 | PGM | 1 | Vanadium | 1 |
| Germanium | 5 | Rare Earths | 2 | Zinc | 1 |
| | | | | Zirconium | 2 |

## TABLE 3: THE MINERAL ASSESSMENT RATINGS

| | PRODUCTION RISKS | | | | | TRANSPORTATION RISKS | |
|---|---|---|---|---|---|---|---|
| | *Existing capacity* | *Labour disputes* | *Violent conflict* | *Range of primary supply sources* | *Time lags for new supplies* | *Primary* | *Secondary* |
| Aluminium | 1 | 2 | 1 | 5 | 8 | 5 | 3 |
| Antimony | 1 | 1 | 1 | 5 | 4 | 1 | 1 |
| Beryllium | 5 | 2 | 1 | 6 | 6 | 3 | 1 |
| Bismuth | 2 | 4 | 1 | 2 | 2 | 1 | 1 |
| Cadmium | 2 | 3 | 1 | 1 | 1 | 2 | 1 |
| Chromium | 3 | 2 | 7 | 9 | 9 | 8 | 2 |
| Cobalt | 1 | 3 | 9 | 8 | 8 | 9 | 2 |
| Columbium/ Niobium | 2 | 3 | 1 | 8 | 5 | 5 | 2 |
| Copper | 2 | 7 | 2 | 6 | 8 | 7 | 2 |
| Diamond | 1 | 2 | 7 | 4 | 6 | 2 | 3 |
| Gallium | 1 | 1 | 1 | 2 | 4 | 1 | 7 |
| Germanium | 5 | 1 | 3 | 8 | 1 | 2 | 1 |
| Gold | 1 | 1 | 10 | 10 | 10 | 1 | 1 |
| Indium | 3 | 1 | 1 | 7 | 7 | 1 | 1 |
| Iron | 1 | 3 | 1 | 1 | 1 | 4 | 4 |
| Lead | 1 | 5 | 1 | 2 | 4 | 2 | 2 |
| Lithium | 1 | 2 | 1 | 2 | 1 | 1 | 1 |
| Magnesium | 1 | 2 | 1 | 3 | 3 | 1 | 1 |
| Manganese | 2 | 1 | 7 | 8 | 6 | 8 | 8 |
| Mercury | 1 | 2 | 1 | 3 | 2 | 2 | 2 |
| Molybdenum | 1 | 6 | 1 | 5 | 8 | 1 | 1 |
| Nickel | 1 | 6 | 2 | 1 | 7 | 8 | 2 |
| Platinum Gp | 1 | 1 | 6 | 8 | 8 | 7 | 1 |
| Rare Earths | 2 | 1 | 1 | 4 | 6 | 1 | 1 |
| Rhenium | 1 | 4 | 3 | 2 | 7 | 1 | 1 |
| Selenium | 1 | 3 | 1 | 1 | 6 | 1 | 1 |
| Silicon | 1 | 1 | 1 | 1 | 1 | 1 | 1 |
| Silver | 1 | 4 | 6 | 1 | 3 | 1 | 1 |
| Tantalum | 1 | 1 | 2 | 7 | 8 | 5 | 1 |
| Tellurium | 1 | 2 | 1 | 3 | 4 | 1 | 1 |
| Tin | 1 | 6 | 3 | 7 | 7 | 6 | 2 |
| Titanium | 1 | 2 | 3 | 7 | 7 | 5 | 1 |
| Tungsten | 1 | 1 | 5 | 6 | 5 | 6 | 1 |
| Uranium | 1 | 2 | 1 | 2 | 4 | 2 | 10 |
| Vanadium | 1 | 2 | 2 | 7 | 9 | 3 | 1 |
| Zinc | 1 | 6 | 1 | 3 | 4 | 6 | 3 |
| Zirconium | 2 | 1 | 1 | 4 | 5 | 2 | 1 |

*Each mineral is assessed under the 13 headings on a scale from 1 (lowest) to 10 (highest) risk. For commentaries on these assessments see the mineral profiles in Section Two.*

**Table 3: The Mineral Assessment Ratings (Cont)**

| *APPLICATION /USE RISKS* | | | | *TRADE RISKS* | |
|---|---|---|---|---|---|
| *Total economic impact* | *Effect on key industries* | *Availability of substitutes* | *Longer term substitutability* | *Collusive price agreements* | *Embargoes* |
| 10 | 9 | 9 | 8 | 8 | 1 |
| 1 | 3 | 3 | 3 | 1 | 1 |
| 1 | 7 | 9 | 9 | 5 | 2 |
| 2 | 3 | 2 | 2 | 1 | 1 |
| 3 | 5 | 4 | 3 | 1 | 1 |
| 7 | 7 | 10 | 8 | 5 | 8 |
| 4 | 9 | 7 | 7 | 7 | 4 |
| | | | | | |
| 4 | 6 | 5 | 5 | 7 | 2 |
| 10 | 7 | 9 | 8 | 5 | 1 |
| 1 | 8 | 7 | 4 | 8 | 2 |
| 1 | 8 | 6 | 6 | 2 | 1 |
| 1 | 9 | 9 | 9 | 2 | 1 |
| 1 | 8 | 6 | 6 | 8 | 3 |
| 1 | 9 | 7 | 6 | 3 | 1 |
| 10 | 3 | 10 | 1 | 3 | 1 |
| 7 | 4 | 5 | 5 | 2 | 1 |
| 2 | 4 | 4 | 3 | 2 | 1 |
| 4 | 7 | 6 | 4 | 3 | 1 |
| 8 | 7 | 9 | 9 | 5 | 5 |
| 2 | 3 | 6 | 5 | 6 | 3 |
| 5 | 9 | 9 | 8 | 3 | 1 |
| 6 | 8 | 6 | 6 | 5 | 1 |
| 2 | 7 | 5 | 6 | 9 | 8 |
| 1 | 8 | 8 | 6 | 1 | 1 |
| 1 | 6 | 1 | 3 | 7 | 2 |
| 2 | 7 | 3 | 4 | 1 | 1 |
| 10 | 2 | 6 | 6 | 1 | 1 |
| 2 | 4 | 5 | 5 | 8 | 1 |
| 1 | 8 | 6 | 7 | 5 | 1 |
| 1 | 5 | 3 | 3 | 1 | 1 |
| 6 | 5 | 3 | 3 | 10 | 1 |
| 6 | 10 | 8 | 8 | 3 | 2 |
| 5 | 7 | 6 | 6 | 3 | 1 |
| 3 | 7 | 7 | 3 | 1 | 9 |
| 4 | 6 | 7 | 6 | 6 | 3 |
| 7 | 5 | 5 | 5 | 5 | 1 |
| 4 | 3 | 5 | 4 | 2 | 1 |

**TABLE 4: MINERAL PRODUCTION RISKS: LABOUR DISPUTES**

| | | | | | |
|---|---|---|---|---|---|
| Aluminium | 2 | Gold | 1 | Rhenium | 4 |
| Antimony | 1 | Indium | 1 | Selenium | 3 |
| Beryllium | 2 | Iron | 3 | Silicon | 1 |
| Bismuth | 4 | Lead | 5 | Silver | 4 |
| Cadmium | 3 | Lithium | 2 | Tantalum | 1 |
| Chromium | 2 | Magnesium | 2 | Tellurium | 2 |
| Cobalt | 3 | Manganese | 1 | Tin | 6 |
| Columbium | 3 | Mercury | 2 | Titanium | 2 |
| Copper | 7 | Molybdenum | 6 | Tungsten | 1 |
| Diamond | 2 | Nickel | 6 | Uranium | 2 |
| Gallium | 1 | PGM | 1 | Vanadium | 2 |
| Germanium | 1 | Rare Earths | 1 | Zinc | 6 |
| | | | | Zirconium | 1 |

are separated out. This means that strikes must exceed several months' duration before their effects are felt. This would be the case for such metals as selenium, cadmium, indium, germanium, and the rare earths.

### 1.1(c) Violent Conflict

What is the likelihood of wars, insurrections and rebellions in the major production countries?

This is chiefly a geopolitical assessment and reference should be made to the section for each country concerned. What is the expected effect of a war in the major production areas?

In the event of a war in Southern Africa, copper, gold, uranium and platinum mines would be targets for destruction in addition to the havoc associated with a state of national emergency in the countries involved. The changing trends of mineral production during World War II provide a useful guide to the possible effects of a future localised or generalised war.

Metals susceptible to supply disruption in the event of a local war include tin, with its preponderance of small, labour-intensive, seasonally-operating mines in South East Asia, and copper, with its large-scale, export-oriented mines in Zambia, Zaire, Chile and Peru. The effect of a war can be considered as affecting not only current production, but the purchasing and investment policies of consumers in the longer term. (Copper output has increased in Chile, at the expense of Central Africa.)

### 1.1(d) Range of Primary Supply Sources

Are reserves and existing mine/refinery capacity concentrated in two or three countries or are they geographically widespread? It is important to determine whether the state controlled economies are major exporters of a metal and whether they contribute significantly to Western world supply. For example, lead and zinc are both produced in over 50 countries on all populated continents and net trade with the Communist bloc is relatively small.

**TABLE 5: MINERAL PRODUCTION RISKS: VIOLENT CONFLICT**

| | | | | | |
|---|---|---|---|---|---|
| Aluminium | 1 | Gold | 10 | Rhenium | 3 |
| Antimony | 1 | Indium | 1 | Selenium | 1 |
| Beryllium | 1 | Iron | 1 | Silicon | 1 |
| Bismuth | 1 | Lead | 1 | Silver | 6 |
| Cadmium | 1 | Lithium | 1 | Tantalum | 2 |
| Chromium | 7 | Magnesium | 1 | Tellurium | 1 |
| Cobalt | 9 | Manganese | 7 | Tin | 3 |
| Columbium | 1 | Mercury | 1 | Titanium | 3 |
| Copper | 2 | Molybdenum | 1 | Tungsten | 5 |
| Diamond | 7 | Nickel | 2 | Uranium | 1 |
| Gallium | 1 | PGM | 6 | Vanadium | 2 |
| Germanium | 3 | Rare Earths | 1 | Zinc | 1 |
| | | | | Zirconium | 1 |

In contrast, the platinum group metals are produced mainly in South Africa and the USSR, with less than six per cent mined elsewhere. A similar, although slightly less critical, situation applies in both manganese and vanadium, where South Africa and the USSR jointly produce over 60 per cent of world mined output.

Any examination of supply sources must now consider the possibility of obtaining the material from previously uneconomic ores in untroubled areas, using national strategic or industrial stockpiles or investing time and effort into improving recycling efficiency.

### 1.1(e) Time Lag for New Supplies

Supposing that alternative sources of supply are available, how long will it take to obtain these supplies?

The development of new orebodies typically involves a timescale of between six and ten years, the upper end of the range for underground mines, for those in remote areas of the world requiring extensive infrastructure and for those in difficult terrain. The expansion of existing mines and plants requires less time as does the use of specialised techniques (e.g. tin gravel pump mining off the coast of Thailand, panning for gold or exploiting the rich surface deposits of gold in Brazil).

The availability of above ground stocks shortens the lead time substantially but their form can vary from stocks of concentrate or slags, such as the tantalite-bearing tin slags now being reworked in South East Asia, through the intermediate stages of partially refined metals, such as blister copper, to metals stocks of the commercial purity.

Industrial specifications have, in many cases, been tightened over the past decade to accommodate the demands of technology. Strategic stockpile qualities have to be constantly reviewed to ensure their ready industrial acceptability.

**TABLE 6: PRODUCTION RISKS: PRIMARY SUPPLY SOURCES**

| | | | | | |
|---|---|---|---|---|---|
| Aluminium | 5 | Gold | 10 | Rhenium | 2 |
| Antimony | 5 | Indium | 7 | Selenium | 1 |
| Beryllium | 6 | Iron | 1 | Silicon | 1 |
| Bismuth | 2 | Lead | 2 | Silver | 1 |
| Cadmium | 1 | Lithium | 2 | Tantalum | 7 |
| Chromium | 9 | Magnesium | 3 | Tellurium | 3 |
| Cobalt | 8 | Manganese | 8 | Tin | 7 |
| Columbium | 8 | Mercury | 3 | Titanium | 7 |
| Copper | 6 | Molybdenum | 5 | Tungsten | 6 |
| Diamond | 4 | Nickel | 1 | Uranium | 2 |
| Gallium | 2 | PGM | 8 | Vanadium | 7 |
| Germanium | 8 | Rare Earths | 4 | Zinc | 3 |
| | | | | Zirconium | 4 |

**TABLE 7: PRODUCTION RISKS: TIME LAGS FOR NEW SUPPLIES**

| | | | | | |
|---|---|---|---|---|---|
| Aluminium | 8 | Gold | 10 | Rhenium | 7 |
| Antimony | 4 | Indium | 7 | Selenium | 6 |
| Beryllium | 6 | Iron | 1 | Silicon | 1 |
| Bismuth | 2 | Lead | 4 | Silver | 3 |
| Cadmium | 1 | Lithium | 1 | Tantalum | 8 |
| Chromium | 9 | Magnesium | 3 | Tellurium | 4 |
| Cobalt | 8 | Manganese | 6 | Tin | 7 |
| Columbium | 5 | Mercury | 2 | Titanium | 7 |
| Copper | 8 | Molybdenum | 8 | Tungsten | 5 |
| Diamond | 6 | Nickel | 7 | Uranium | 4 |
| Gallium | 4 | PGM | 8 | Vanadium | 9 |
| Germanium | 1 | Rare Earths | 6 | Zinc | 4 |
| | | | | Zirconium | 5 |

**TABLE 8: MINERAL PRODUCTION RISK ASSESSMENT**
(Average of the five supply disruption factors to a scale of 1-10)

| | | | | | |
|---|---|---|---|---|---|
| Aluminium | 3.4 | Gold | 6.4 | Rhenium | 3.4 |
| Antimony | 2.4 | Indium | 3.8 | Selenium | 2.4 |
| Beryllium | 4.0 | Iron | 1.4 | Silicon | 1.0 |
| Bismuth | 2.2 | Lead | 2.6 | Silver | 3.0 |
| Cadmium | 1.6 | Lithium | 1.4 | Tantalum | 3.8 |
| Chromium | 6.0 | Magnesium | 2.0 | Tellurium | 2.2 |
| Cobalt | 5.8 | Manganese | 4.8 | Tin | 4.8 |
| Columbium | 3.8 | Mercury | 1.8 | Titanium | 4.0 |
| Copper | 5.0 | Molybdenum | 4.2 | Tungsten | 3.6 |
| Diamond | 4.0 | Nickel | 3.4 | Uranium | 2.0 |
| Gallium | 1.8 | PGM | 4.8 | Vanadium | 4.2 |
| Germanium | 3.6 | Rare Earths | 2.8 | Zinc | 3.0 |
| | | | | Zirconium | 2.6 |

Recycling provides only a partial solution given the extended lifetime of vehicles (three to six years in the USA, Europe and Japan), buildings (40 years) and durable consumer goods (five to ten years). A possible exception is aluminium used in cans with an effective life of three months or the precious metals where jewellery and coins can be requisitioned.

## 1.2 TRANSPORTATION RISKS

These can be divided into two categories, primary risks involving the movement of raw materials and secondary risks concerned with the transportation of finished and semi-finished metals.

### 1.2(a) Primary

What is the likelihood of disruption in transport links vital to the movement of ore and metal from the site of production to that of fabrication? The movement of minerals is normally undertaken by rail, road, ship, air and pipelines.

There are several risks attendant upon transport by ship; dock strikes, weather and other natural forces and the length and direction of the route to be followed. Dock strikes still occur with monotonous regularity in the USA, Canada and Australia although their severity has declined in line with the health of the shipping industry.

The established concept of freedom of the seas is in some dispute, given the growth in territorial claims which affect the passage and safety of ships passing in many areas. World shipping route choke points, where problems may be expected, include the Panama Canal, the English Channel, the Straits of Gibraltar, the Horn of Africa, the Straits of Hormuz and the Red Sea entrance, the Cape of Good Hope, the Malacca Straits, the Sri Lankan/Indian seas, the area surrounding Singapore and the Tasman Straits.

Transport by rail can be disrupted by energy shortages, strikes, a shortage of rolling stock as occurs frequently in Southern Africa, natural disasters such as avalanches, subsidence etc or violence. This last factor embraces guerilla attacks such as those which have affected the railway lines between the Central African copper belt and the sea.

Transport by road is less liable to disruption since lorries and trucks are smaller and can be flexible enough to travel over difficult terrain or damaged roads if necessary. Air transport is often used as a temporary solution rather than as the main type of transport except for such high-value, low-volume commodities as diamonds, the platinum-group metals and gold. Air freight is expensive and can only cope with limited weights, but it is flexible and reliable and has been used in extreme circumstances, eg to airlift cobalt out of Zaire in the 1978 Shaba invasion, and was considered during the 1981-82 tin market manipulation.

Transport of ore and intermediate-stage metal still tends to follow traditional colonial patterns with the bulk going from the Third World southern hemisphere plus Australia to the industrialised countries of the northern hemisphere.

### 1.2(b) Secondary

What are the risks in transporting fabricated products containing the particular metal to its final consumer?

Here the trade tends to be between industrialised countries, rather than from the Third World to the industrialised world, and this difference reduces the risks other than those relating to labour problems. Exceptions include the environmentally-related problems attaching to the shipment of processed uranium, radio-active waste, sulphur, mercury and other toxics. The risk of damage applies to delicate, low-volume, high-value electronics, and the risk of theft to precious metals and gems.

## 1.3 APPLICATION/USE RISKS

In contrast to the production and transportation risks this family of risks is concerned essentially with the costs of possible disruption. As set out in Table 1 the four sub-components fall within this category.

### 1.3(a) Total Economic Impact

This sub-component provides an overall volume weighting.

Is the metal or mineral used in large weights or volumes worldwide? What is the market size? On this basis it is not surprising that iron, copper, and aluminium score highly due to the large tonnages used in almost every type of industry, whilst zirconium, the rare earths, gallium and others score low.

### 1.3(b) Effect on Key Industries

What is the end-use pattern of demand for the material? Is it vital in the growth industries of defence, aerospace and electronics or is it merely price related? Is consumption of the metal actually increasing in these industries or does each technological development within a sector reduce the unit requirements for the metal?

There has been a pronounced move in the developed countries towards the service industries and towards closer control of material consumption within each industry. There is a declining trend in the income elasticity of demand for many high-volume metals. This suggests that a one per cent increase in the GDP of a country would require, perhaps, a 0.5 per cent increase in lead or copper demand in that country as opposed to a one per cent increase ten years ago.

This effect has been seen for lead, zinc, copper, aluminium, steel, chromium and many other metals and is a function of a drive towards smaller and/or lighter and overall cost-competitive products in all sectors of manufacturing industry.

Within the developed countries, few metal-consuming growth industries remain: the electronics industry has grown phenomenally over the last decade and the growth in units sold has offset the effects of continuing miniaturisation. Continuing research and development with different materials ensures that the continuing predominance of any one metal cannot be assured for any application in the longer term.

### 1.3(c) Availability of Substitutes

Are substitutes available? Do they come from the same area or mine as the metal whose supplies have been interrupted? What is the cost of changing to the substitute and the cost of the substitute itself? Is a reformulation of the final

**TABLE 9: TRANSPORT RISK ASSESSMENT**

| | | | | | |
|---|---|---|---|---|---|
| Aluminium | 4.0 | Gold | 1.0 | Rhenium | 1.0 |
| Antimony | 1.0 | Indium | 1.0 | Selenium | 1.0 |
| Beryllium | 2.0 | Iron | 4.0 | Silicon | 1.0 |
| Bismuth | 1.0 | Lead | 2.0 | Silver | 1.0 |
| Cadmium | 1.5 | Lithium | 1.0 | Tantalum | 3.0 |
| Chromium | 5.0 | Magnesium | 1.0 | Tellurium | 1.0 |
| Cobalt | 5.5 | Manganese | 8.0 | Tin | 4.0 |
| Columbium | 3.5 | Mercury | 2.0 | Titanium | 3.0 |
| Copper | 4.5 | Molybdenum | 1.0 | Tungsten | 3.5 |
| Diamond | 2.5 | Nickel | 5.0 | Uranium | 6.0 |
| Gallium | 4.0 | PGM | 4.0 | Vanadium | 2.0 |
| Germanium | 1.5 | Rare Earths | 1.0 | Zinc | 4.5 |
| | | | | Zirconium | 1.5 |

**TABLE 10: STRATEGIC APPLICATION RISK ASSESSMENT**

| | | | | | |
|---|---|---|---|---|---|
| Aluminium | 9.0 | Gold | 5.3 | Rhenium | 2.8 |
| Antimony | 2.5 | Indium | 5.8 | Selenium | 4.0 |
| Beryllium | 6.5 | Iron | 6.0 | Silicon | 6.0 |
| Bismuth | 2.3 | Lead | 5.3 | Silver | 4.0 |
| Cadmium | 3.8 | Lithium | 3.3 | Tantalum | 5.5 |
| Chromium | 8.0 | Magnesium | 5.3 | Tellurium | 3.0 |
| Cobalt | 6.8 | Manganese | 8.3 | Tin | 4.3 |
| Columbium | 5.0 | Mercury | 4.0 | Titanium | 8.0 |
| Copper | 8.5 | Molybdenum | 7.8 | Tungsten | 6.0 |
| Diamond | 5.0 | Nickel | 6.5 | Uranium | 5.0 |
| Gallium | 5.3 | PGM | 5.0 | Vanadium | 5.8 |
| Germanium | 7.0 | Rare Earths | 5.8 | Zinc | 5.5 |
| | | | | Zirconium | 4.0 |

fabricated product needed or merely a simple change of process? How long would it take to adapt to using the substitute and how closely does the possible substitute resemble what it is replacing: metal for metal or plastic–glass–ceramic for metal? How would the consumer feel about the change in terms of cost, appearance and performance?

These complex questions must all be considered and the overall substitutability tends to be a compromise between cost and fitness for purpose; whether this is measured in terms of strength, density, melting point, conductivity, corrosion resistance, hardness, colour, lustre, catalytic properties, toughness etc or a combination of three or four such properties.

### 1.3(d) Longer Term Substitutability

If the disruption of supplies of a particular mineral is regarded temporary policy makers may not consider it worthwhile to obtain and adapt to substitutes. However if the disruption is lengthy then a substitute may prove to be so effective and economical that it permanently retains a significant section of the market. Cobalt lost a large section of its permanent magnet market following the 1978 price peak and tantalum lost 15 per cent of its tantalum pentoxide capacitor powder market after the record price highs of May 1980.

Other factors which affect longer term substitutability include energy costs and environmental requirements. Energy costs cover both the energy required in extracting and fabricating the metal plus that consumed during the lifetime of the fabricated component. Comparisons can be drawn between the different effects of the 1973-74 oil crisis on aluminium and on zinc. Aluminium consumption increased steeply in the automobile and aircraft industry due to its low density/strength ratio. In contrast, zinc lost its prime position as chief ingredient for cast automobile components as producers moved to smaller, lighter cars using magnesium or aluminium alloys.

Environmental legislation has become tighter since 1970, affecting the usage of certain cadmium compounds in Sweden and Japan, the lowering of permissible tetra-ethyl lead-in-petrol levels in the USA, Europe and Japan and the stagnation of growth in all mercury's markets.

## 1.4 TRADE RISKS

Two trade risk sub-components were identified in Table 1, both falling under the umbrella of disruption probability.

### 1.4(a) Collusive Price Agreements

Are the prices for a metal and its ores determined by a free market or are they fixed by agreement, be it a formal international association or just an informal arrangement between companies ostensibly in competition? Such price agreements vary from the formal cartels such as CIPEC or the International Bauxite Association (comprising mainly Third World producers) through the gamut of international organisations including producers and consumers, such as the International Tin Council and other UN-sponsored groups, to informal arrangements such as that rumoured to exist between Russia and South Africa on gold and platinum sales.

Third World organisations have seen little success due to the lack of economic power wielded by their members, who depend heavily on the revenues from raw mineral sales. The moderate ITC, combining producer and consumer interests, has fared better in its aims of stabilising prices and ensuring an approximate supply/demand balance. Other UN-proposed organisations, including those for tungsten and manganese, have met with no enthusiasm from consumers and rarely moved beyond the committee stage.

Informal cartels include those between the Canadian producers, Inco and Falconbridge, for nickel, between Russia and Impala and Rustenburg of South Africa for the platinum group metals and between European zinc toll smelters. These private arrangements violate EEC regulations and US anti-trust laws, but remain.

In several cases, one company effectively dominates Western world supplies of a particular metal e.g. Amax with 57 per cent of world molybdenum production in 1981, De Beers and the Central Selling Organisation with a monopoly on diamond marketing worldwide and Codelco Chile with over 70 per cent of Western rhenium output. This often allows the group almost total freedom of price fixing when demand is buoyant. However, during a recession producer price wars are common and the free market mechanism grows in importance.

Terminal free markets such as the London Metal Exchange and the New York Commodity Exchange provide the vehicles and are much less open to individual manipulation although notable exceptions include silver in 1980-81 and the tin market manipulation (thought by many to have been masterminded by the Malaysians) in 1981-82. A merchant free market operates for most metals and published prices and records are available in most cases to provide at least a semblance of competitiveness.

**1.4(b) Embargoes**

Are trade patterns or prices likely to be affected politically? There have been instances of governments imposing total or partial bans on the export of particular minerals and for various reasons. Under a Labour administration Australia stopped the export of titanium sands in the 1970s in an attempt to influence prices. Canada halted uranium sales selectively to countries which did not comply with nuclear non-proliferation demands.

The imposition of an embargo affects both parties adversely by disrupting trade and tends to be a tactic of last resort. A pessimistic analysis suggests that Third World countries could at some future stage halt mineral exports as a form of economic blackmail. Fears have prompted France to tend its old colonial connections, Japan to build up a widely-based strategic stockpile and the USA to expand its own huge strategic inventories. Even without these safeguards, embargoes are unlikely to establish any lasting effect on trade in the foreseeable future.

## 1.5 COMPARATIVE STRATEGIC RATINGS

It was noted earlier in this section that the sub-component ratings, as listed and quantified in Table 3, form the basis for the overall assessment for each mineral. There are two steps in this process.

Firstly the sub-components are grouped under the two risk factor categories of probability or likelihood of supply disruption and the economic cost

**TABLE 11: TRADE RISK ASSESSMENT**

| | | | | | |
|---|---|---|---|---|---|
| Aluminium | 4.5 | Gold | 5.5 | Rhenium | 4.5 |
| Antimony | 1.0 | Indium | 2.0 | Selenium | 1.0 |
| Beryllium | 3.5 | Iron | 2.0 | Silicon | 1.0 |
| Bismuth | 1.0 | Lead | 1.5 | Silver | 4.5 |
| Cadmium | 1.0 | Lithium | 1.5 | Tantalum | 3.0 |
| Chromium | 6.5 | Magnesium | 2.0 | Tellurium | 1.0 |
| Cobalt | 5.5 | Manganese | 5.0 | Tin | 5.5 |
| Columbium | 4.5 | Mercury | 4.5 | Titanium | 2.5 |
| Copper | 3.0 | Molybdenum | 2.0 | Tungsten | 2.0 |
| Diamond | 5.0 | Nickel | 3.0 | Uranium | 5.0 |
| Gallium | 1.5 | PGM | 8.5 | Vanadium | 4.5 |
| Germanium | 1.5 | Rare Earths | 1.0 | Zinc | 3.0 |
| | | | | Zirconium | 1.5 |

**TABLE 12: TOTAL STRATEGIC RATINGS**

| | | | | | |
|---|---|---|---|---|---|
| Aluminium | 23.0 | Gold | 26.4 | Rhenium | 9.9 |
| Antimony | 3.2 | Indium | 10.4 | Selenium | 4.6 |
| Beryllium | 17.6 | Iron | 9.5 | Silicon | 4.3 |
| Bismuth | 3.7 | Lead | 9.0 | Silver | 11.6 |
| Cadmium | 4.5 | Lithium | 3.5 | Tantalum | 13.6 |
| Chromium | 41.5 | Magnesium | 5.9 | Tellurium | 3.8 |
| Cobalt | 35.3 | Manganese | 36.7 | Tin | 21.8 |
| Columbium | 22.3 | Mercury | 8.8 | Titanium | 17.7 |
| Copper | 28.8 | Molybdenum | 16.1 | Tungsten | 13.9 |
| Diamond | 19.0 | Nickel | 18.8 | Uranium | 14.2 |
| Gallium | 7.7 | PGM | 28.8 | Vanadium | 17.6 |
| Germanium | 15.5 | Rare Earths | 6.6 | Zinc | 14.9 |
| | | | | Zirconium | 5.9 |

of that disruption. An average is calculated for each factor (from 1.0 to 10.0) and these numbers appear at the beginning of each mineral profile in Section Two.

The two factor assessments are then multiplied together to give the overall strategic assessment for each mineral. The assessment ranges therefore from 1 to 100. This multiplicative approach results in most minerals appearing towards the lower end of the scale as shown in Table 12.

It can be seen from this table that antimony emerges with the lowest risk rating at 3.2 whilst chromium at 41.5 has the highest rating. In broad terms it can be concluded that minerals with overall ratings lower than 10.0 can be considered low risk. There are 15 of these as listed in Table 12. The medium risk range (from 10.0 to 25.0) accounts for a further 16 minerals including uranium, the backbone of the world nuclear power industry, diamond and tin.

Table 13 lists the six minerals which emerge from this survey with the highest strategic risk. It can be seen from this table that all have higher cost factor ratings underlining their present importance in applications and in some cases, e.g. copper, the important effect any disruption would have on the world economy.

**TABLE 13: THE HIGHEST RISK MINERALS**

| | *Strategic Rating* | *Likelihood of Disruption Factor* | *Cost of Disruption Factor* |
|---|---|---|---|
| Chromium | 41.5 | 5.0 | 8.3 |
| Manganese | 36.7 | 4.7 | 7.8 |
| Cobalt | 35.3 | 4.9 | 7.2 |
| Copper | 28.8 | 3.6 | 8.0 |
| PGM | 28.8 | 4.8 | 6.0 |
| Gold | 26.4 | 4.0 | 6.8 |

# COUNTRY RISK ASSESSMENT

A total of 34 countries have been selected for inclusion in Section Three. The basis for this selection has been:

- –the largest raw material producers: Australia, Canada, South Africa, USA and USSR
- –the second rank of producers: which make available substantial quantities on the world markets: Bolivia, Brazil, Chile, Mexico, Peru, Philippines, Zaire and Zambia
- –smaller producers with a wide range of minerals: Cuba, Finland, Indonesia, Malaysia, Papua New Guinea, Spain, Thailand and Zimbabwe
- –minor producers with at least one world-ranking product: Botswana, Cuba, Eire, Gabon, Guyana, India, Italy, Jamaica and Namibia
- –major refiners and consumers: France, Germany, Japan and the United Kingdom.

In the case of the last group there is clearly overlap as the larger countries are both producers and consumers but their relative exposure to risk is analysed in the strategic rating table.

The risk assessment for each country is based on an examination of seven primary components. These together with their sub-components are listed in Table 14.

Table 15 highlights the component ratings determined for each country. The sub-component ratings can be found in the risk assessment tables which appear within each country profile in Section Three.

## 2.1 LABOUR RISKS

Manpower represents the most volatile risk factor in minerals production, impinging upon every sector and being for the most part unpredictable. Labour disruptions have an immediate effect and are likely to occur in the most simple and in the most advanced of economies, throughout the political spectrum. The occurrence of a labour dispute can be the end product of any one of a number of problems; political, financial, locational or organisational.

Table 15 shows that the advanced economies, such as Australia and Canada, can be highly vulnerable alongside much poorer developing countries such as Guyana and Zambia. The Polish experience of 1980 onwards has similarly shown that the centrally planned economies have problems of no lesser intensity than the West. The USSR is also known to have major labour difficulties and low productivity. The sub components of labour risk clearly overlap but their importance, as noted below, varies significantly country to country.

### 2.1(a) Incidence of Strikes

Assessments for this sub component by necessity run fairly closely to 2.1(c) General Unrest. Examples would include tin production in Bolivia and copper and cobalt production in Zaire.

But a high incidence of strikes is not necessarily more damaging than less frequent strikes of longer duration. In North America, where the three-year industrial labour contract is well established, the onset of a prolonged strike around contract-negotiation time is often regarded as inevitable, and has damaging consequences. The copper and nickel industries suffered particularly badly in this way between 1979 and 1981.

Australia has been particularly prone to wild-cat strikes, in mining, transport and dock areas since the 1960s when its large tonnage iron ore, coal and bauxite operations began to expand. Reasons cited include harsh environmental conditions, e.g. the iron ore regions of Western Australia and the influx of union-orientated immigrants.

The incidence of strikes in the big three mining countries of South America–Chile, Peru and Bolivia–associated with a mixture of political unrest and poor conditions periodically threaten a significant proportion of the world's copper, molybdenum, silver and lead.

**TABLE 14: COMPONENTS OF THE COUNTRY RISK ASSESSMENTS**

| *Components* | *Sub Components* |
|---|---|
| 1. Labour Risks | (a) Incidence of Strikes<br>(b) Quality of Labour<br>(c) General Unrest |
| 2. Political Risks | (a) History<br>(b) Stability<br>(c) External Dangers |
| 3. Location Risks | (a) Hostile Borders<br>(b) Critical Land/Sea Routes |
| 4. Mineral Resource Risks | (a) Adequacy of Reserves<br>(b) Costs of Production<br>(c) Development Appeal |
| 5. Financial and Economic Risks | (a) Currency Performance<br>(b) Vulnerability to Manipulation<br>(c) External Indebtedness<br>(d) Domestic Productivity<br>(e) Foreign Ownerships |
| 6. Energy and Other Natural Resources Risks | (a) Domestic Energy Sources<br>(b) Domestic Energy Production<br>(c) Sources of Outside Supply |
| 7. Dependence on Foreign Natural Resources | |

The presence or absence of trade unionism affects the intensity of labour disputes rather than the frequency. Where unionism has reached a high degree of sophistication as in the OECD area its influence spreads into every facet of the industry, often including the direction of investment, manning levels and such. In the younger economies such as the post-colonial independent states, a strong political-industrial overlap occurs which, in countries such as Guyana and Zambia, has ultimately led to the manipulation of labour disputes and a higher strike incidence.

Metals and countries considered most susceptible to the effects of strikes include:

—bauxite in Australia, Guyana and Jamaica

—copper in Chile, Peru, the Philippines, Zaire and Zambia

—tin in Bolivia.

**2.1(b) Quality of Labour**

The most highly trained labour is concentrated in the OECD area but the cost/efficiency advantages derived are to some extent offset by higher living standards, and a greater degree of environmental control, the reverse of that pertaining in parts of Africa and South America. A mid-point situation was sustained successfully in many developing countries for much of the 1920-1970 period when most of the big mining companies were owned by European or American entities and staffed from a fairly junior level by expatriates. The dismantling of this system in the post 1960 period has led to serious man-

**TABLE 15: THE COUNTRY ASSESSMENT RATINGS**

| *Country* | *Components of Risk Assessment* | | | | | | |
|---|---|---|---|---|---|---|---|
| | *Labour* | *Political* | *Location* | *Mineral Reserves* | *Financial & Economic* | *Energy Reserves* | *Dependence on Foreign Natural Resources* |
| Australia | 4.0 | 4.0 | 3.5 | 5.3 | 6.0 | 5.0 | 2.0 |
| Bolivia | 6.7 | 7.3 | 6.0 | 8.3 | 7.2 | 4.0 | 5.0 |
| Botswana | 4.3 | 5.3 | 7.5 | 5.7 | 6.2 | 5.3 | 9.0 |
| Brazil | 5.7 | 5.0 | 5.0 | 5.0 | 7.4 | 5.7 | 5.0 |
| Canada | 5.3 | 5.7 | 1.5 | 5.0 | 6.8 | 1.3 | 2.0 |
| Chile | 6.3 | 6.3 | 5.5 | 5.0 | 5.0 | 6.3 | 6.0 |
| Finland | 3.7 | 7.0 | 7.0 | 7.7 | 4.4 | 6.6 | 7.0 |
| France | 5.3 | 7.0 | 5.0 | 7.0 | 5.6 | 7.7 | 8.0 |
| Gabon | 5.3 | 5.7 | 4.0 | 5.3 | 5.8 | 4.0 | 5.0 |
| Germany (FR) | 4.3 | 4.7 | 6.5 | 7.0 | 3.8 | 7.0 | 8.0 |
| Guyana | 7.7 | 7.3 | 6.0 | 7.3 | 7.2 | 7.7 | 7.0 |
| India | 5.0 | 7.0 | 6.5 | 7.3 | 6.3 | 6.7 | 9.0 |
| Indonesia | 4.7 | 7.3 | 7.5 | 5.7 | 5.2 | 2.3 | 8.0 |
| Ireland | 6.7 | 5.7 | 1.5 | 7.3 | 7.4 | 6.3 | 8.0 |
| Italy | 5.0 | 7.3 | 4.0 | 7.0 | 5.0 | 6.0 | 7.0 |
| Jamaica | 7.0 | 7.0 | 5.0 | 7.7 | 6.8 | 7.7 | 9.0 |
| Japan | 2.7 | 2.7 | 5.5 | 7.7 | 3.8 | 7.7 | 7.0 |
| Malaysia | 6.0 | 6.3 | 7.0 | 6.3 | 5.6 | 4.3 | 7.0 |
| Mexico | 5.7 | 4.7 | 3.0 | 5.7 | 7.2 | 2.3 | 4.0 |
| Namibia | 7.7 | 8.0 | 7.5 | 5.7 | 5.2 | 7.7 | 6.0 |
| Papua New Guinea | 6.3 | 5.3 | 5.5 | 4.7 | 5.2 | 5.0 | 7.0 |
| Peru | 7.3 | 5.7 | 5.0 | 7.7 | 7.0 | 3.0 | 5.0 |
| Philippines | 6.0 | 6.0 | 7.0 | 5.3 | 7.0 | 6.0 | 7.0 |
| South Africa | 5.3 | 7.3 | 7.0 | 3.3 | 5.4 | 4.3 | 4.0 |
| Spain | 4.7 | 5.7 | 2.5 | 4.7 | 6.6 | 6.3 | 6.0 |
| Thailand | 5.3 | 7.3 | 8.0 | 5.3 | 5.8 | 4.0 | 8.0 |
| United Kingdom | 4.3 | 3.3 | 4.0 | 7.3 | 6.4 | 2.3 | 8.0 |
| USA | 3.0 | 2.7 | 1.5 | 6.3 | 5.6 | 4.0 | 7.0 |
| Zaire | 7.0 | 7.3 | 7.5 | 6.0 | 7.0 | 5.7 | 7.0 |
| Zambia | 7.7 | 5.3 | 6.0 | 6.3 | 6.8 | 7.7 | 8.0 |
| Zimbabwe | 7.3 | 4.7 | 6.5 | 5.7 | 6.2 | 6.7 | 2.0 |

***Notes:***
***Whilst profiles appear on the USSR, China and Cuba in Section Three risk assessments cannot be established as they are inappropriate given their political regimes.***

***Each country is assessed under the seven component headings on a scale from 1.0 (lowest) to 10.0 (highest) risks. For the sub-component ratings and commentaries see the country profiles in Section Three.***

power difficulties particularly affecting bauxite production in Guyana and Jamaica and copper and cobalt production in Zaire and Zambia.

### 2.1(c) General Unrest

Labour difficulties of this nature normally occur as a result of factors beyond the immediate influence of a labour force. These factors include political, ethnic, regional and union divisions within a country or between one country and another. Political divisions affect the democratically governed multi-party states of Canada, USA, the EEC, Australia and Japan differently from the dictatorships or single party systems which control much of Africa and South America. In the former grouping day-to-day politics have little influence on the minerals industry. Overall attitudes change when the party seen as more representative of workers interests is in power.

A more permanent state of unrest obtains amongst the one-party systems of the developing countries of Africa, the Caribbean and South America where frustrations cannot be vented on the ballot box. In Africa they are largely the teething problems following the rapid transition from colonialism superimposed on tribalism to independence based loosely upon Western democratic constitution. Their effects are felt in:

—Zambia: union-led dissent against one party state, falling living standards. Minerals affected—cobalt, copper

—Zaire: general unrest following many years dictatorial leadership, poor living standards, unresolved tribal problems and admitted corruption. Minerals affected—copper, cobalt, germanium, diamonds

—Angola: revolutionary state, Cuban troops in occupation, major internal dissent and guerilla problems within and without its borders. Minerals affected—diamonds, oil

—Namibia: disputed, South African administered territory with major ethnic and political problems. Minerals affected—diamonds, uranium, base metals

—Zimbabwe: underlying tribal differences and a Socialist-led regime paying lip service to a Westminster-style constitution. Minerals affected—chrome, nickel, gold, lithium.

The labour risk component within a country or region is critical to the ability to attract investment from abroad or to prevent domestic capital flowing overseas. In the developing world investment decisions by foreign parties are based less on the quality of labour and effect of unionism than they are on the state of general unrest and financial viability. Certain developed countries have lost marks since the 1960s for their deteriorating labour performance, a rising incidence of strikes being aggravated by rising costs and falling or stagnant productivity.

Australia is a particular example, its poor record affecting plans for the further expansion of coal, iron ore, bauxite and alumina and almost certainly curtailing its ambition to become the world's number two aluminium producer by 1990. Canada is also afflicted with a poor labour record, particularly in the far north where there is a growing concentration of nickel, oil, uranium and gold production.

## 2.2 POLITICAL RISKS

Metals have been used as a political weapon, or as a reason to begin hostilities, for centuries. Most wars have territorial gain as their incentive and the value of territory is greatly enhanced by the presence of economic minerals. The massive colonisation of the Americas was largely based on precious metals, as was the later development of Africa and Australia. The present relationship between politics and metals supply owes much to the historic background. The continuity of supply depends upon political stability and the outlook, both short and long term, is strongly influenced by the presence of external political dangers. These three sub components are important elements in the risk assessment tables for each major country. In some cases they can prove critical. For example Canada's political rating of 5.7 includes a below-average 6 for stability because of the danger of fragmentation of federal government through the increasing power and political differences at provincial level (see the Canada profile in Section Three). Zambia's stability factor is one better at 5 because of the continuation of leadership under

peninsula and selected countries of Western Europe such as Italy, Spain and Portugal.

### 2.2(c) External Dangers

This risk sub component measures the likelihood of a nation's or a region's minerals production being disrupted by outside forces. This factor has changed considerably since 1960 in many parts of the world.

It is the fulcrum of East-West ideology having presented both sides with opportunities to gain politico-commercial advantages worldwide. These opportunities arose from a combination of events previously referred to: decolonisation, the rapid growth of minerals demand and the need to secure finance for expansion. The zenith of political activity in minerals production probably came in the 1965-75 period. It focused on the extreme state of unrest in central Africa involving the Rhodesian UDI, the overthrow of the Portuguese in Angola and Mozambique, and the Congolese campaign.

A second arena developed in the Caribbean, involving Cuba's transition to Soviet communism, Jamaica's decisive swing in the same direction and latterly a more widespread dissent amongst several of the Central American countries.

Another old battleground, the South East Asia region, swung firmly pro-West on a wave of commodity-related affluence in the 1960-80 period. Indonesia moved into oil, copper and nickel, Malaysia benefitted from tin, oil, rubber, Thailand from tin. Yet Vietnam and Cambodia turned to communism and by 1982, the commodity boom clearly over, the prospect of renewed unrest returned.

All these events have contributed to the 'external dangers' rating, which is tempered by the obvious unwillingness of the Superpowers, since 1975, to provoke a direct confrontation in a delicate area. Table 17 highlights those metals which could be at risk from countries whose external dangers factor exceeds 6 on the assessment scale.

The political risk component is probably the most critical of all the components covered in this exercise given that it is the most volatile. A mine or mineral production source can only be established where a suitable deposit exists and this is invariably a lengthy and costly operation. The commercial risk can be reasonably assessed and changes within it can be accommodated to a degree. The labour risk tends to change only slowly.

The political risk can, and often does, change rapidly, totally and without any opportunity of redress. It is the factor now receiving much more attention by investors worldwide in their choice of a minerals production location. The USA remains a firm favourite in spite of high operating costs, South America receives preference over Africa, France and Spain have been derated. Australia and Canada are no longer automatic choices. The political factor in minerals supply is being seen more aggressively in marketing where government agencies are tightening their grip e.g. in Zaire, Zimbabwe and Zambia.

An increasing political risk factor for a country indicates that the immediate availability of its major metals is in doubt and the maintenance or expansion of its long term facilities is in danger. Minerals in this category include: bauxite, cobalt, chromite, gold, platinum and tin.

## 2.3 LOCATION RISKS

These risks attach to the supply of a mineral by virtue of a hostile location and not with the economics of production from that location; the latter is a matter for the marketplace. The two ingredients are the presence or absence of hostile borders and the necessity to use transport routes which could be easily disrupted by hostile acts or natural disasters.

Typical of the location risks attaching to a mineral are its production in either one single or very few locations worldwide. Gold, with over 60 per cent produced in South Africa and cobalt, over 50 per cent from Zaire, are examples. Where within a country production is further restricted to only one or two locations the risk is measurably increased.

A history of hostilities between neighbouring countries further increases the location risk.

Since 1980 the inter-African risk has all but dissolved with the exception of the border terrorist activities against South Africa. South America's problems are largely confined to a minor territorial dispute between Argentina and Chile and a long standing one between Guyana and Venezuela. The Canadian, USA, Mexican alliance appears set fair in spite of Mexico's financial problems and Europe's East-West demarcation is not seriously in dispute. Even in the Caribbean there are no border problems of world significance but South East Asia once more presents a real threat.

Commercial problems of location attach to the high mountain regions of the Andes (Bolivia's tin, Chile's copper and molybdenum), the harsh climatic regions of the Arctic (Canada's oil, uranium and gold in part, the Soviet Union's platinum, nickel, oil and gas) and the long road and rail routes connecting mines and ports in Central Africa, South Africa and parts of Australia. In some cases these spill over into military significance. Certain lanes are recognised as potential choke-points and have been the subject of close attention by the world's naval powers. They include the Cape of Good Hope, the Suez/Red Sea route, the Indian Ocean, the Caribbean and much of the eastern Pacific area.

### 2.3(a) Hostile Borders

In general terms this consideration in minerals supply has been assuming a reduced role with the conclusion of the sweeping-up of the last major colonial problems of Africa—South Africa excepted—in the 1975-79 period. This coincided with world economic downturn and a corresponding change of attitude by the Superpowers. Even the Iraq-Iran war which began in 1979 and dragged on into 1983, with its attendent loss of oil exports and dangerous proximity to the Soviet Union made little real impact on the minerals world. The assessment of hostile borders recognises the following groupings, which have a strong political undertone:

—Europe: inherent stability in the Western bloc, capable of accommodating some internal communist influence, e.g. Italy and France, but having a common border with the Soviet Bloc

—Africa south of the Sahara: no obvious international disputes other than the presence of Cuban troops in Angola, which is internally confused, the Namibian administration dispute and the long term unacceptability of a white minority government in South Africa

—South America: a long standing consensus of right-wing politics and what border disputes exist are of a minor nature

—Central America: serious political problems but few involving border disputes. The presence of Soviet dominated Cuba is seen as a serious threat by the USA

—North America: a maintained alliance between Canada, USA and Mexico

—The Middle East: potentially very dangerous but less so imminently with the apparent acceptance of the revolutionary regime in Iran. Fears for the long term stability of Saudi Arabia and the settlement of the Israel/PLO problem are, however, major factors

—South East Asia: a very important area in minerals, including the production of tin, copper, nickel, oil, chromite, gold, having a history of communist influence on the Malaysian Peninsular mainland, in Cambodia, Vietnam, Indonesia and the Philippines.

**TABLE 18: MINERALS AND COUNTRIES CONSIDERED OF ABOVE AVERAGE RISK FROM HOSTILE BORDER PROBLEMS**

| *Mineral* | *Country* | *Percentage of Western World Output* |
|---|---|---|
| Bauxite | Guyana, Indonesia, Malaysia | 5 |
| Cobalt | Zaire, Botswana | 58 |
| Copper | Zaire, Philippines, Namibia, South Africa | 18 |
| Chromite | South Africa | 58 |
| Gold | South Africa | 68 |
| Tin | Indonesia, Thailand | 33 |

**2.3(b) Critical Land/Sea Routes**

These have been the subject of detailed study but as a risk factor transport is governed by one overriding consideration, namely since 1960 it has been recognised that the industrialised world will continue to rely increasingly upon the developing world for supplies of raw materials. Transport links will become more vulnerable; choke points will arise; and the possibility of disruption, by natural or artificial means, will grow.

Outside the assumed safety of a producing country, commercial transport risks and natural disasters can be recreated by hostilities. For example copper from landlocked Zambia, about 600,000 tonnes per year or 10 per cent of world supplies, flows to port along one of three single track railway systems via neighbouring countries The bulk of South Africa's mass exports—coal, iron ore, manganese, chromite, asbestos—flow through two ports, Richards Bay on the Indian Ocean and Saldanha Bay on the Atlantic, which also handle a substantial tonnage of exports from neighbouring Zambia, Zimbabwe and Botswana.

The Red Sea-Suez Canal route has resumed as the major oil artery from the Middle East to Europe and presents a glaring bottleneck. The vast flow of raw materials to Japan from Australia northwards, from North and South America westwards and from Africa eastwards bunches in the Western Pacific.

These and the sealanes around Britain and continental Europe are obvious military targets. The Soviet Union's massive naval expansion since 1962, the year of the Cuban missile crisis, is well documented and has to be more than a coincident event in the growth of ocean transport and the political awareness of raw materials supply.

The interraction between land and sea based transport systems is critical to the entire risk assessment. It is ironic that several countries use the South Arican railway and port network to bring their minerals to the world market and in doing so have to run the gauntlet of two critical transport choke points, one internal, one external.

But, as noted previously, mines can only be built where ore bodies exist. The location factor is therefore practically a constant whose effect on the market is a variable. A persistently high location risk rating can deter investment but only to a degree. For example Chile could not be replaced as the producer of one sixth of the world's copper in less than five years, whatever the impetus. Long term its reserves are critical to the whole balance. Other minerals are even more vulnerable, the extreme example being gold in South Africa, probably accounting for over 80 per cent of known reserves. Under the circumstances it is understandable that the location factor, in particular the transport element, has received the closest attention of the strategic minerals specialists.

## 2.4 MINERAL RESOURCES RISKS

When the post-World War Two commodity boom was at its height during the 1960s and the annual demand for the major minerals was increasing at rates of anything between 5 and 10 per cent, it was a popular belief that world resources were in danger of total depletion. It is now obvious they are not. The USBM gives a broad resources base which includes in years of total world consumption at 1981 rates:

> bauxite (260 years), chromite (374 years), cobalt (116 years), copper (65 years), ilmenite (138 years), iron ore (308 years), manganese (186 years), molybdenum (97 years), nickel (84 years), platinum (175 years), tin (40 years)

What is more relevant strategically and economically is the world distribution, the cost of production and the development appeal.

**2.4(a) Adequacy of Reserves**

Although exceptions abound, mineral deposits tend to occur in geo-deposits such as the copper-molybdenums of the Rocky Mountains-Andes chain, the lead-zinc-silvers of the Canadian Shield, the igneous Bushveld complex of South Africa and the Middle East oil geosyncline. The discovery of these and similar occurrences has led to the development of the major mining regions of the world and to the serious international competition which exists in the marketplace. In addition to these well defined and developed deposits more sporadic reserves have been located including such sub-economic occurrences as sea bed nodules of manganese-nickel-cobalt, titanium as a residue of bauxite and aluminium as a residue of coal ash. In particular circumstances these are, or could be, exploited where local economics permit or strategic requirements dictate. Table 19 sets out the broad reserve base for world minerals.

The sub-economic deposits of a wide range of minerals could become a measurable factor if

## TABLE 19: MAJOR MINERAL RESERVES

| *Mineral* | *Reserve Base (million tonnes)* | *Years Supply at 1981 Consumption* | *Major Location of Reserves* |
|---|---|---|---|
| Antimony | 4.5 | 70 | Bolivia, South Africa, Mexico |
| Bauxite | 20,300 | 260 | Guinea, Australia, Brazil, Jamaica |
| Beryllium | N/A | N/A | Brazil |
| Bismuth | 0.1 | 30 | Australia, Bolivia, Canada |
| Cadmium | 0.7 | 39 | Canada, USA, Australia |
| Chromium | 3,350 | 374 | South Africa, Zimbabwe, Finland |
| Cobalt | 3.1 | 116 | Zaire, Zambia, Morocco |
| Columbium | 3,450 (tonnes) | 206 | Brazil, Canada, Nigeria |
| Copper | 505 | 65 | Chile, USA, Zambia, Canada, USSR |
| Diamond | 620 (m carats) | 20 | Zaire, Botswana, Australia |
| Gallium | 1 | 50 | Contained in Bauxite (see above) |
| Germanium | very large | very large | Zaire, USA |
| Gold | 37,000 (tonnes) | 30 | South Africa, USSR, USA |
| Hafnium† | 0.5 | N/A | Contained in Zircon (see below) |
| Ilmenite* | 660 | 138 | Norway, South Africa, Canda, Finland |
| Indium | 1,500 (tonnes) | 36 | Canada, USA, Peru, USSR |
| Iron Ore | 108,000 | 410 | USSR, Brazil, Australia, India |
| Lead | 165 | 48 | USA, Australia, Canada |
| Lithium | 2.22 | large | Chile, USA, Zaire, Canada |
| Manganese | 5,000 | 186 | USSR, South Africa, Australia |
| Mercury | 1.5 | 23 | Spain, USSR, Algeria |
| Molybdenum | 9.8 | 97 | USA, Chile, Canada |
| Nickel | 54 | 76 | New Caledonia, Canada, Cuba |
| Platinum Group | 37,000 (tonnes) | 176 | South Africa, USSR, Zimbabwe |
| Rare Earths | N/A | large | USA, India, Australia |
| Rhenium | 3.175 (tonnes) | 254 | USA, Chile, USSR, Peru |
| Rutile* | 131 | 337 | Brazil, Australia, India |
| Selenium | 0.1 | 84 | USA, Canada, Chile, Peru |
| Silver | 262,000 (tonnes) | 24 | USA, Canada, Mexico |
| Tantalum | 22,000 (tonnes) | 49 | Thailand, Australia, Canada, Zaire |
| Tellurium | 32,000 (tonnes) | 156 | USA, Canada, Peru |
| Tin | 10 | 40 | Malaysia, Indonesia, Thailand, China |
| Tungsten | 2.9 | 56 | China, Canada, USA, Korea |
| Vanadium | 19 | 520 | South Africa, USSR, China |
| Yttrium+ | 44,000 (tonnes) | large | India, Australia, USA |
| Zinc | 240 | 41 | Canada, USA, Australia |
| Zirconium | 43 | 83 | Australia, USA, South Africa |

** for details in Section Two see under Titanium*
*† for details in Section Two see under Zirconium*
*+ for details in Section Two see under Rare-Earth Minerals*

ever consumption suddenly outstripped available supply. This it did in the 1950-60 period when industrial demand for nickel, aluminium and titanium threatened to continue increasing at double figure annual percentage rates. Such deposits include:

- —Bauxite: clays, coal wastes and oil shales containing alumina
- —Bismuth: large, unmeasured deposits occur in the districts listed in the table above
- —Cadmium: large resources occur in the zinc-bearing coals of the USA
- —Chromium: very low grade chromite occurs in Stillwater, Montana, USA
- —Cobalt: Western USA, seabed nodules, lateritic nickel-irons
- —Copper: seabed nodules
- —Gallium: some zinc ores contain recoverable gallium
- —Germanium: coal ash and flue dusts
- —Gold: minor values occur worldwide
- —Manganese: seabed nodules
- —Nickel: seabed nodules, very large low grade deposits worldwide
- —Platinum: Stillwater Complex, USA
- —Rare Earths: spent uranium leach solutions
- —Selenium: currently uneconomic concentrations are present in coal ash
- —Tellurium: also occurs in lead ores at about one quarter that found in copper ores
- —Vanadium: also present in bauxite, crude oil, coal ash, tar sands
- —Zinc: possible by-product of some US coals and perhaps others.

The total reserve of most metals is such that few fears of long term supply inadequacy need be entertained, but the factors which do give rise to concern are:

- —the diminishing of reserves in specific areas
- —the lack of development of reserves in critical regions
- —the overdevelopment of reserves of particular minerals to the detriment of an orderly market.

Diminishing reserves in established areas affect traditional marketing and investment relationships as well as the strategic and economic importance of producing regions. Zambia has been a vital source of copper since the 1920s and latterly of cobalt. A deteriorating domestic economic climate and the development of more attractive deposits elsewhere will probably cause Zambia's output to fall from its 600,000 tonnes a year base, to be replaced progressively as a source by Chile, Philippines and elsewhere. Other examples of locally diminishing reserves are gold in South Africa and titanium sands in Australia.

Lack of development of reserves in a region often follows a loss of political or economic confidence, the discovery of more attractive resources elsewhere or a prolonged downturn in demand. Gold output worldwide fell progressively between 1970 and 1980 from 1,273 to 943 tonnes a reduction of 26 per cent. The largest reductions were registered in South Africa (32.5 per cent), USA (49 per cent), Canada (35 per cent). Copper production in Chile declined in the early 1970s following nationalisation whilst zinc output in the USA almost halved between 1970 and 1979 as demand fell and environmental pressures grew.

Over development of reserves results in long term excess capacity of which the 1970-83 casualties have been aluminium and nickel. In the 1960s Canada still produced over half the world's nickel but an over-reaction to a healthy demand profile left overcapacity estimated at almost 50 per cent by 1983. Aluminium moved into similar difficulties by development away from traditional areas.

### 2.4(b) Costs of Production

These vary greatly across the world, even within similar types of orebody, but they are more directly affected by the richness of the ore worked and the method of extraction employed. The major elements in the production cost factor are now described.

Operating Costs are clearly of major importance. As equipment costs vary only marginally across the world, the two key variables are labour and energy. In the industrialised countries the employment of labour, per capita, is usually more expensive than elsewhere. This is often offset by greater skill levels and a higher degree of mechanisation. The overuse of labour in developing countries is often defended on political and humanitarian grounds.

Power Costs have assumed critical proportions since the 1973 energy crisis and these are dealt with separately as a separate risk component below.

The grade of ore mined also has a major influence on cost. The grade in the ground is determined by geological forces of formation but varies greatly from deposit to deposit, within the same metal group, across the world. The grade which can be economically mined depends on a multitude of local and international factors and has changed progressively over time. In 1900 the world average grade for copper was over four per cent but by 1980 it had fallen to less than one per cent. The major reasons were increasing demand raising real prices and the discovery and mechanisation of very large, low grade ore bodies.

Mining methods must also be considered. Even where ore grades are favourable, factors such as the size, thickness and depth of an orebody can determine the mining method to be employed and the average mining cost.

Location is also of great importance. The low grade (30 per cent) iron ores of the northern USA are close to the Great Lakes transportation system whereas the high grade (60 per cent) iron ores of Western Australia are far inland.

The ultimate determinant of production cost is the market place since a world price basis exists for practically all metals, even where the producers effect a strong degree of control. Regional cost differences of a high order are still sustained as a result of rapid market changes, particularly as occurred in the major growth era of 1950-75. Production is also maintained in some cases as a matter of government policy, notwithstanding adverse economics, and particularly where hard currency earnings are at stake. Typical examples include the copper industry in Zambia and the tin industry in Bolivia.

### 2.4(c) Development Appeal

The decision to produce a certain mineral in a certain place, is measured as the sum of all the strategic factors but each is a variable whose weighting depends on both near and long term considerations. The most important factors include: the choice of mineral, the choice of region/country and the availability of finance.

Since the time scale from discovery of an orebody to production of metal can be up to ten years, there is no guarantee that a mineral with a healthy price-demand profile today will retain it indefinitely. Nickel, aluminium and titanium are examples of metals which lost much of their development appeal through over-development in the 1960-80 period.

With respect to the region/country selected a healthy profile includes positive labour relations and energy costs, a stable political profile, established infrastructure and an availability of local skills. On these bases South Africa scores an attractive 5 since only the underlying political risk is negative, whereas Guyana registers basement 9 because of the presence of almost every adverse factor.

Methods of financing mineral ventures have swung from the high equity-low debt formula to the reverse, particularly where the development of foreign ventures is involved. Arbitrary nationalisations, compulsory purchases and outright expropriations, which became a feature of the post-colonial 1960-70 period, have lowered the development appeal of many regions, particularly the Caribbean, South America and many parts of Africa. Political risks apart, the aids to finance which affect the choice of a location include:

—presence of a major group

—forward sales contracts

—long term borrowings

—currency and interest rate protection

—a mixture of equity and debt

—government or other official sponsorship.

The major problem associated with development appeal is that it can change quite rapidly, whilst the commitment to a minerals' venture is permanet, irreversible and capital intensive. Australia's development appeal hit a low during the 1970's Labour administration of Gough Witlam, recovered on a change of government and in the light of massive mining expansion but turned down again by 1980 as demand fell and labour problems rose. Other countries whose development appeal is in question include: Botswana (high infrastructure costs), Brazil (poor economic performance), Canada (labour and economic problems), Indonesia (tighter foreign investment rules), and Namibia (long term political uncertainty).

The adequacy of reserves is obviously a prerequisite to development but in a market which has become distorted by the effects of a high growth era ending with an abrupt recession, production costs and overall development appeal dictate the direction of investment. Developing nations have lost some of their investment attraction but the established lenders, who are also the major commodity consumers, have recognised that unless the emerging nations are encouraged to produce a high level of raw materials output they will be unable to sustain themselves. The problem would then be thrown back at the establishment: trade or aid. The mineral resource component does not consider this aspect; it confines itself to the existing mining investment attractions, balanced between the availability of resources and the ability to exploit them efficiently.

## 2.5 FINANCIAL AND ECONOMIC RISKS

Although the causes and effects will long remain a source of debate, there can be no argument that inflation and general monetary and economic instability closely followed the 1973 OPEC oil crisis. The commodity consumer boom halted, raw materials price control began to break down and the rational analysis of investment in minerals production was rendered more difficult. It was not a simple case of reassessing the situation in the light of new circumstances. The minerals industry was geared to expansion and had incurred a heavy borrowing commitment.

The succession of financial and economic traumas afflicting the 1973-82 period has been well documented but Table 20 lists the principal events of consequence to the international minerals market. The sum result of these has been the state of total uncertainty which surrounded the industry in 1983.

By 1983 it was reluctantly accepted that prospects of a sustained world economic recovery were slender. Overspending by developing nations, encouraged by over-eager bank lending, was the most serious threat to the economic order. Extreme volatility in all sectors—currency parities, interest rates, commodity prices—was inevitable.

In attempting to determine the financial risk-rating of the mineral-producing nations in the aftermath of these events the following sub components have been isolated:

—currency performance

—vulnerability to manipulation

—external indebtedness

—productivity

—foreign ownership.

### 2.5(a) Currency Performance

This is a relative, not an overall measure. To a particular country the value of its currency is really of importance only in comparison with that of its trading partners. Most traded minerals are priced as a legacy of history, in pounds sterling or, as a measure of marketability, in US dollars: the other major trading currencies being the German mark, the Japanese yen, the Swiss and the French franc.

As currencies themselves are also traded their relative levels are affected by the same forces as commodities: interest rates, the strength of economic activity and the resultant supply-demand balance. Market actions since World War Two, either in the form of arbitrary, periodic realignments or the more recent system of floating rates have determined that in the hard currency area, the yen and D-mark have strengthened considerably, the Swiss franc has remained stable and the pound and dollar suffered extreme pressure. The French franc enjoyed a long period of stability from 1962 onwards but came under pressure with the 1982 election of a Socialist government in the midst of recession.

The soft currency areas present a more complex problem. Successive devaluations against the major currencies can ease internal hardships but achieve only short term benefits. They raise the cost of imports, decrease foreign earnings and exacerbate long term debt problems. Most developing countries score badly on the currency risk factor. A recent exception is Chile which, under strict military control, has brought inflation down into single figure and reduced its foreign debt to measurable levels.

### 2.5(b) Vulnerability to Manipulation

This affects hard currency regions where daily changes in financial conditions—balance of payments, interest rates, external debts, etc—are immediately registered in parity levels. The pound sterling, US dollar and French franc have all suffered in this way. Countries with a high dependence on mineral earnings are vulnerable to movements in the commodity markets. This particularly affects single or major product

**TABLE 20: KEY EVENTS 1973-82 AND THEIR IMPACT ON THE INTERNATIONAL MINERALS MARKET**

| | |
|---|---|
| November 1973: | The OPEC Group arbitrarily quadrupled the basis price for crude oil from circa \$2.50 to circa \$10 per barrel. |
| December 1973-April 1974: | Inflationary fears gripped the commodity markets. LME copper registered an all-time high of £1,400 per tonne, over three times the 1972 average; zinc hit £938, six times its previous all-time high; tin and lead more than doubled. Coal and uranium prices reacted to the OPEC initiative. |
| May 1974-February 1975: | The base metals boom collapsed, inventories rose to record levels and a serious market recession set in, affecting all normal investment sectors. Precious metals benefitted as a barometer of fear—gold peaked at over \$200 per ounce. Metals related to the electronics industry—gallium, cadmium, germanium, indium, selenium, and silicon—all recorded new highs amidst fears of acute shortage. |
| 1975: | Massive OPEC cash surpluses led to a redistribution of international liquidity. Serious financial problems in non-oil countries contrasted with major capital spending programmes within OPEC. |
| 1976-77: | Adjustment to new oil prices promoted a mini-recovery in economic confidence. The oil price rose further. Inflation became a major problem. |
| 1978-79: strategic | A single guerrilla incursion into Southern Zaire temporarily cut off supplies of over 60 per cent of the world's cobalt, causing the free market price to increase ten fold from \$4 to \$40 per pound. This brought about a renewed awareness of the vulnerability of the West to the supply of key materials. The 'strategic' metals concept was born. Western world inflation and interest rates began to move out of control. |
| 1980-82: | A serious collapse in the world economy dropped metal prices to their lowest levels for several years, ended growth expectations and put oil prices under pressure. Non-OPEC oil output exceeded OPEC production for the first time. Inflation turned down and interest rates fell sharply. Several major nations including Poland, Mexico, Brazil and Argentina came to the edge of bankruptcy. |

economies such as Zambia, Zaire, Chile and the Philippines, all heavily dependent on copper exports.

### 2.5(c) External Indebtedness

This subject has received well deserved attention since 1980. The financial distortions which followed OPEC encouraged an increased level of lending by industrial nations, private banks and the IMF, to developing countries, particularly for the funding of natural resource projects. As these loans were sunk into infrastructure on long-term developments the interest rate burden began to mount before the means of servicing it came on stream. Then product demand fell and commodity prices collapsed.

Recent experiences have led to the development of a solvency indicator, the proportion of a country's export earnings required to service its external debts. It has thrown up some frightening numbers as shown in Table 21. Significantly, three of the countries in this table are major oil producers: Mexico, Venezuela and Ecuador.

**TABLE 21: DEBT FIGURES FOR KEY BORROWERS**

| | Total Debt* Sbn | Borrowing from Banks at end 1981 Sbn† | Debt Service in 1982+ as Percentage of Export of Goods and Services | | |
|---|---|---|---|---|---|
| | | | Total | Interest | Principal |
| Mexico | 80 | 56.9 | 129 | 37 | 92 |
| Brazil | 75 | 52.7 | 122 | 45 | 77 |
| Argentina | 37 | 24.8 | 179 | 44 | 135 |
| South Korea | 32.5 | 19.9 | 53 | 11 | 43 |
| Venezuela | 18.5 | 26.2 | 95 | 14 | 81 |
| Yugoslavia | 18 | 10.7 | 46 | 14 | 32 |
| Philippines | 15 | 10.2 | 91 | 18 | 74 |
| Chile | 15 | 10.5 | 116 | 40 | 76 |
| Ecuador | 6.6 | 4.5 | 122 | 30 | 92 |

**Latest available official estimates where available, otherwise bankers estimates*
*†Source: Bank for International Settlements*
*+Source: Morgan Guaranty*

### 2.5(d) Domestic Productivity

This indicator is used as a measure of a country's ability to compete financially, given that the market price for its product equates worldwide. It is also a measure of the degree of efficiency of its minerals industry generally in the light of its own domestic circumstances. Thus South Africa scores well at 4, in spite of its mining industry being labour intensive. This reflects a high degree of efficiency and technology within local circumstances. In contrast Bolivia emerges badly at 8 because of the scattered nature of its mining activities and a generally low level of productivity throughout its industry.

### 2.5(e) Foreign Ownerships

These are regarded as detrimental to the long-term reliability of supply of a mineral from a particular location, since control rests with outside parties. This is the reverse of a mineral being produced simply because its derived revenues are vital to a nation to the exclusion of market economics or demand. Major changes in mineral ownerships have taken place worldwide since 1960 and the process is far from complete. The most significant of these changes have been:

- —outright expropriation: Chilean copper mines, Guyana's bauxite mines
- —partial nationalisation: Zambia's copper mines
- —major takeovers involving foreign assets: oil company purchases of other natural resource groups.

The effect on the supply of a mineral can be profound. RTZ Limited, one of the world's largest groups, owns copper mines in South Africa, Canada and Papua New Guinea. It controls operations in Namibia, Australia, Canada and the USA. It could retain, curtail or expand output selectively amongst all these sources. Countries which have shown a tendency to change their foreign ownership rules arbitrarily and politically score badly in this sector. They include Canada and Australia. Namibia also scores poorly because its disputed sovereignty leaves the ownerships of its established industries in extreme doubt.

With five separate sub-factors, this sector has the most important strategic weighting, representing 20 per cent of the total. In the greatly changed circumstances of the 1980s, countries with a poor rating in this sector are finding it increasingly difficult to attract investment funds. In some cases, however, it will be impossible to avoid such investment.

Several vulnerable countries produce an overriding proportion of vital strategic minerals as listed in Table 22. The loss of output from any of the countries shown in this table would affect the market immediately. Moreover if such a loss became long term or investment confidence ebbed it would result in a serious geographical shift in investment direction.

## 2.6 ENERGY AND OTHER NATURAL RESOURCES RISKS

In a Western energy market dictated purely by cost and availability the Middle East would

**TABLE 22: COUNTRIES WITH ABOVE AVERAGE RATINGS FOR FINANCIAL AND ECONOMIC RISK**

| *Country* | *Financial Risk Factor* | *Mineral and Percentage of Western World Output* |
|---|---|---|
| Bolivia | 7.2 | Tin (15) |
| Brazil | 7.4 | Iron ore (19), manganese (15), tin (4), gold (4) |
| Guyana | 7.2 | Bauxite (3) |
| Mexico | 7.2 | Silver (18), copper (4) |
| Philippines | 7.0 | Copper (5), chromite (10), cobalt (6), gold (2) |
| Zaire | 7.0 | Copper (8), cobalt (60), diamonds (30) |

produce most of the oil for international consumption, Australia, USA and South Africa the coal and uranium whilst hydro-electricity would be generated wherever possible. Until 1973 this was not far from reality since low production costs dictated price levels.

Once OPEC emerged as not only a commercial but a political force the direction changed radically. Higher oil prices encouraged higher energy costs generally. These combined with strategic uncertainty to persuade the non-OPEC world to seek energy self sufficiency. This has been substantially successful. By 1982, non-OPEC oil output exceeded OPEC production, US imports had halved from their peak levels, Mexico and the UK had established themselves in the top six of world producers and Japan had made a substantial switch away from oil.

Outside factors assisted. Within OPEC, Iran and Iraq were at war whilst world recession cut overall demand by over five per cent. Overspending by oil producers (see 2.5 Financial and Economic Risks) demanded that they maintain output. The non-oil developing nations suffered heavily in the energy price explosion and some have incurred untenable liabilities. Not only have existing minerals production facilities in some areas been rendered uneconomic almost wholly on fuel-cost considerations but future expansion has been jeopardised over a much wider area.

Three sub components have been identified as affecting the energy rating of a country and ultimately the supply of minerals, namely domestic sources, domestic production and the nature of outside supply sources. The potential of synthetic oil or chemicals derived from coal has not been ignored but their high cost and long lead-times have been recognised.

The supply of oil is critical to every serious economy and in spite of a weakening of the OPEC position, still has political connotations. Canada has an excellent energy profile (rating 1.3) since it contains major reserves of coal, oil, gas and hydro-electricity, expanding production and not net import reliance. Namibia and Jamaica are highly vulnerable. The UK, thanks to North Sea oil, has a rating of 2.3, whereas before achieving self-sufficiency the rating would have been closer to 7.0. South Africa occupies a unique position. Although without domestic natural oil sources as yet, it has expanded its coal industry into synthetic fuel production which accounts for 50 per cent of its domestic oil requirement.

### 2.6(a) Domestic Energy Sources

World consumption of water power and nuclear energy is still a fraction of the total demand for energy. Hydro power is limited by geography and is unlikely to become strategic in other than local circumstances. Nuclear power will grow, but is constrained by high capital costs and environmental risk. It is also clear that sources of uranium are widespread and abundant. Coal, gas and oil are the truly strategic fuel minerals and their reserve locations are listed in Table 23.

Long term demands present obvious imbalances on the figures based on present knowledge. Known oil reserves represent only 12 per cent of total fossil fuels yet over 36 per cent of production; coal supplies only 32 per cent of power requirements from 79 per cent of reserves. Yet even oil reserves will last 40 years on current figures and have begun to climb as new fields are discovered. If an effective method of recovering the hydrocarbon content of tar sands and oil shales could be commercialised the potential of

**TABLE 23: FOSSIL FUEL RESERVES**
**(Million tonnes of oil equivalents 1981)**

| *Region* | *Oil* | *Percentage World Share* | *Natural Gas* | *Percentage World Share* | *Coal* | *Percentage World Share* | *Percentage Total Fossil Fuel* |
|---|---|---|---|---|---|---|---|
| USA | 4,700 | 5.4 | 5,050 | 6.8 | 148,700 | 23.6 | 19.9 |
| Canada | 1,000 | 1.2 | 2,300 | 3.1 | 4,000 | 0.6 | 1.0 |
| Latin America | 11,900 | 12.5 | 4,500 | 6.1 | 3,300 | 0.5 | 2.5 |
| Western Europe | 3,400 | 3.7 | 3,900 | 5.2 | 84,700 | 13.5 | 11.6 |
| Middle East | 49,300 | 53.5 | 19,400 | 26.2 | 0 | 0 | 8.6 |
| Africa | 7,500 | 8.3 | 5,400 | 7.3 | 22,000 | 3.5 | 4.4 |
| USSR | 8,600 | 9.3 | 29,600 | 39.8 | 155,300 | 24.7 | 24.3 |
| Eastern Europe | 400 | 0.4 | 220 | 0.3 | 52,700 | 8.4 | 6.7 |
| China | 2,700 | 2.9 | 620 | 0.8 | 66,000 | 10.6 | 8.7 |
| Other Eastern Hem | 2,600 | 2.8 | 3,250 | 4.4 | 92,000* | 14.6 | 12.3 |
| World | 92,100 | 100.0 | 74,280 | 100.0 | 628,700 | 100.0 | 100.0 |

**Includes Australia, India*
*Source: British Petroleum*

Canada's reserves alone would dwarf those of the Middle East yet still be a fraction of those of the USA. Recent discoveries of all forms of fossil fuel have made it obvious that the possibility of reserves being substantially increased in spite of consumption is considerable.

The problem is a strategic one. Over half the known oil reserves lie in the Middle East geosyncline, 19 per cent in the Americas and 9 per cent in the Soviet Union. Three important power blocs: Africa, China and Eastern Europe, are critically short.

Natural gas is in many respects a substitute for oil and whilst the two are often found in association, the global distribution in gas favours the Soviet Union 40 per cent, Middle East 26 per cent, and the Americas 16 per cent. Coal is widespread commercially but the reserve base massively favours the Soviet Union 25 per cent, USA 24 per cent, and Western Europe 13 per cent. Many of the more important mineral producers lack economic sources of energy:

—Chile: copper, molybdenum

—Guyana: bauxite

—Jamaica: bauxite

—Namibia: uranium, diamonds

—Zambia: copper, cobalt

—Zimbabwe: chromite, nickel

**2.6(b) Domestic Energy Production**

Production is more immediately critical to a nation and its mineral output for three basic reasons: commercial security of supply, political security and control of cost. The Arab-Israeli conflicts which heightened in the 1970s highlighted the political dangers when major importers including Germany, the Netherlands and Japan were forced to concede to the anti-Israel lobby on pain of losing their oil supplies.

Most oil is still produced, even under the most difficult conditions, at a price well below the official level. Producing countries, particularly newer ones such as the UK, Indonesia and Mexico, take advantage of this by charging out at the posted rate—both domestically and for export—and levying a substantial tax charge. This allows a self-sufficient country to subsidise its other sectors. The non-oil minerals producers hold no such advantage.

The energy mix is also crucial. South Africa has excellent coal reserves and production capacity but no domestic oil or gas. Its political disadvantage is such that in spite of a world oil surplus having built up since 1979 it still employs restrictions in usage tantamount to rationing. It has also taken a world lead in the production of synthetic oil from coal. Australia has a similar coal/oil imbalance whilst several nations have an over-abundance of gas. The OPEC turnaround of late 1973 served as a serious stimulus to fuel production generally, coal and gas gaining a secondary benefit from the increased price of oil. How this pattern has changed is shown in Table 24.

The national approach to an energy production policy produces the widest risk-rating spread of any strategic factor. Canada scores well (2) on the basis of an established industry covering every major energy sector, no net import demand but a good export base. Jamaica and Guyana score badly on a high import demand with little immediate prospect of relief.

**2.6(c) Sources of Outside Supply**

Dependence on energy imports brings consequences of both a commercial and economic nature. The commercial aspect is simply one of arranging an import source with respect to price, delivery, life of contract and method of payment. Payment has become the most complex consideration in that it may involve the raising of hard currency, barter or switch trading and maintaining a balance of trade between countries.

Imbalances are sometimes inevitable. One half of Western Europe's annual oil imports of 500

**TABLE 24: FOSSIL FUEL PRODUCTION 1971-81**
**(Million tonnes oil equivalent)**

| *Region* | *Oil Production 1981* | *Percentage Change 1971-81* | *Gas Production 1981* | *Percentage Change 1971-81* | *Coal Production 1981* | *Percentage Change 1971-81* | *Total Fossil Production (Percentage Change 1971-81)* |
|---|---|---|---|---|---|---|---|
| USA | 482 | -9 | 499 | -11 | 406 | +28 | -1.4 |
| Canada | 74 | -4 | 68 | +17 | 23 | +44 | +9.0 |
| Mexico | 128 | +433 | 32 | +167 | 0 | 0 | +344.0 |
| Venezuela | 112 | -40 | 16 | +78 | 0 | 0 | -35.0 |
| Total Western Hemisphere | 874 | 0 | 634 | -3 | 446 | +30 | +4.5 |
| UK | 89 | – | 32 | +100 | 70 | -16 | +93.0 |
| Norway | 25 | – | 23 | | 1 | 0 | – |
| Total Western Europe | 134 | +509 | 158 | +93 | 264 | 0 | +50.2 |
| Iran | 66 | -71 | 6 | -57 | 0 | 0 | -70.1 |
| Iraq | 44 | -48 | 0 | 0 | 0 | 0 | -47.6 |
| Saudi Arabia | 492 | +119 | 13 | +550 | 0 | 0 | +122.5 |
| Total Middle East | 788 | -20 | 38 | +46 | 0 | 0 | -1.0 |
| Libya | 54 | -59 | 3 | +200 | 0 | 0 | -57.4 |
| Nigeria | 71 | -5 | 1 | | 0 | 0 | -4.0 |
| Total Africa | 233 | -17 | 25 | +525 | 67 | +52 | -1.2 |
| SE Asia | 100 | +82 | 13 | +117 | 57 | +68 | +78.9 |
| USSR | 609 | +62 | 412 | +216 | 337 | +12 | +56.1 |
| China | 101 | +173 | 12 | +200 | 394 | +44 | +61.5 |
| World | 2,890 | +16 | 1,378 | +37 | 2,007 | +23 | +22.2 |

million tonnes originates from the Middle East. A large part of Japan's raw minerals import demand is met from Canada, Australia and Africa yet its major markets for finished goods lie in Europe. As oil products are priced in dollars, hard currency problems arise and barter and switching deals grow.

A further development has been the growth of the free market in oil; a spot market has become well established in Rotterdam whilst futures markets in oil products trade in both the USA and Europe.

The commercial pricing practices of oil contracts has led to serious disagreements both within and outside OPEC. The OPEC reference price, relating to Saudi Arabian light crude with additional quality differentials, has been subjected to local manipulation to combat falling sales since 1979. Libya, Nigeria and Iran have all been suspected of dealing well below the line whilst non-OPEC producers, particularly Canada, Mexico and the UK have happily accepted the role of a non-union employee, enjoying the price and output controls of the OPEC bloc without having to subject themselves to the rules. This situation has come under increasing pressure and is likely to last only as long as any semblance of unity prevails within OPEC.

Political considerations in energy supply weigh heavily, even though muted in recession. In oil, the prime example has been the use of supply boycotts to bolster the anti-Israeli stance by the Arab nations.

In gas, the Soviet pipeline debate has proved equally perplexing. The US government strongly objected in 1982 to the continued supply of Western technology and manufactured goods to the construction of a pipeline intended to transport Siberian gas to Western Europe. The basis of objection was the probable political consequence of the Soviet Union being the source of a growing percentage of Europe's energy needs. As shown in Table 14 the Soviet share of world energy reserves, particularly gas, is very high and its desire to export not only understandable, but critical.

The transportation of energy also has political undertones. A terrorist raid on the Zimbabwe oil terminal at Beira on the Mozambique, Indian Ocean coast in December 1982 demonstrated the dangers to supply faced by landlocked nations. Hence Zambia, Zimbabwe and Zaire score badly in the energy sources rating alongside Namibia and South Africa. The UK, Indonesia and Mexico, with few political or commercial problems, score well.

The energy resources risk has changed considerably since 1973, moving from complacency, through fear, political manipulation, and acute shortage into recession and surplus. It will undoubtedly become acute again since it contains so many obvious potential problems—political, geographical and commercial. The major exporters, USSR and the OPEC bloc, are financially and therefore politically less strong than in 1973, having the greater need to sustain their economies by export sales.

## 2.7 DEPENDENCE ON FOREIGN NATURAL RESOURCES

This factor is clearly more important to the established industrial nations and the rapidly developing regions than to the lower range of underdeveloped countries. But to each individual country, the factor is measured as its dependence upon supplies not found or produced domestically. The UK does not score well at 8, since other than in energy and some iron ore and tin it is import dependent. South Africa emerges favourably with a rating of 4 because of its wide range of domestic resources, but the USA, whose own mineral output is falling and import demand growing, scores 7.

This final factor encapsulates the whole question of the strategic minerals policies of nations. Is it satisfactory to secure outside sources of supply if the economic and political conditions surrounding them are acceptable? As a matter of prudence should a sovereign nation develop at least partial supply domestically? Should it stockpile? Should it develop substitutes?

A great deal of work has been carried out on this problem in the USA, and with some interest in Europe and Japan. The US strategic stockpile is under constant review as to both content and volume. The foreign policy of the USA bends heavily towards the security of supply of raw materials from abroad. The development of even sub-economic domestic resources of critical materials is also under constant investigation (see 2.4 Mineral Resources Risks). As a major trading nation the USA has to walk a tightrope between national security and commercial reality. The USSR suffers a degree of the same problem. It is deficient in such key minerals as bauxite, tin, cobalt, nickel and fluorspar and its trading, if not its political activities, veer towards the countries of origin of those minerals.

The approach to natural resource dependency can be commercial, strategic, or a combination of both. The basis of the US strategic stockpile was an inventory equivalent to three years consumption in the event of hostilities cutting off outside supplies. The initial calculations took into account recycling, commandeering of non-commercial domestic supplies as well as the rapid development of sub-economic reserves. It resulted in the accumulation of a very large reserve of tin, whose supplies from South East Asia were jeopardised by World War Two, the Korean and the Vietnam Wars, silver, chrome, manganese and many other materials. Over a thirty year period, world conditions have changed and with them the nature and requirements of the stockpile. Tin and silver have become less critical, manganese and chromite sources less dependable and cobalt, platinum and a range of electronics-related minerals even more so.

In 1983, the UK government's plans for a £40-45 million strategic metals stockpile were made clear. Purchases were effected by Brandeis Intsel in the 1982-83 financial year and included 20,000 tonnes high-carbon charge ferrochrome, 40,000 tonnes ferromanganese and unspecified amounts of cobalt and vanadium pentoxide. The stockpile is not planned for the traditional war scenario but appears to provide a short-term hedge against a disruption of mineral supplies from Southern Africa.

Japan is to implement purchases during 1983 for its official strategic metals stockpile. The stockpile will cover only cobalt, nickel, chromium, tungsten, molybdenum, manganese and vanadium, a downward revision from the original plan which covered 13 metals. The government's energy agency aims initially to accumulate a five-day stockpile of these metals, which will supplement a private stockpile equivalent to 10 days of consumption of cobalt, nickel, chromium, molybdenum and tungsten currently being accumulated with government assistance in the form of interest rate subsidies.

France began a strategic metals stockpile in 1975 and its initial purchases, then worth FFr 250 million, have increased markedly in value. The detailed contents of the stockpile have been kept secret but include cobalt from Zaire and an estimated 40,000 tonnes of ferrochrome from Zimbabwe. The purchases have been centrally controlled by the government, acting on advice from BRGM (Bureau de Recherches Geologiques et Minieres).

The focus on financial and economic problems rather than immediate military and political threats has characterised the early 1980s to the detriment of the resource-dependency issue. It would only need a single serious disruption in a critical minerals supply region to swing attention back to the awareness of a growing danger: that the supply of most vital minerals is increasingly capable of immediate disruption with a degree of relative ease.

## 2.8 COMPARATIVE STRATEGIC RATINGS

It might be considered inevitable that the mineral-rich developed western democracies—the USA, Canada, Australia—should achieve the highest strategic ratings. Yet the controls established by the 20 sub-factors ensure that fair weighting is given to every aspect which impinges on the vulnerability of a nation in respect of key minerals.

Thus Japan, which is largely deficient in domestic minerals, scores a better than average 5.0 because of its financial record and its industrial discipline. Zaire, blessed with excellent geology, scores badly because of its shocking economic conditions. Many of the underdeveloped nations are bunched together in the 6.0—6.6 range because they share common problems: ample natural resources but a lack of capital funding, mounting debts and a market in surplus.

The entire range is fairly closely bunched between 4.30 (USA) and 7.30 (Guyana). This results from the smoothing effect of allotting a total 200 points (maximum 10 per factor x 20 factors). Thus if Chile's economy suffered a serious setback, losing it 10 points in the overall economic sectors, its rating would only fall by 0.2. Its longer term attractions of geology, reserves and industrial competitiveness would remain.

The ratings generally would only undergo a serious and rapid revision if the world economy took a decisive swing away from the period of low growth and inflation control which is characterising the 1980's. Financial constraints would recede in favour of acute supply shortages and political considerations would become less onerous. There is little to indicate that this is about to happen. Just as the economic turndown of 1978-79 was really the peaking of the consumer book of 1960-1975, the early 1980s must first soak up the previous excesses before determining their own rate of progress.

# SECTION TWO:

# THE MINERALS

# ALUMINIUM (Al)

| | |
|---|---|
| FACTOR FOR LIKELIHOOD OF SUPPLY DISRUPTION | 2.8 |
| FACTOR FOR COST OF SUCH A DISRUPTION | 8.2 |
| TOTAL STRATEGIC FACTOR | 23.0 |

*The factors for supply disruption and cost of such a disruption are assessed on a scale from 1.0 (lowest) to 10.0 (highest) risk. Multiplied together these factors give the total strategic factor rating. For components of these assessments see the table later in this section.*

Aluminium has a moderately high strategic factor due to its vital importance in all aspects of modern technology, from cans to supersonic jet aircraft. Energy requirements and long haul transport remain problems but the ore is mainly mined in politically stable countries. Substitutes for the metal and its ore are difficult to exploit and currently non-viable or inferior in quality. Aluminium is a high volume metal with particular involvement in growth industries due to its unique combination of properties.

**Physical Characteristics**

Low specific density (2.7); relatively low melting point (660$^{o}$C); strong, tough, corrosion resistant; good conductor of heat and electricity.

## OCCURRENCE AND PRODUCTION

Aluminium is present in many ores of which the most commonly exploited is bauxite containing aluminium oxide ($Al_2O_3$).

**World Reserves**

These are estimated at 22 billion tonnes (bauxite), which is sufficient for over 275 years of consumption at current levels. Reserves are concentrated in the tropics or in areas which once had hot climates. Communist countries control less than one per cent of world bauxite reserves but possess alternative ores which are exploited to a limited extent.

**Production Characteristics**

| | |
|---|---|
| *Mining Methods:* | Open cast extraction of ores typically containing 45-65 per cent bauxite |
| *Processing:* | Conversion to alumina then electrolytic reduction to aluminium |
| *Energy Requirements:* | High: at 205-215MJ/kg aluminium ingot in ratio 1:11:1:41 mining: conversion: transportation: electrolysis |
| *By-Products:* | Gallium; possibly titanium and iron |
| *Environmental Factors:* | Disposal problems of residues from alumina production; disposal of poisonous fluorine gas during electrolysis |
| *Development Lead Time:* | Up to ten years for green field projects. |

**The Major Producers**

The production figures demonstrate the concentration of the final processing stages in the developed economies of Western Europe, North America and Japan. Brazil and Australia are the two bauxite producers moving fastest towards vertical integration, aided by foreign investment capital attracted by their local power supplies and the latter's stable political base.

The six multinational integrated aluminium producers have tended to control bauxite mining operations in Third World countries but recent efforts by the host countries have been aimed towards increased financial control of their resources, either in the form of nationalisation, higher unit taxes or complex reciprocal trade agreements. During late 1981-82 savage aluminium production cutbacks were effected in the USA and in Japan in an attempt to counterbalance a world surplus. Japan with its preponderance of oil-fired smelters has been particularly hard hit with 1982 output likely to be 30 per cent down on 1981.

**Technological Developments**

Oil-fired smelters are now proving uneconomic and there is a marked move towards the use of alternative fuels where these can now be supplied at lower cost, notably hydro-electricity, coal, natural gas and occasionally nuclear power. The Japanese have developed a direct reduction process from bauxite which, it is claimed, produces aluminium at substantially lower cost than the traditional two-stage process.

## COMPONENTS OF RISK ASSESSMENT

| *Factor* | *Rating* | *Comment* |
|---|---|---|
| **Production risks** | | |
| Existing capacity | 1 | Current surplus capacity; mothballed plants could be brought bac on stream should demand warrant it. |
| Labour disputes | 2 | Most likely to occur (if at all) in the later production stages e.g during electrolytic reduction or fabrication processes largely under taken in the developed countries. |
| Violent conflict | 1 | Unlikely in most of the important supplier countries. |
| Range of primary supply sources | 5 | Reasonably broad range of bauxite producers although five countrie control 75 per cent of world output. |
| Time lags for new supplies | 8 | Supply lines and lead times are long in almost all cases. |
| **Transportation Risks** | | |
| Primary | 5 | Interruptions feasible considering the extended shipping routes involved which incorporate several choke-points. |
| Secondary | 3 | Transportation of aluminium metal and fabricated products is generally undertaken between industrialised stable countries. History of strikes at Canadian ports. |
| **Application/Use Risks** | | |
| Total Economic Impact | 10 | Large volume market with world consumption around 16 million tonnes per annum. |
| Effect on key industries | 9 | High technology base with bias towards aerospace, improved performance vehicles etc. |
| Availability of substitutes | 9 | Alternative aluminium ores are available but currently non-viable for bulk output. Alternative end-use materials often available but with cost, density or other disadvantages |
| Longer-term substitutability | 8 | Adequate long-term substitutes for aluminium unlikely. |
| **Trade Risks** | | |
| Collusive price agreements | 8 | Likelihood of the International Bauxite Association toughening up on prices and producer country revenues within the next decade, although not immediately. Host government involvement in projects initiated by multi-national corporations is growing |
| Embargoes | 1 | Unlikely given the dependence of many countries on the income derived from bauxite and alumina output. |

*These components are assessed on a scale from 1 (lowest) to 10 (highest) risk. Grouped together, they form the basis for the assessments of the factors for likelihood of supply disruption, and for cost of such disruption which appear in the first table in this section.*

## WESTERN WORLD BAUXITE RESOURCES

Total = 21.2 Billion Tonnes

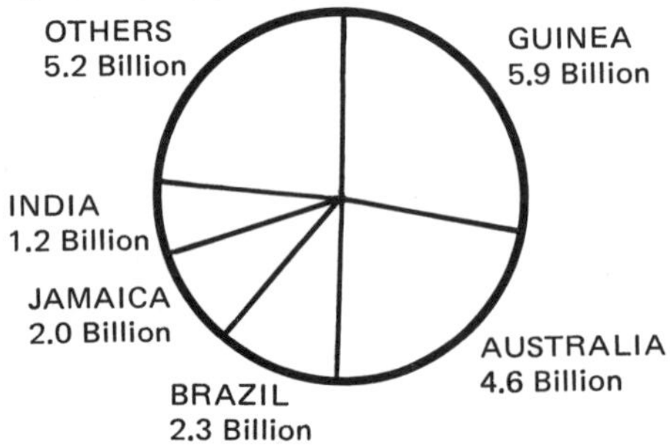


*Source: US Bureau of Mines, 1982*

## US ALUMINIUM CONSUMPTION

By Market Sector 1981

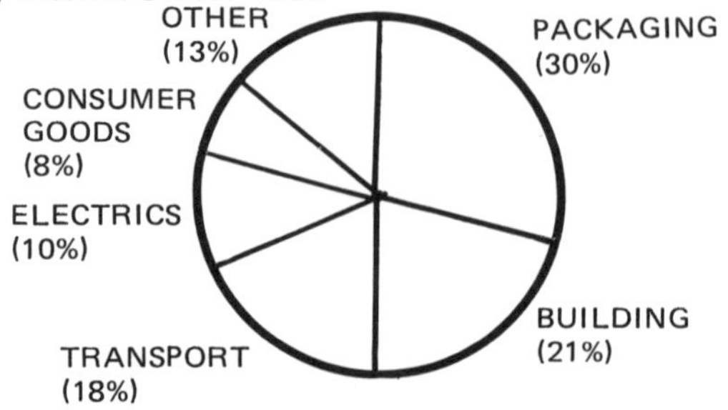


*Source: US Bureau of Mines,*
*Mineral Commodities Summaries 1982*

## *ALUMINIUM*

### The Major Producers

Bauxite

| *Countries* | *Output 1980 (Million Tonnes)* | *Percentage of Western World Output* |
|---|---|---|
| Greece | 3.3 | 4.2 |
| Yugoslavia | 3.2 | 4.0 |
| Brazil | 4.2 | 5.3 |
| Guyana | 3.1 | 3.9 |
| Jamaica | 12.1 | 15.3 |
| Surinam | 4.9 | 6.2 |
| Guinea | 11.8 | 14.9 |
| Australia | 27.2 | 34.3 |
| Others | 9.5 | 11.9 |
| WESTERN WORLD TOTAL | 79.3 | 100.0 |

(Estimated output from the Communist bloc 11.8 million tonnes.)

Alumina

| *Countries* | *Output 1981 (Thousand Tonnes)* | *Percentage of Western World Output* |
|---|---|---|
| Europe | 4,182 | 15.7 |
| Africa | 678 | 2.5 |
| Asia | 2,764 | 10.4 |
| America | 10,746 | 40.3 |
| Australasia | 7,087 | 26.6 |
| Other | 1,186 | 4.5 |
| WESTERN WORLD TOTAL | 26,643 | 100.0 |

Aluminium

| *Countries* | *Output 1981 (Thousand Tonnes)* | *Percentage of Western World Output* |
|---|---|---|
| France | 436 | 3.5 |
| W Germany | 729 | 5.8 |
| Norway | 633 | 5.1 |
| Japan | 771 | 6.2 |
| Canada | 1,125 | 9.0 |
| USA | 4,491 | 36.0 |
| Africa | 439 | 3.5 |
| Australasia | 536 | 4.3 |
| Other | 3,309 | 26.6 |
| WESTERN WORLD TOTAL | 12,469 | 100.0 |

(Estimated output from the Communist bloc 3,290,000 tonnes.)

*Source: World Bureau of Metal Statistics, May 1982.*

## TRANSPORTATION

Transportation is an important link in the bauxite-alumina-aluminium metal-fabricated aluminium products chain. Since bauxite largely occurs in the Southern Hemisphere or in Third World countries, the movement of this ore to refineries is of major interest. The bauxite producers have tended to be relatively stable with labour disruptions the main problems. Australian dock unions tend to be militant which has often caused loading delays at port, similarly in Canada with respect to bauxite-alumina imports and aluminium exports. Japan's shipping schedules are well organised but the main bauxite-alumina route from Australia to Japan is long and expensive. Rotterdam is the main clearing port in Europe. It also houses some of the LME warehouses. This port is often congested or strikebound, but in the main the transport risks for this mineral are largely those of delay rather than stoppage.

## APPLICATIONS

### Consumption Trends

The rapid growth rates of the last two decades, when the increase in aluminium demand averaged between six and eight per cent per annum, have ended. Western world consumption levelled off at around 12-12.5 million tonnes per annum during the recession but is expected to advance again by between two to four per cent per annum if and when recovery gets underway.

### The Principal Markets

Demand growth from the end-use industries is based on the need for light, strong alloys for the aerospace sector, vehicle engine components, aluminium or composite beverage cans and aluminium foil packaging.

A pie chart showing the end-use markets in the USA is given on page 38. Western Europe follows much the same pattern with slightly more emphasis on transport due to a drive towards fuel conservation. Japan uses up to 30 per cent of its aluminium for construction: in window frames, sliding partitions etc., whilst in the developing countries, such as Brazil and India, around 50 per cent is used in the electrical industry for overhead cables.

Growth in the packaging industry is expected at a rate exceeding four per cent per annum as aluminium retains the beverage can market and makes headway into the food can and laminated foil packaging sector. The electrical sector in the developing Third World countries could see 10 per cent per annum demand growth over the medium term. Use in construction will show a moderate increase due to ease of application, further increase in mobile homes, double glazing etc. Use in the automobile industry is increasing at the expense of steel and the weight of aluminium per vehicle could double during the decade.

### Recycling

The recycling process requires only five per cent of the energy needed to produce virgin metal. Problems include the separation of various aluminium types and the prevention of chlorine emissions during the process. The use of recycled aluminium scrap in 1981 is estimated to have provided 15 per cent of world aluminium consumption. In the USA recycled scrap accounted for 15 per cent of the country's aluminium consumption with almost two-thirds of the scrap coming from recycled cans.

### Substitution

Copper can be used instead of aluminium in electrical and heat conduction applications but, at a higher cost per unit, can only be justified in special circumstances, for example in underground power transmission lines. Magnesium and titanium can also be substituted for many structural and weight-saving applications but again at a higher cost and these minerals are normally limited to aerospace and satellite applications.

Steel may be used where weight considerations are not important, for example in static consumer durables. Steel composites and wood compete, both functionally and aesthetically in the building trade. Steel, glass, plastics and paper compete with aluminium in the packaging sector although aluminium's share of this market is increasing.

## INTERNATIONAL TRADE AND WORLD PRICES

### Supply Arrangements

90 per cent of all aluminium sales are effected directly by the producer, often within different levels of an integrated multinational. The big six companies (see below) publish official virgin ingot prices but, in times of glut, unofficial discounts or credit lines may be offered. Smaller producers and the Communist bloc tend to supply the free market which includes the London Metal Exchange aluminium contract started in October 1978. (The contract unit is for 25 tonnes of 99.5 per cent ingots).

Bauxite prices are loosely correlated worldwide by the level of international trade. The basis was reorganised in 1974 by the impact of the International Bauxite Association linking the raw material to the price of aluminium ingots, but this has largely lost its momentum in a falling market. Composed largely of Third World countries, the International Bauxite Association has so far proved to be an ineffective cartel.

### Companies

Six multinational companies, all vertically integrated, control between 40 and 50 per cent of world aluminium supplies. They are: Alcan (Canada), Alcoa (USA), Reynolds Metals (USA), Kaiser Aluminium (USA), Pechiney Ugine Kuhlmann (France) and Alusuisse (Switzerland). The balance of supplies is provided by smaller companies and state-owned agencies in the Communist and Third World countries.

### Prices

Free market prices over the last decade have been volatile for this effectively dollar-based commodity. During times of glut, as in 1981-83, the six major producers have been forced to abandon their list prices (fixed on the basis of costs plus a return) and adhere more closely to the free market. Due partly to the energy-intensiveness of aluminium's production process and the increase in demand in the early to mid 1970s, free market aluminium prices tripled over the decade but this performance is unlikely to be repeated in the absence of a purely speculative boom.

**Aluminium Ingots (99.5%) Prices—Annual Averages, Highs and Lows (Free Market)**

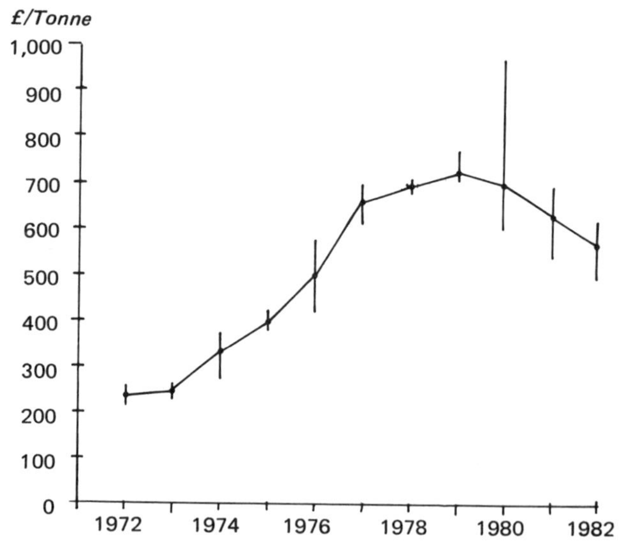

# ANTIMONY (Sb)

| | |
|---|---|
| **FACTOR FOR LIKELIHOOD OF SUPPLY DISRUPTION** | **1.0** |
| **FACTOR FOR COST OF SUCH A DISRUPTION** | **3.2** |
| **TOTAL STRATEGIC FACTOR** | **3.2** |

*The factors for supply disruption and cost of such a disruption are assessed on a scale from 1.0 (lowest) to 10.0 (highest) risk. Multiplied together these factors give the total strategic factor rating. For components of these assessments see the table later in this section.*

Antimony has a relatively low strategic factor due to its moderate importance in military applications and in flame-proofing paints and batteries. Reserves are plentiful and antimony has retained its competitive edge in many applications due to its traditional low cost and ample supply. Substitutes are available and viable for many of its uses. It is, however, largely a by-product metal, leading to a measure of supply uncertainty.

### Physical Characteristics

Silvery-white, hard, brittle metal; melting point of 630°C, suitably elevated for use in flame proofing; forms a wide series of alloys with useful properties.

## OCCURRENCE AND PRODUCTION

Principal antimony ore is stibnite $Sb_2S_3$ but lower-grade oxide ores are also worked.

### World Reserves

These are estimated at 4.5 million tonnes antimony content, sufficient for over 65 years of consumption at current levels. Reserves are concentrated in mainland China with much of the remainder in South Africa and the Comecon bloc.

### Production Characteristics

| | |
|---|---|
| *Mining Methods:* | Underground or open-cast depending on ore type |
| *Processing:* | Stibnite ore/by-product concentrates are roasted to the oxide, then reduced to the metal |
| *Energy Requirements:* | Low-moderate |
| *Co-Products:* | Mainly produced as by-product of silver/lead or lead/zinc ores |
| *Environmental Factors:* | Problems minimised by strict control of emissions and effluents at processing plants |
| *Development Lead Time:* | Eight to ten years for by-product projects, less for small primary product mines. |

### The Major Producers

| *Countries* | *1981 (Tonnes)* | *Percentage of Western World Output* |
|---|---|---|
| Yugoslavia | 1,455 | 3.8 |
| South Africa | 9,747 | 25.4 |
| Thailand | 1,750 | 4.6 |
| Turkey | 835 | 2.2 |
| Canada | 1,670 | 4.4 |
| Bolivia | 15,296 | 39.8 |
| Mexico | 1,800 | 4.7 |
| Australia | 1,130 | 2.9 |
| Others | 4,703 | 12.2 |
| WESTERN WORLD TOTAL | 38,386 | 100.0 |

(Communist bloc production is currently estimated at 16,100 tonnes and production in China 9,000 tonnes per annum.)

*Source: World Bureau of Metal Statistics, August 1982*

A handful of large companies dominate mined output in South Africa and the Americas whilst in many of the Asian countries, small mines are worked on an intermittent basis by independent groups. Producers in Bolivia, Mexico and South Africa have traditionally shipped their concentrates to the USA, UK, Europe and Japan for refining but they are beginning to follow the example of China and Yugoslavia in processing their own output. Bolivian output has been cut sharply over the last 18 months due to voluntary controls in the face of a weak, glutted market.

### Technological Developments

Stable production technology.

## COMPONENTS OF RISK ASSESSMENT

| *Factor* | *Rating* | *Comment* |
|---|---|---|
| **Production Risks** | | |
| Existing capacity | 1 | Current surplus capacity; mothballed operations in Bolivia could be brought back onstream and production cutbacks revoked by Consolidated Murchison |
| Labour disputes | 1 | Unlikely to affect supplies |
| Violent conflict | 1 | Possibility in South Africa |
| Range of Primary supply sources | 5 | Dependent on three countries (South Africa, China and Bolivia) for over half of world supplies |
| Time lags for new supplies | 4 | Availability of recycled antimony reduces supply uncertainty. By-product output depends on lead/silver or lead/zinc production |
| **Transportation Risks** | | |
| Primary | 1 | Low tonnages; shipped as low-value concentrate or metal |
| Secondary | 1 | Converted to trioxide or alloy and then consumed largely in situ with little further transportation |
| **Application/Use Risks** | | |
| Total economic impact | 1 | Low volume market with annual world consumption around 60-65,000 tonnes |
| Effect on key industries | 3 | Employed in growth area of flame retardants, but use in automobile batteries declining |
| Availability of substitutes | 3 | Available for all applications but often less cost effective |
| Longer term substitutability | 3 | Available for all applications but antimony has the cost and flexibility to retain competitive edge |
| **Trade Risks** | | |
| Collusive Price Agreements | 1 | Embryonic Antimony Producer's Association with little power |
| Embargoes | 1 | Unlikely since it provides only a small fraction of export earnings for the countries where it is mined. |

***These components are assessed on a scale from 1 (lowest) to 10 (highest) risk. Grouped together, they form the basis for the assessments of the factors for likelihood of supply disruption, and for cost of such disruption which appear in the first table in this section.***

## WORLD ANTIMONY RESOURCES

Total = 4.5 Million Tonnes

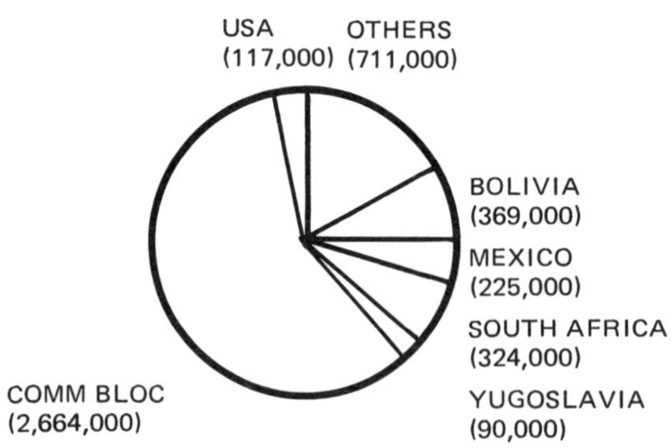


*Source: US Bureau of Mines, 1982*

## US ANTIMONY CONSUMPTION

By Market Sector 1981

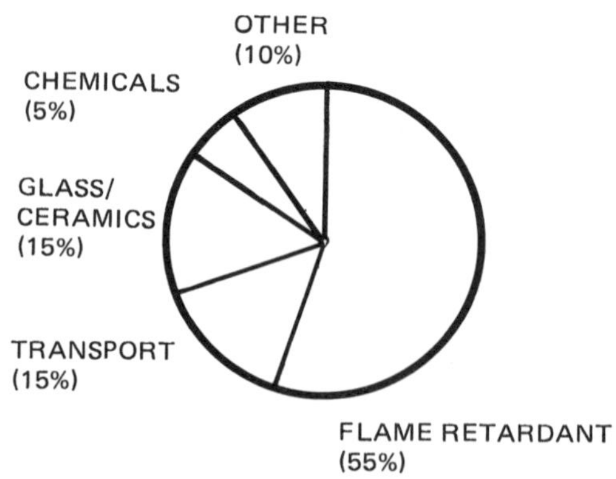


*Source: US Bureau of Mines, 1982*

## TRANSPORTATION

Antimony is a relatively low-volume metal and is generally transported as a low unit-value concentrate, not a prime target for supply disruption. Supplies from China have tended to arrive erratically, but this has been more a function of internal disruption than a result of delays in shipping.

## APPLICATIONS

### Consumption Trends

Antimony has shown a stable level of consumption over the past decade. The US Bureau of Mines predicts a 3.1 per cent a year growth rate in consumption during the period to 1990. Consumption particularly affected by the recession due to its dependence on the hard-hit transportation and construction industries.

### The Principal Markets

Demand is based on the flame-proof properties of antimony trioxide and on the ability of antimony to confer hardness and strength to the alloys in which it is present as a minor component.

The principal end uses to which antimony is put in the USA are shown in the pie-chart. These vary little throughout the industrialised West. Antimony trioxide confers flame-proof properties on a wide variety of materials including paints, plastics, rubber, adhesives, textiles and paper for use in the automobile, building, clothing, and mining industries. This sector is expected to show a higher growth rate than that envisaged for the antimony market in total. The use of lead batteries, containing four per cent antimony, is on a declining trend with the move towards smaller, lighter batteries and competition from calcium/lead maintenance free batteries. Other antimonial alloys are employed as solders, type metal and armour piercing shells. Antimony compounds are used in fireworks and smokescreen chemicals, a small but stable sector.

### Recycling

Between 45 and 50 per cent of Western consumption derives from old scrap and 85-90 per cent of this comes from antimonial lead battery plates. The remainder is recovered from type metal, bearing metal and alloy scrap. This 'above ground mine' acts to reduce supply instability caused by erratic Communist sales.

### Substitution

Selected organic compounds and aluminium oxide are widely accepted alternative materials in flame retardant systems. Titanium, zinc, chromium, tin and zirconium are suitable substitutes for antimony chemicals in paints and enamels. Calcium, strontium, selenium and cadmium are all alternatives to antimony for the alloying of lead for battery plates. Antimony has retained its markets due to its general versatility and low price.

## INTERNATIONAL TRADE AND WORLD PRICES

### Supply Arrangements

The larger Western refiners tend to have long-term supply contracts with producers, but there is also an active non-terminal free market in the powdered ore and in 99.6 per cent purity regulus ingots. The smaller producers together with the Communist bloc countries tend to be the main suppliers to the free market via specialised merchants. Over the past decade this market has aroused much speculative interest. But major fundamental influences, including China's policy of restricting supplies to maintain prices and Japan's increasing reliance on imported metal for trioxide production, have remained paramount.

### Companies

Mine production in the West is dominated by South Africa's Consolidated Murchison, National Lead of the USA and Mexico and C Tennant in Bolivia. Smaller mines in Bolivia fall within the aegis of the Banco Minero de Bolivia. Scrap is recycled by a wide range of smaller companies scattered over the USA, Europe and Japan and their operations are largely controlled by prevailing lead market conditions.

### Prices

Antimony has a volatile price history and 1982 saw both metal and trioxide prices at eight to ten year lows. This price decline in real terms has forced several producers, including Consolidated Murchison, to cut back sharply on output in an attempt to retain some measure of profitability.

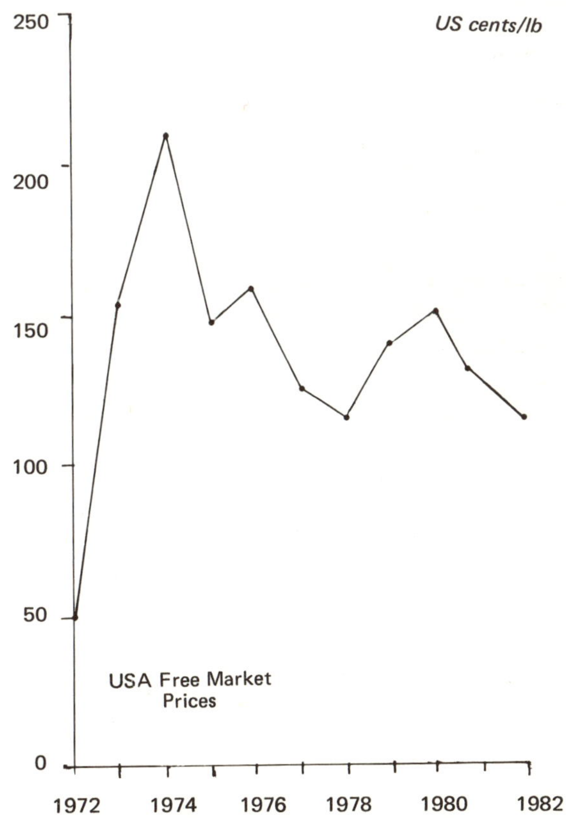


**Antimony Metal (99.5-99.6%) Prices—Annual Averages**

# BERYLLIUM (Be)

| | |
|---|---|
| **FACTOR FOR LIKELIHOOD OF SUPPLY DISRUPTION** | **2.8** |
| **FACTOR FOR COST OF SUCH A DISRUPTION** | **6.3** |
| **TOTAL STRATEGIC FACTOR** | **17.6** |

*The factors for supply disruption and cost of such a disruption are assessed on a scale from 1.0 (lowest) to 10.0 (highest) risk. Multiplied together these factors give the total strategic factor rating. For components of these assessments see the table later in this section.*

Beryllium with its consumption bias towards electronics, aerospace and nuclear applications has a moderately high strategic factor. Refined production in the Western world is dominated by two US companies and Brazil possesses the major part of Western reserves. The metal is a low volume market, partly due to health hazards and working difficulties, but its unique properties assure it a useful place in modern technology.

**Physical Characteristics**

High strength, high stiffness, low density (1.85) metal with excellent corrosion resistance; good heat dissipation ability plus useful microwave characteristics, nuclear moderating and reflecting properties; high melting point (1289°C).

## OCCURRENCE AND PRODUCTION

Beryllium forms a constituent of 40-50 minerals of which beryl ($3BeO: Al_2O_3: 6SiO_2$) and bertrandite ($Be_4Si_2O_7(OH)_2$) are viably exploited.

**World Reserves**

These have so far been inadequately delineated although Brazil is thought to contain the major part of beryl reserves. Resources are found throughout the Americas and Africa and the USA has bertrandite reserves containing 25,000 tonnes of beryllium. Currently known reserves are considered ample to meet likely future demand.

**Production Characteristics**

*Mining Methods:* Mechanised open-cast mining on outcrop deposits

*Processing:* Beryl converted to beryllium oxide and then to the metallic form by chemical means via the fluoride. Powder technology used in metal fabrication

*Energy Requirements:* High due to difficulty of reduction to the metal

*Co-Products:* Usually feldspar, mica and tantalum ores

*Environmental Factors:* Dust and fumes cause beryliosis and the metal is thought to be carcinogenic. Health hazards can be controlled by suitable standards of industrial hygiene

*Development Lead Time:* Generally under five years due to relatively small scale of extraction.

**The Major Producers**

| *Countries* | *1981 Mined Output (tons beryl)* | *Percentage of Western World Output* |
|---|---|---|
| USA (estimate) | 2,000 | 68.9 |
| Brazil | 600 | 20.7 |
| Rwanda | 100 | 3.4 |
| South Africa | 110 | 3.8 |
| Others | 93 | 3.2 |
| WESTERN WORLD TOTAL | 2,903 | 100.0 |

(Communist bloc output estimated at 80-100 tons metal a year.)

*Source: US Dept. of Interior, Minerals Yearbook 1981.*

Refined production in the Western world is concentrated in two plants in the USA owned by two companies. There are new potential sources of beryllium in France and Canada. The metal is rarely used in pure form but is usually converted to the copper/four per cent beryllium master alloy (Be 4%/Cu) or to a ceramic for further use. The conversion improves ease of handling and reduces health risks. Communist production figures are given as estimates only, with Communist net exports totalling approximately half annual output.

## COMPONENTS OF RISK ASSESSMENT

| *Factor* | *Rating* | *Comment* |
|---|---|---|
| **Production Risks** | | |
| Existing capacity | 5 | Able to cope with current demand |
| Labour disputes | 2 | Unlikely in the USA, possible at the southern African mine operations |
| Violent conflict | 1 | Unlikely to affect output of USA, the chief refined metal supplier |
| Range of primary supply sources | 6 | Mined supply sources limited to USA, Zimbabwe, Southern Africa, Brazil and the USSR. Refined output concentrated in USA and Japan |
| Time lags for new supplies | 6 | Up to five years for open cast mines but refining plants are sophisticated and need a long lead time |
| **Transportation Risks** | | |
| Primary | 3 | High value commodity, generally transported by ship in the beryl state |
| Secondary | 1 | Disruptions in deliveries of higher value-added beryllium metals and compounds would be costly but rarely occur |
| **Application/Use Risks** | | |
| Total economic impact | 1 | Low volume market of under 600 tonnes a year beryllium metal |
| Effect on key industries | 7 | Consumption bias towards electronics, aerospace and nuclear applications. Growth possibilities in defence sector |
| Availability of substitutes | 9 | Substitutes are available but tend to be inferior, particularly in microwave and nuclear applications |
| Longer term substitutability | 9 | Likely to retain its markets due to low density and excellent physical properties |
| **Trade Risks** | | |
| Collusive price agreements | 5 | Feasible considering the small handful of producers at mining and refining stage. However several producers depend too heavily on beryllium sales to risk losing the market by charging too high prices |
| Embargoes | 2 | Unlikely except as part of a ban on high technology exports to Communist bloc or Third World countries |

*These components are assessed on a scale from 1 (lowest) to 10 (highest) risk. Grouped together, they form the basis for the assessments of the factors for likelihood of supply disruption, and for cost of such disruption which appear in the first table in this section.*

## US BERYLLIUM CONSUMPTION

By Market Sector 1981

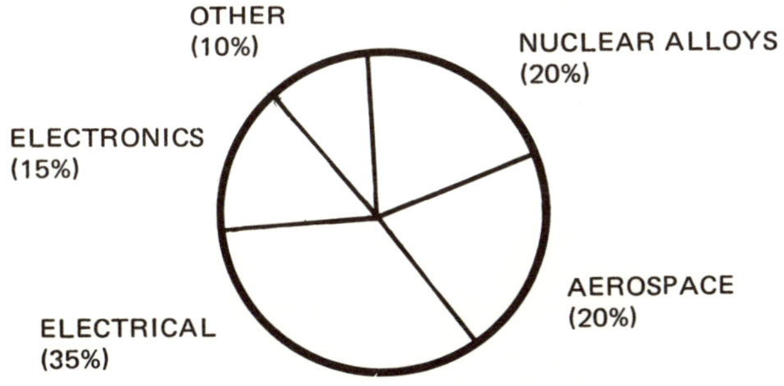


*Source: US Bureau of Mines, 1982*

### Technological Developments

Brush Wellman is currently modifying processing plant to treat ores grading down to seven per cent beryllium oxide instead of the current ten per cent limit.

## TRANSPORTATION

Hand-sorted beryl, exported from the main mining countries to the USA, tends to be transported by ship with relatively large distances involved. Beryllium metal, alloys and compounds are exported from the USA to Western Europe and particularly to Japan. Disruptions would be more costly to these higher value-added forms, but are considered to have a lower incidence of occurrence.

## APPLICATIONS

### Consumption Trends

About 75 per cent of beryllium is consumed as the Be 4%/Cu alloy, with 15 per cent as the ceramic oxide and 10 per cent in other alloys. Consumption has been on a moderate uptrend over the past five years reflecting increased defence and aerospace spending plus the electronics boom. The USBM see a 0.3 per cent a year increase in US consumption until 1990 from a 1979 base of 303 tonnes.

### The Principal Markets

Beryllium compounds find useful application as nuclear fuel containers and as a neutron moderator. Growth in this sector will be limited by a reduction in the growth expectations of the Western world's energy consumption plus environmentalist pressure against nuclear energy. The use of beryllium alloys for precision components in civil and military aircraft, however, could treble over the next two decades, with increased NATO emphasis on preparedness in the air. Electrical applications employ copper strengthened with four per cent beryllium to provide reliable springs and contacts. The major growth area is for beryllium-based ceramics as heat-sinks and housings in electronic circuitry. Beryllium oxides also provide strong, micro-wave permeable radomes and windows for military equipment and installations. A small stable sector is provided by the use of beryllium in alloys for plastics moulds and heavy-duty machine tools.

### Recycling

Old beryllium scrap is not recycled due to the technical difficulties and cost of sorting and processing to give virgin material of a suitable standard. New scrap is usually recycled in situ by the companies that generate it.

### Substitution

Steel, titanium or graphite composites may be substituted for beryllium metal in some alloys and phosphor-bronze for beryllium/copper alloys but with substantial loss of performance. No substitutes are available for beryllium oxide for microwave-permeable applications. Substitutes for beryllium ceramics in electronic applications include aluminium-based ceramics but they tend to give inferior thermal dissipation properties.

## INTERNATIONAL TRADE AND WORLD PRICES

### Supply Arrangements

Beryllium ore, alloys and compounds are mainly traded on a producer price basis with minimal merchant activity. It is traded as beryllium ore with prices quoted per unit of beryllium oxide contained, or as the Be 4%/Cu alloy with a price quoted as an aggregate of contained copper and beryllium prices. Pure metal is rarely traded due to its carcinogenicity. The USA regards beryllium to be of strategic importance with a stockpile goal slightly greater than their current inventory of beryl ore, Be 4%/Cu alloy and pure beryllium metal.

### Companies

Brush Wellman Inc of the USA is the only Western producer of beryllium metal whilst Cabot Berylco stopped production of the metal in 1979. Both companies remain integrated producers of beryllium alloys. They tend to control prices for their products on a sensible and ordered basis, despite their virtual monopoly. Japan relies on oxide imported from the USA and converts it domestically to the required alloy/ceramic forms.

### Prices

Beryllium metal, alloy and oxide prices have been maintained on a gentle upward incline by the restraint of the small number of producers. This controlled state of affairs is expected to continue over the next decade, with no new major producers on the horizon to break the effective monopoly.

**Beryllium Metal and Ore Prices—Annual Averages**

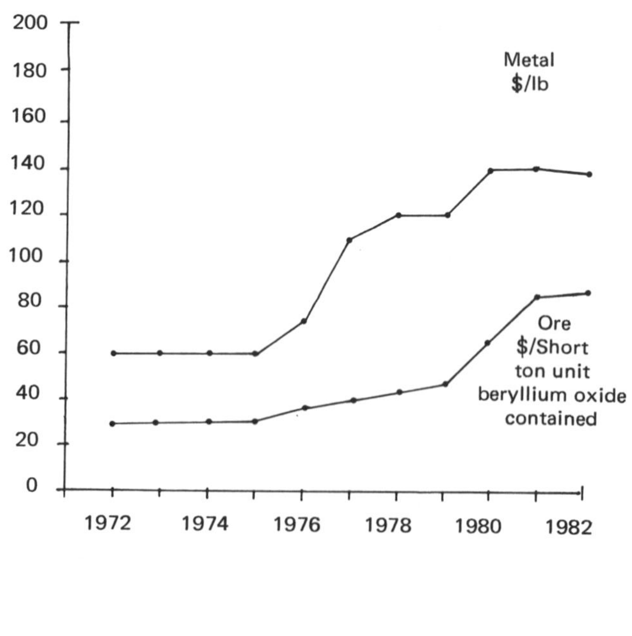

# BISMUTH (Bi)

| | |
|---|---|
| **FACTOR FOR LIKELIHOOD OF SUPPLY DISRUPTION** | **1.7** |
| **FACTOR FOR COST OF SUCH A DISRUPTION** | **2.2** |
| **TOTAL STRATEGIC FACTOR** | **3.7** |

*The factors for supply disruption and cost of such a disruption are assessed on a scale from 1.0 (lowest) to 10.0 (highest) risk. Multiplied together these factors give the total strategic factor rating. For components of these assessments see the table later in this section.*

Bismuth has a low strategic factor since supply disruption or manipulation is unlikely despite its status as a byproduct of copper and lead. Moreover it can only be considered essential in very few applications. Convenience and cost advantages maintain its market shares in the pharmaceutical, industrial chemicals and fusible alloys fields but little growth in consumption is expected.

**Physical Characteristics**

Dense, soft, silvery white metal tinged with red. Relatively low melting point (271$^{o}$C), low thermal and electrical conductivities. Recent medical research relates encephalopathy to bismuth consumption.

## OCCURRENCE AND PRODUCTION

Bismite ($Bi_2O_3$) and bismuthinite ($Bi_2S_3$), commonly found in base metal ores (lead, zinc, copper, molybdenum), are the main viable sources of bismuth.

**World Reserves**

These are estimated at 222m lbs bismuth content, sufficient for 30 years of consumption at current levels. Land based resources are huge and well scattered throughout all the continents except Africa. Bismuth is also present in concentrations of up to 24 ppm in oceanic manganese nodules.

Bolivia, however, is the only place where bismuth is found in sufficient concentrations to justify mining for bismuth alone and in Korea it is a product of tungsten recovery. The Communist bloc, notably China, possesses almost 15 per cent of known reserves.

**Production Characteristics**

| | |
|---|---|
| *Mining Methods:* | Depends on parent metal; lead, copper, tungsten, etc |
| *Processing:* | Recovered from anodic slimes of copper extractive process and in pyrometallurgical extraction from lead bullion |
| *Energy Requirements:* | Largely irrelevant due to by product status and necessity of removing bismuth as an impurity from lead |
| *By-products:* | Lead, tin, copper, molybdenum, tungsten |
| *Environmental Factors:* | None unique to bismuth |
| *Development Lead Time:* | Depends on parent metal. |

**The Major Producers**

| *Countries* | *1981 Tonnes (Bi content)* | *Percentage of Western World Output* |
|---|---|---|
| USA* | 225 | 7.3 |
| Australia | 905 | 29.3 |
| Canada | 180 | 5.8 |
| Japan | 315 | 10.2 |
| Korea | 135 | 4.4 |
| Mexico | 725 | 23.4 |
| Peru | 545 | 17.6 |
| Others | 63 | 2.0 |
| WESTERN WORLD TOTAL | 3,093 | 100.0 |

(Communist bloc output is currently estimated at 480 tonnes a year.)

*Figure withheld by USBM, estimated from data for imports and consumption.

*Source: USBM, Mineral Commodity Summaries, 1982.*

Mined output dominated by the major lead and copper producers such as the USA, Mexico, Peru, Australia, Canada and Japan. The Missouri lead belt area of the USA contains no known bismuth reserves so US bismuth output is likely to decline with the production fall from lead mines in the western states of the USA. The Communist bloc produces around 15 per cent of total world output, mainly from China.

## COMPONENTS OF RISK ASSESSMENT

| *Factor* | *Rating* | *Comment* |
|---|---|---|
| **Production Risks** | | |
| Existing capacity | 2 | Current excess capacity both of bismuth and of its main parent metals—copper and lead |
| Labour disputes | 4 | Possible considering labour histories in Canada, Bolivia, Australia etc but due to by-product status, any hiatus in production would take time to feed through to the consumer and allow alternative arrangements to be made |
| Violent conflict | 1 | Unlikely |
| Range of primary supply sources | 2 | Producers are spread throughout all continents, except Africa |
| Time lags for new supplies | 2 | Stocks usually maintained at sufficiently high level to enable demand peaks to be met short term |
| **Transportation Risks** | | |
| Primary | 1 | Transported as impurity in lead bullion or electrode slimes from copper processing. These forms are sufficiently dilute to prevent supply interruptions if one route is blocked |
| Secondary | 1 | Bismuth metal has few transportation problems |
| **Application/Use Risks** | | |
| Total economic impact | 2 | Total market volume of approximately nine million lbs a year |
| Effect on key industries | 3 | Slight |
| Availability of substitutes | 2 | Substitutes available in all applications, inconvenience and cost would pose minor problems |
| Longer term substitutability | 2 | Possible in all, except its limited electronic applications |
| **Trade Risks** | | |
| Collusive price agreements | 1 | The free market is sufficiently large to hinder such agreements and for the size and the value of the market it would not be worthwhile |
| Embargoes | 1 | Unlikely |

*These components are assessed on a scale from 1 (lowest) to 10 (highest) risk. Grouped together, they form the basis for the assessments of the factors for likelihood of supply disruption, and for cost of such disruption which appear in the first table in this section.*

## WESTERN WORLD BISMUTH RESERVES

Total = 206 Million lbs Bismuth content

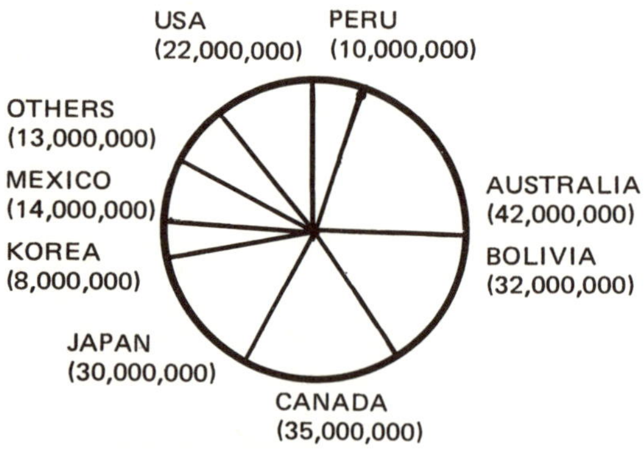


*Source: US Bureau of Mines, 1982*

## US BISMUTH CONSUMPTION

By Market Sector 1981

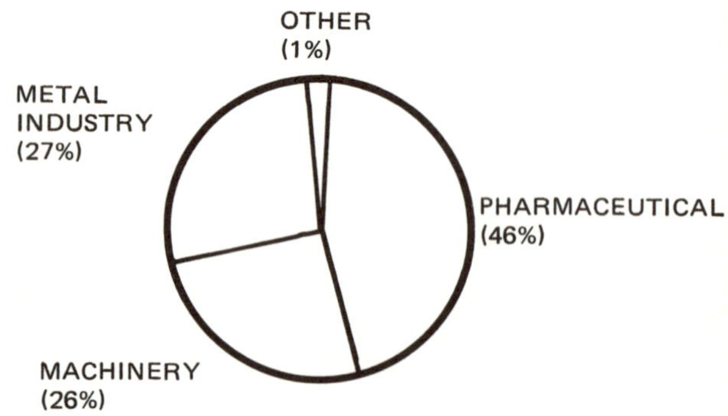


*Source: US Bureau of Mines, 1982*

### Technological Developments

Largely stable production technology. Possibility of realising new supply sources from certain types of coal.

## TRANSPORTATION

Not a major link in the chain. Mainly transported in intermediate metallurgical products such as lead and copper concentrates and then extracted to be used relatively locally.

## APPLICATIONS

### Consumption Trends

Demand for bismuth has never experienced spectacular growth and it is not expected to start showing such characteristics now. Demand over the next decade is expected to grow by 1.5-2.0 per cent a year, displaying a decreasing demand per unit of gross national product ratio in most Western countries. The USBM expect US consumption of bismuth in 1982 at 2.4m lbs.

### The Principal Markets

Bismuth has a broadly based market with around half of all consumption in the pharmaceutical and chemical industries. This sector has a strong regional bias with a noted popularity for bismuth-based antacid stomach preparations in France. The luminescent quality of bismuth-containing cosmetics has made them appreciated in Western Europe. Bismuth is a component of several industrial catalysts, including that for acrylonitrile production, but there is a constant drive to reduce the bismuth content of these catalysts.

Bismuth is an important minor additive in certain primary metal industries; in cast irons to improve ductility and strength and in aluminium alloys to give free machining properties. In 1981, several major steel companies gave some thought to the idea of using bismuth instead of lead in free machining grades but it never took off. Another useful application is in low melting point alloys necessary for the fabrication of machinery components. The demand for bismuth-based temperature control devices which rely on low melting point alloys is also increasing. Other applications include the use of bismuth in rubber vulcanising, in dental alloys and in the manufacture of selected electronic devices.

### Recycling

Only a very small quantity of old bismuth scrap is recycled since the pharmaceutical, chemical and metallurgical additive sectors are dissipative.

### Substitution

Antibiotics, magnesia and alumina can replace bismuth in pharmaceutical uses, mica and fishscales in cosmetics. Iron, phosphorus and potassium are used as catalysts for the production of acrylonitrile. Tellurium and selenium are suitable additives to give free machining steels whilst plastics can replace low melting point alloys in holding devices and jigs.

## INTERNATIONAL TRADE AND WORLD PRICES

### Supply Arrangements

Australia, Bolivia, Japan, Mexico and Peru continue as the major exporters of bismuth and its source materials. The free market is limited to 25 per cent of total supply and this relatively high percentage has meant that prices have shown a great degree of volatility in the past decade, despite the relatively stable nature of many of its applications. Bismuth is generally traded as a metal of 99.97 per cent purity with a detailed impurity content. Prices are quoted in $/lb for tonne lots in a specified warehouse.

### Companies

Asarco is the only US primary bismuth producer. Other major producers include the state-owned Industrias Penoles SA de Cv, Mexico, Peko-Wallsend of Australia, Mitsui Mining & Smelting of Japan. Bismuth production statistics are difficult to acquire and in the USA, such information is regarded as proprietary. However bismuth has never been regarded as a political pawn, mainly because of its modest contribution to company balance sheets, so embargoes and cartels are unlikely.

### Prices

Bismuth free market prices dipped below $1.50/lb in 1982 as a result of the recession. This continued a downward price trend from the 1974 peaks, coinciding with the price peaks of its parent metals; copper and lead. A fundamental price recovery is unlikely, even in a buoyant economy, unless a significant new use is developed.

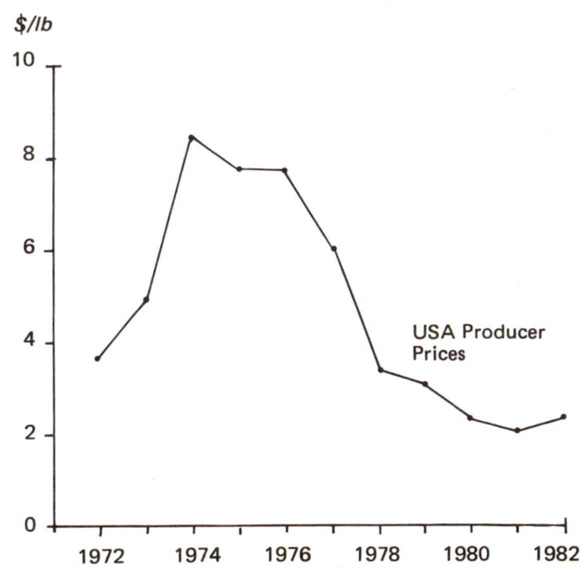


**Bismuth Prices—Annual Averages**

# CADMIUM (Cd)

| | |
|---|---|
| **FACTOR FOR LIKELIHOOD OF SUPPLY DISRUPTION** | **1.6** |
| **FACTOR FOR COST OF SUCH A DISRUPTION** | **2.8** |
| **TOTAL STRATEGIC FACTOR** | **4.5** |

*The factors for supply disruption and cost of such a disruption are assessed on a scale from 1.0 (lowest) to 10.0 (highest) risk. Multiplied together these factors give the total strategic factor rating. For components of these assessments see the table later in this section.*

Cadmium has a relatively low strategic factor due to its abundance and moderate cost. It has a broad market base, spanning pigments, plastics additives, batteries and plating, although suitable substitutes are available in most applications. Cadmium is a by-product of zinc whilst the recycling element is small, so supply disruptions are feasible. Cadmium's main disadvantage is its toxicity to mammals. World reserves are large and widespread.

**Physical Characteristics**

Silver-white, soft malleable metal with fairly high specific density (8.7) and a very low melting point of 321°C; it causes organ dysfunction in mammals.

## OCCURRENCE AND PRODUCTION

Independent minerals include the sulphides, hawleyite and greenockite but neither is sufficiently abundant to be exploited on a viable basis. All commercial sources are of cadmium sulphide (CdS) in geochemical association with zinc sulphide ores.

**World Reserves**

Estimated at 680,000 tonnes contained cadmium, sufficient for 40 years of consumption. World resources of cadmium, particularly in the zinc-bearing carboniferous-age coals, are very large, estimated at up to 50 million tonnes cadmium. The Communist bloc controls 11 per cent of world reserves with the remainder spread between the continents, with Canada and Australia holding around 16 per cent each.

**Production Characteristics**

*Mining Methods:* Depends on the zinc ore body being exploited

*Processing:* Collected from dust and sludge during the roasting of zinc concentrates and the electrolytic method of producing zinc. Electrolytically refined to give pure cadmium metal. Usually 1.5-6.0 kg cadmium recovered per ton of slab zinc

*Energy Requirements:* Largely irrelevant since the zinc extraction proceeds in any case with little extra energy needed to refine the waste dust/sludge to cadmium

*By-Products:* Zinc, silver

*Environmental Factors:* Cadmium dissipation at primary smelters and industrial operations causes ecological problems and research is currently proceeding to minimise them

*Development Lead Time:* Depends on zinc ore body. Also there is a time lag of 3-6 months before a change in mined zinc output feeds into a similar move in cadmium output since the cadmium-rich sludges/dusts are stockpiled to give sufficient volumes prior to treatment.

**The Major Producers**

| *Countries* | *1981 Metal Output (Tonnes)* | | *Percentage of Western World Output* | |
|---|---|---|---|---|
| EUROPE | | 5,334 | | 40.0 |
| Belgium | 1,065 | | 8.0 | |
| West Germany | 1,073 | | 8.1 | |
| AFRICA | | 420 | | 3.2 |
| ASIA | | 2,519 | | 18.9 |
| Japan | 2,036 | | 15.3 | |
| AMERICA | | 4,015 | | 30.2 |
| Canada | 1,293 | | 9.7 | |
| USA | 1,871 | | 14.1 | |
| AUSTRALIA | | 1,027 | | 7.7 |
| WESTERN WORLD TOTAL | | 13,315 | | 100.0 |

(Communist bloc output is estimated at 4,250 tonnes a year.)

*Source: World Bureau of Metal Statistics, June 1982*

## COMPONENTS OF RISK ASSESSMENT

| *Factor* | *Rating* | *Comment* |
|---|---|---|
| **Production Risks** | | |
| Existing capacity | 2 | Current surplus capacity which is basically a function of zinc over production |
| Labour disputes | 3 | Most likely in Canada and the Latin American countries which produce a large proportion of zinc and hence cadmium output |
| Violent conflict | 1 | Unlikely |
| Range of primary supply sources | 1 | Produced in over 25 countries in the Western world, with 24 per cent from Canada and the USA |
| Time lags for new supplies | 1 | Usually a six month lag to zinc output, but few problems in establishing new sources |
| **Transportation Risks** | | |
| Primary | 2 | Few supply lines for cadmium concentrates are particularly long or hazardous |
| Secondary | 1 | Problems unlikely |
| **Application/Use Risks** | | |
| Total economic impact | 3 | Small-to-medium volume market with total world consumption around 17-18,000 tonnes a year |
| Effect on key industries | 5 | Broad market base with important applications in plastics, batteries, pigments and plating |
| Availability of substitutes | 4 | Suitable substitutes available for most applications but cadmium's advantages include cost and convenience |
| Longer term substitutability | 3 | Total substitutability feasible in certain applications if necessary, e.g. its use as plastics stabilisers and for plating in several EEC countries and Japan has been banned by law. |
| **Trade Risks** | | |
| Collusive price agreements | 1 | Unlikely since mainly sold on free market basis, except in the USA |
| Embargoes | 1 | An unrealistic proposition |

***These components are assessed on a scale from 1 (lowest) to 10 (highest) risk. Grouped together, they form the basis for the assessments of the factors for likelihood of supply disruption, and for cost of such disruption which appear in the first table in this section.***

## WORLD CADMIUM RESERVES

Total = 680,000 Tonnes

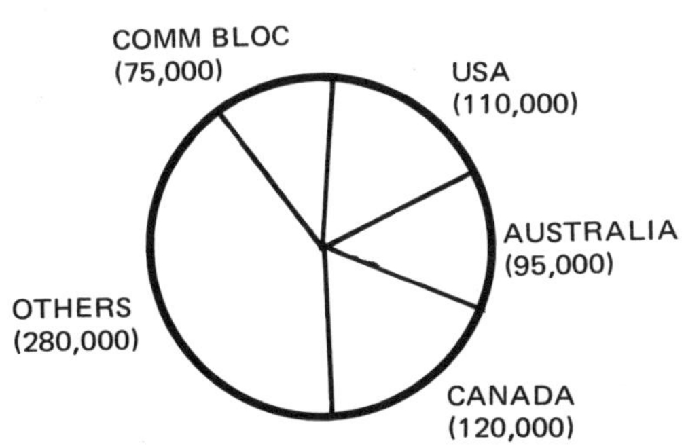


*Source: US Bureau of Mines, 1982*

## US CADMIUM CONSUMPTION

By Market Sector 1981

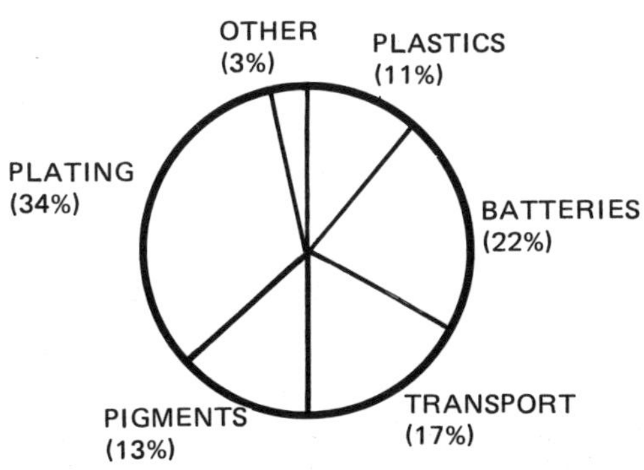


*Source: US Bureau of Mines, 1982*

Cadmium is generally refined in the country of origin of the ore, with Belgium being a major exception. Belgium's toll zinc smelters refine imported concentrates and produce cadmium as a by-product. The revenue derived from cadmium sales is merely considered as a sweetener in the zinc extraction equation and hence a true cost of production for cadmium does not exist. Thus cadmium producers are unlikely to show restraint even in a weak, over-supplied market since the cadmium-bearing sludges/dusts cannot be dumped and any price realised for cadmium sales is a bonus.

### Technological Developments

Mainly aimed towards general pollution control techniques and the recovery of cadmium from primary and secondary materials.

## TRANSPORTATION

Supply sources are numerous so transport distances for concentrates tend to be short and often confined to intra-continent and often intra-country journeys. Routes tend to follow those of zinc including Mexico and Canada to the USA and Northern Ireland; Peru to Western Europe; and Australia to Japan. In 1981, the USA had a 63 per cent import dependence for cadmium supplies. Japan traditionally exports 1,000 tonnes a year cadmium metal to Western Europe and this is an important trading route.

## APPLICATIONS

### Consumption Trends

Consumption is concentrated in Western Europe and the USA, with Japan using slightly less than Belgium. Cadmium consumption in the USA is expected to grow by 1.8 per cent a year until 1990 from a 1978 base. However it is almost impossible to predict the effect of the environmentalist lobby in the Western world which has already successfully fought against cadmium's use in plating in several countries, including Japan.

### The Principal Markets

Plating applications are growing moderately in volume terms in the USA where cadmium's cost and corrosion advantages are realised. The main growth area is in nickel/cadmium and, to a smaller extent in silver or mercury/cadmium batteries to provide standby, portable power cells with long life, good performance under a wide temperature range, low operating costs and extended shelf life. The long-awaited electric car or state subsidies for wind/solar electric power storage could herald a major boom in nickel/cadmium batteries.

Cadmium usage is also a sensitive function of the health of the plastics industry since its compounds are used both as pigments in the yellow/red range and also as stabilisers to inhibit degradation by heat and ultraviolet light.

Miscellaneous applications employ cadmium in bearing, brazing and other low melting point alloys and as the sulphide in television picture tubes and visual data display terminals. A research application that will become very important, if proven economic on a large scale, will be for cadmium compounds in solar energy cells.

### Recycling

Cadmium use is largely dissipative and recovery is only profitable from nickel-cadmium batteries plus some alloys and industrial uses.

### Substitution

Technically suitable substitutes for cadmium are available in almost all its applications but cadmium often has the advantages of cost and convenience. Coatings of zinc or vapour deposited aluminium can be used instead of cadmium for many plating applications whilst the traditional lead acid battery provides a low cost alternative to the nickel/cadmium batteries. Ferric oxides and various chromates can be used in place of cadmium pigments but often at the risks of reduced heat stability and brilliance. Organotins provide a substitute for use as plastics stabilisers.

## INTERNATIONAL TRADE AND WORLD PRICES

### Supply Arrangements

In the main, producers sell their cadmium in free market conditions except in North America where producers publish a fixed price for domestic buyers and for export. Non-integrated zinc concentrate producers receive credits from the smelters to whom they sell their material on the basis of the cadmium content. These credits depend on the state of the free market for cadmium and the level of stocks. Cadmium metal is traded in a variety of forms and purities depending on the end-use requirements. These include balls of two inch diameter for plating, sticks for chemical applications and ingots to make alloys. Prices are quoted in US$/lb of metal of 99.95 per cent purity, although a premium will be charged for the 99.99 per cent purity required in some cases.

The cadmium market has been noted for its volatility over the past two decades and has recently attracted the attention of speculators and investors.

## Companies

The major cadmium producers tend to be synonymous with the major zinc producers including Cominco of Canada, Asarco of the USA, Preussag of West Germany and others. Supply sources are well dispersed, and include the state owned operations of China, North Korea and the USSR whose erratic offerings to the West total some 500-700 tonnes a year.

## Prices

Free market cadmium prices fell in 1982 to their lowest level for over a decade. Cadmium's free market price saw a high of $5.05/lb in 1974 in the general boom following the first oil crisis. Since then, speculators and investors have enhanced its price volatility by active trading, although in real terms its value has fallen by over 50 per cent in the last 15 years. Market manipulation by the Chinese has also been an important, if sporadic, factor.

**Cadmium Prices—Annual Averages**

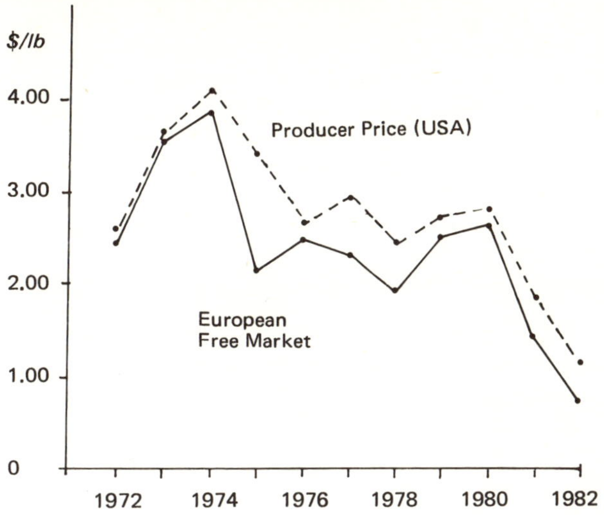

# CHROMIUM (Cr)

1983

| | |
|---|---|
| **FACTOR FOR LIKELIHOOD OF SUPPLY DISRUPTION** | **5.0** |
| **FACTOR FOR COST OF SUCH A DISRUPTION** | **8.3** |
| **TOTAL STRATEGIC FACTOR** | **41.5** |

*The factors for supply disruption and cost of such a disruption are assessed on a scale from 1.0 (lowest) to 10.0 (highest) risk. Multiplied together these factors give the total strategic factor rating. For components of these assessments see the table later in this section.*

Chromium has a high strategic factor due to its wide range of uses in the metallurgical, chemical and refractory industries, and its absolute necessity during the manufacture of stainless steels. The bulk of production is centred in South Africa, Zimbabwe, Turkey and the Communist bloc and there is a long history of supply interruptions during times of civil and political unrest.

Chromium's properties of strength, toughness, corrosion resistance and moderate cost make it one of industry's most versatile metals.

### Physical Characteristics

Shiny, steel-grey metal with specific density close to that of iron (7.19); high melting point (1,857$^{o}$C); high corrosion resistance due to the formation of a thin but coherent oxide skin; good physical and mechanical properties plus virtually zero mammalian toxicity.

## OCCURRENCE AND PRODUCTION

The main viable ore is chromite ($FeCr_2O_4$) which occurs with varying percentages of chromium, iron, aluminium and magnesium oxides and is classified into three general grades depending on the content.

### World Reserves

These are estimated at 3.7 billion tons chromite, sufficient for 375 years of consumption at current levels. Reserves are concentrated in South Africa and metallurgical grade chromite derives almost wholly from this area. There are low-grade resources in the USA, Canada and Greenland.

### Production Characteristics

*Mining Methods:* Both surface and underground methods depending on ore body shape. Mechanised underground mines in Zimbabwe and South Africa

*Processing:* (i) Metallurgical grade ore smelted with fluxes and reducing agents to ferro chromium. Pure chromium may be produced electrolytically from ferro chromium
(ii) Chemical grade chromite treated hydrometallurgically to give sodium dichromate
(iii) Refractory grade chromite, over 20 per cent aluminium oxide content, crushed with magnesite to give bricks

*Energy Requirements:* Moderately high, depending on application

*By-Products:* Aluminium, magnesium and iron are present in chromite but virtually never extracted separately from the ore

*Environmental Factors:* Disposal of dust and slag from processing plants remains a problem

*Development Lead Time:* Below three years for small, non-mechanised operations in some developing countries, but six to ten years for major underground mines.

### The Major Producers

| *Countries* | *1981 Output (Thousand Tons)* | *Percentage of Western World Output* |
|---|---|---|
| Finland | 180 | 2.9 |
| Philippines | 600 | 9.7 |
| South Africa | 3,400 | 54.8 |
| Turkey | 410 | 6.6 |
| Zimbabwe | 580 | 9.4 |
| Others | 1,030 | 16.6 |
| WESTERN WORLD TOTAL | 6,200 | 100.0 |

(Communist bloc production is currently estimated at 3,700,000 tons of which production in the USSR is 2,600,000 tons a year.)

*Source: US Bureau of Mines, 1982.*

## COMPONENTS OF RISK ASSESSMENT

| *Factor* | *Rating* | *Comment* |
|---|---|---|
| **Production Risks** | | |
| Existing capacity | 3 | Currently in surplus due to depressed state of world steel industry |
| Labour disputes | 2 | Extensive labour problems unlikely in South Africa and other main producers |
| Violent conflict | 7 | The possibility of political problems leading to violence in South Africa, Turkey and Zimbabwe |
| Range of primary supply sources | 9 | Almost 25 per cent of world output concentrated in USSR, with another 40 per cent in South Africa and Zimbabwe |
| Time lags for new supplies | 9 | Metallurgical grade ore mainly comes from the African continent and is mined in underground operations with a long lead time |
| **Transportation Risks** | | |
| Primary | 8 | Extended transport lines for chromite and ferrochromium. High bulk, high volume product which depends on cheap, efficient transport for competitiveness |
| Secondary | 2 | Little risk associated with stainless steel and alloy shipments |
| **Application/Use Risks** | | |
| Total economic impact | 7 | High volume market with world consumption around 10 million tons a year chromite |
| Effect on key industries | 7 | Wide range of uses in the metallurgical, chemical and refractory industries although of these only the metallurgical applications have healthy growth potential |
| Availability of substitutes | 10 | No adequate substitutes for stainless steel production and in other areas, substitutes tend to be inferior on grounds of performance and cost |
| Longer term substitutability | 8 | Functionally acceptable substitutes could be made more available but at great cost |
| **Trade Risks** | | |
| Collusive price agreements | 5 | The South African producers control to a great extent Western world producer powers and could break the market by undercutting other producers due to their low costs |
| Embargoes | 8 | Political action of this nature is feasible but most producers are currently too dependent on the foreign exchange generated by mineral exports |

***These components are assessed on a scale from 1 (lowest) to 10 (highest) risk. Grouped together, they form the basis for the assessments of the factors for likelihood of supply disruption, and for cost of such disruption which appear in the first table in this section.***

### WORLD CHROMITE RESERVES

Total = 3.7 Billion Tons Gross Weight

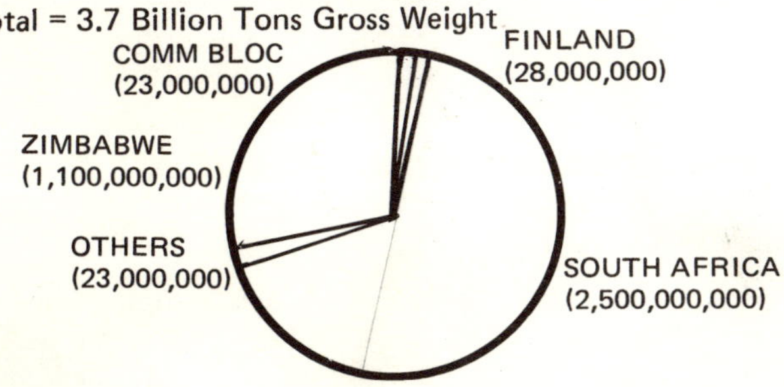


*Source: US Bureau of Mines, 1982*

### US CHROMIUM CONSUMPTION

By Market Sector 1981

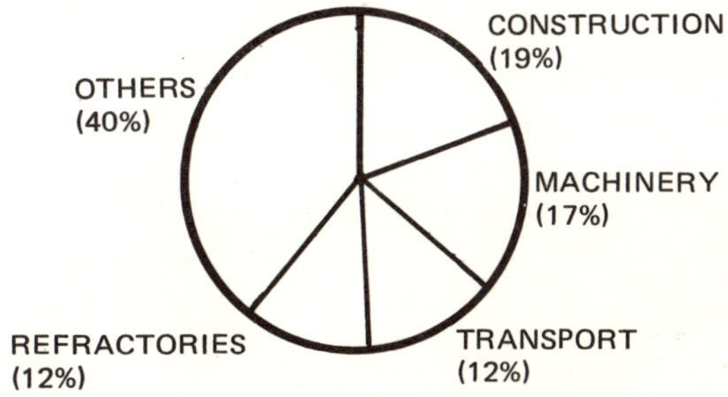


*Source: US Bureau of Mines, 1982*

Chromite is mined in over 20 countries on all five continents. South Africa is by far the largest producer followed by the USSR with Zimbabwe and Turkey vying for third place. It is a bulky ore and transport costs make it economic for processing centres to be located near the major mines. Historically the most valuable grade of ore was that containing minimum 46 per cent $CrO_2$ (chromic oxide) and a high Cr:Fe ratio and this metallurgical grade originates chiefly in southern Africa. A recent development has been the increased beneficiation of ordinary ore taking place to upgrade the chromic oxide content to over 40 per cent so it can be used to produce ferrochromium which is then added to steels. Ferrochrome production is concentrated in Zimbabwe, South Africa and Turkey plus such countries as the USA, Japan and West Germany which import beneficiated chromite. South Africa has the advantages of possessing suitable ore, indigenous coal, cheap labour and efficient transport facilities. These help consolidate the country's dominance in the ferrochromium market and allow it to move into stainless steel production.

### Technological Developments

The introduction of the argon-oxygen-decarburising vessel into stainless steel production has made the use of high-carbon, and hence lower grade, charge chromium viable and over the past decade use of high-carbon ferrochromium has increased by over 65 per cent in the USA, 35 per cent in the UK and 35 per cent in Japan. This is at the expense of the previously favoured low-carbon ferrochromium.

## TRANSPORTATION

These aspects are particularly important for the EEC, Japan and the USA. The last country is 90 per cent import dependent for chromium. All import chromium at the chromite, ferrochromium or steel product stages. Transport lines are extended while those from Zimbabwe are subject to periodic guerilla disruptions. Supplies from the USSR are erratic, despite long-term contracts, due to shipping difficulties and the problems of adequately equipping port loading facilities. With chromium being a high bulk, high volume product, transport is a vulnerable link in the supply chain through to steel products.

## APPLICATIONS

### Consumption Trends

Chromium has three main markets: the refractory, chemical and metallurgical industries, in which it is used in the volume ratio 1:1:2. The USBM estimates that chromium demand will increase at an annual rate of 3-3.5 per cent until 1990, from a 1978 base. This is, not coincidentally, the same growth rate predicted for the stainless steel industry which consumes some 70 per cent of all chromium metal/alloys.

### The Principal Markets

The metallurgical industry uses chromium to enhance the hardenability, strength, corrosion resistance and wear resistance of any alloy to which it is added. 70 per cent of all chromium metal is employed in stainless steels and 25 per cent in specialised alloy steels for the chemical, aerospace and power generating industries. In the chemical industry chromium compounds provide yellow, green and orange pigments as the dichromate, or are used for leather tanning, catalysts, corrosion inhibitors and wood treatments. Chromite refractory bricks are used in iron and steel processing, non-ferrous alloy refining, glass making and cement processing due to its corrosion and thermal shock resistance. Chromite sand is also an important moulding material in foundries.

A pie chart showing the end-use markets in the USA is given on page 55 and the same broad usage pattern is seen through the industrialised world. Use in mass transportation is likely to grow at a faster rate than use in refractories etc, due to experimentation on rustless steel car bodies but the specialist and super-alloy sectors could see a growth rate of five per cent a year albeit from a lower base.

### Recycling

The chemical and refractory industry applications are largely dissipative whilst in the metallurgical sector, chromium is one of the lower cost alloy additives and is only recycled when it is viable to recover and reuse the alloy as such. Around 10 per cent of chromium consumed is recycled in the form of old stainless steel scrap as under present conditions it is not worth extracting it from the steel alloy.

### Substitution

According to a 1978 US National Materials Advisory Board Study, about 30 per cent of chromium usage could be saved by functionally acceptable chromium-free substitutes currently available. Cost and consumer appeal, however, have so far retained chromium its markets. There is no viable substitute for chromium in the bulk of stainless steel applications but other possibilities are: nickel, zinc or cadmium for corrosion protection of iron and steel; aluminium and plastics for automotive decorative trim; nickel, cobalt, molybdenum or vanadium for alloying iron and steel; titanium for chemical processing equipment; cadmium in some yellow pigments and magnesite and zircon refractories for furnace bricks and foundry moulds.

Despite these possibilities, the USA has set up its own stockpile of chromium source materials, Japan has declared its intention of stockpiling chromium along with four other strategic metals and the UK Government bought unspecified amounts of chromium source materials in 1983.

## INTERNATIONAL TRADE AND WORLD PRICES

### Supply Arrangements

There is a very wide range, if not a bewildering number, of chromium-bearing materials which can be traded, mainly on a producer basis. The free market is small and effectively limited to inter-merchant trading of chromium metal flake. The balance of the market is distorted by the fact that the USA, which consumes around 15 per cent of world annual production, has no indigenous production. Concern over supply is two pronged: apart from the threat of supply disruptions, there is the prospect of South Africa flooding the market with cheap ferrochromium or stainless steel. A selection of the specifications traded is detailed below:

—electrolytic chromium of minimum 99 per cent purity

—chromite with variable, but specified, oxide contents from a stated country of origin since this tends to define the grade and impurities

—ferrochrome or silico-chrome master alloys with specified contents, particularly of carbon.

**Companies**

Major chromite producers include Samancor; Rand Mines and Transvaal Consolidated Land and Exploration of South Africa, Etibank (state-owned) of Turkey, Rio Tinto Mining (Zimbabwe) Ltd, and the state-owned operations of the USSR and Algeria. There are over 35 major ferrochromium producers in the world and prices tend to be set on an individually negotiated basis within the country or region. This segmentation of the market is aided by the tariff system applying to steel in many countries.

**Prices**

The dominance of several major producers has acted to minimise the importance of the free market whilst maintaining prices on an upward trend in nominal terms until the late 1970s, when prices plateaued.

**Chromium Ore Prices—Annual Averages (44% $Cr_2O_3$, Transvaal)**

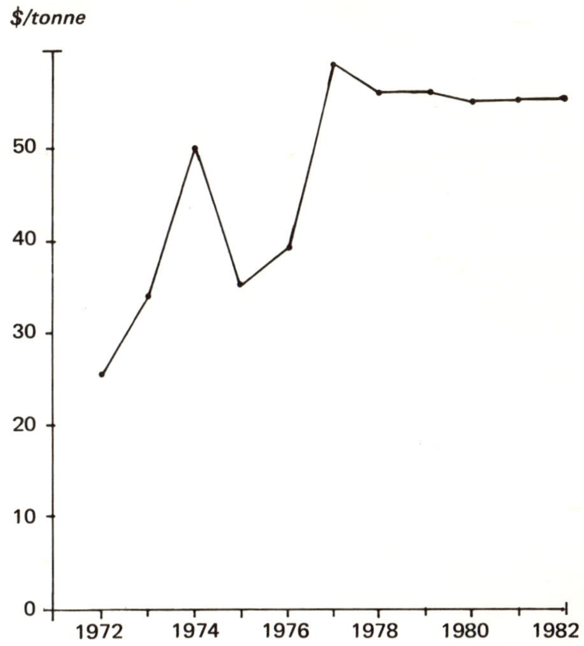

# COBALT (Co)

| | |
|---|---|
| **FACTOR FOR LIKELIHOOD OF SUPPLY DISRUPTION** | **4.9** |
| **FACTOR FOR COST OF SUCH A DISRUPTION** | **7.2** |
| **TOTAL STRATEGIC FACTOR** | **35.3** |

*The factors for supply disruption and cost of such a disruption are assessed on a scale from 1.0 (lowest) to 10.0 (highest) risk. Multiplied together these factors give the total strategic factor rating. For components of these assessments see the table later in this section.*

Cobalt has a high strategic factor due to its importance in the electrical, aerospace and machine tool sectors and fears that the supply disruptions of 1978 could be repeated. Over 60 per cent of all cobalt derives from southern Africa and it is largely a by-product of nickel or copper. Substitutes are usually of inferior quality and not suitable for cobalt's performance-sensitive markets. Viable land-based reserves are relatively limited.

### Physical Characteristics

Silvery-grey metal with a high melting point (1,495$^{o}$C); moderately high specific density (8.9) and a high Curie point (ie the temperature at which its magnetic properties are lost); hard, corrosion resistant and strong; an essential feed additive for many mammals.

## OCCURRENCE AND PRODUCTION

Generally present in copper or nickel ores as the arsenide, sulphide or oxide.

### World Reserves

These are estimated at 3.4 million tons land-based reserves, sufficient for 115 years consumption at current levels. In addition, offshore resources in the form of nodules on the seabed are immense. The nodules contain 0.35 per cent cobalt and are concentrated on the bed of the Pacific Ocean. Technical and legal barriers to exploitation are such that nodules are not expected to become an important source of the metal this century. Reserves are concentrated in Zaire and Zambia. Morocco's relatively meagre reserves of 50,000 tonnes cobalt are however of note as they grade at 1.2 per cent. This is probably the only area in which deposits are mined solely for their cobalt content.

### Production Characteristics

| | |
|---|---|
| *Mining Methods:* | Depending on the type of ore. Grades 0.25-0.45 per cent in Zaire, elsewhere below 0.15 per cent (except Morocco) |
| *Processing:* | Roasted, leached and electrolysed, followed by purifying to required specifications |
| *Energy Requirements:* | Largely irrelevant due to by-product status |
| *Co-Products:* | Copper, nickel, platinum |
| *Environmental Factors:* | Dust and sulphide emission problems due to copper and nickel output |
| *Development Lead Time:* | As for copper/nickel mines plus time to accumulate sufficient cobalt bearing concentrate for viable extraction. |

### The Major Producers

| *Countries* | *1981 Output (Tons)* | *Percentage of Western World Output (Independently Rounded)* |
|---|---|---|
| Australia | 1,700 | 6.2 |
| Botswana | 300 | 1.1 |
| Canada | 1,500 | 5.5 |
| Cuba | 1,900 | 6.9 |
| Finland | 1,355 | 4.9 |
| Morocco | 1,000 | 3.6 |
| New Caledonia | 200 | 0.7 |
| Philippines | 1,500 | 5.5 |
| Zaire | 15,000 | 54.7 |
| Zambia | 2,833 | 10.3 |
| Zimbabwe | 120 | 0.4 |
| WESTERN WORLD TOTAL | 27,408 | 100.0 |

(Production in the USSR is currently estimated at 2,250 tonnes a year.) (Independently rounded)

*Source: US Bureau of Mines, 1982, and others.*

The production figures demonstrate the dependence of the Western world on cobalt from Zambia, Zaire and Botswana. Output from the Sudbury nickel deposits in Canada are of importance particularly for Europe and the

## COMPONENTS OF RISK ASSESSMENT

| *Factor* | *Rating* | *Comment* |
|---|---|---|
| **Production Risks** | | |
| Existing capacity | 1 | Current surplus capacity both at mining and processing stages |
| Labour disputes | 3 | Mineworkers in Southern Africa have a poor strike record, as do those in the Sudbury area of Canada, which is however a less important source |
| Violent conflict | 9 | Southern Africa has never been considered a stable area and future coups must be expected |
| Range of primary supply sources | 8 | Two thirds of annual Western output derives from Zambia, Zaire, Zimbabwe and Botswana |
| Time lags for new supplies | 8 | Supply lines are long and lead times depend on nickel/copper project plans |
| **Transportation Risks** | | |
| Primary | 9 | The main producers are landlocked with poor links to ports. Extended shipping lines stretch halfway round the world in many cases |
| Secondary | 2 | Transportation of cobalt metal and compounds is relatively risk-free |
| **Application/Use Risks** | | |
| Total economic impact | 4 | Medium volume market with consumption at around 30,000 tonnes a year |
| Effect on key industries | 9 | Bias towards important growth industries of aerospace, military equipment and power generating machinery |
| Availability of substitutes | 7 | Substitutes viable for magnetic applications but often inferior on performance basis in superalloys, paints etc |
| Longer term substitutability | 7 | This would involve the development of a whole series of alternative alloys which is currently unlikely |
| **Trade Risks** | | |
| Collusive price agreements | 7 | Zairean and Zambian producers attempt to coordinate prices but are often under-cut by Canadian and USSR producers |
| Embargoes | 4 | Unlikely given the dependence of most producers on the income generated from cobalt sales to sweeten the revenue from nickel/copper operations |

***These components are assessed on a scale from 1 (lowest) to 10 (highest) risk. Grouped together, they form the basis for the assessments of the factors for likelihood of supply disruption, and for cost of such disruption which appear in the first table in this section.***

## WORLD COBALT RESERVES

Total = 3.4 Million Tons Cobalt Content

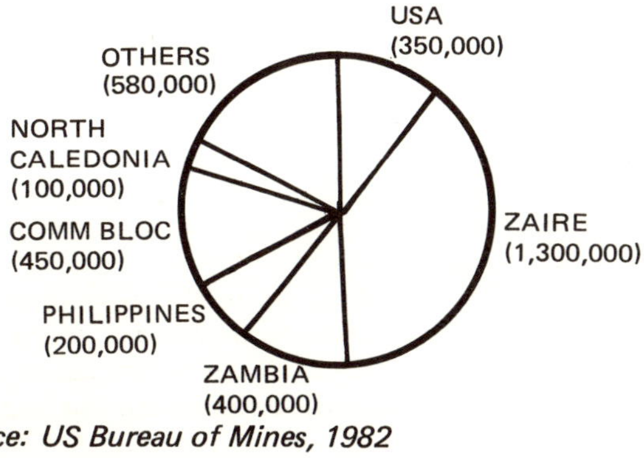


*Source: US Bureau of Mines, 1982*

## US COBALT CONSUMPTION

By Market Sector 1981

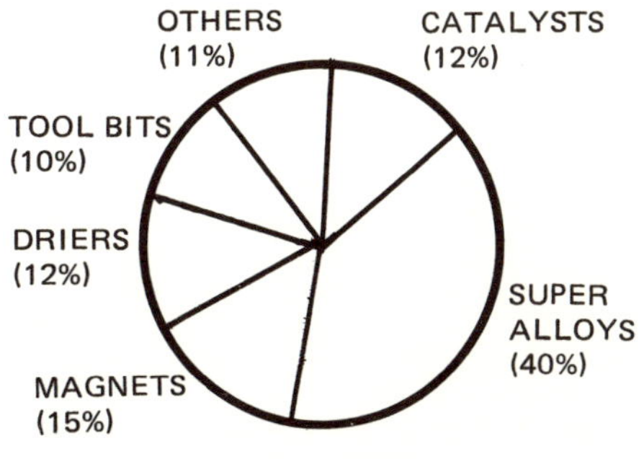


*Source: US Bureau of Mines, 1982*

USA which have no viable, indigenous reserves. The US government is currently encouraging new domestic cobalt producing ventures in Idaho (Blackbird) and California (Gasquet Mountain) but with little success, as yet. There are a number of European and Japanese cobalt refiners but they still depend on supplies from Africa. The Communist bloc currently produces approximately 14 per cent of total world output. Some analysts consider that China has the potential to become a major producer and exporter by the end of the century. There is a trend towards state control of the mining and marketing of cobalt concentrates in the African countries plus a desire to possess their own cobalt refineries.

### Technological Developments

These include the development of more efficient leaching processes for the exploitation of lateritic nickel deposits which contain a low cobalt concentration. Several major international consortia are currently exploring the possibility of deep sea mining whilst UN Law of the Sea discussions continue in the background.

## TRANSPORTATION

Transport disruptions have proved troublesome in the past since Zambia is land locked and Zaire has only limited port facilities. The Tanzania-Zambia railway has rarely functioned efficiently, due to poor maintainance compounded by frequent sabotage. Often one or more months' copper/cobalt output has been stranded along this route. During the Shabah invasion of Zaire it proved viable to airlift cobalt concentrates out for refining in Western Europe but under normal circumstances the concentrates are shipped via South African ports or from the port of Dar es Salaam.

## APPLICATIONS

### Consumption Trends

Cobalt saw high demand growth rates in the 1960s. But over the 1980s the USBM predicts a growth rate of 2.5 per cent a year with the obvious anomalous years of 1981 and 1982.

### The Principal Markets

Cobalt's most strategic application is as an integral part, from two to 65 per cent, of the super-alloys which have the strength, corrosion resistance and reliability to function at elevated temperatures in gas or high speed industrial turbine applications. Use of cobalt in magnets is declining in both total and relative terms mainly due to the price gyrations of 1978-79 and the development of alternative magnetic alloy systems. Cobalt as an additive to machine tool steels or as an abrasion resisting coating for machinery and tool tips represents a growth sector. Around 25 per cent of all cobalt is used in the non-metallic form; as a desulphurising and cracking catalyst in crude oil processing, whilst the oxide and organic compounds are used in paints and ceramics as dyes, driers and pigments. Minor, but stable uses include cobalt as a crop feed or as the isotope Co-60 for medical purposes.

### Recycling

Scrap recovery currently provides 10 per cent of cobalt consumption in the USA, mainly as metal contained in superalloys or magnet alloys. A new oxygen lancing technique has been developed to recover cobalt directly from super-alloy scrap and prevent downgrading of the scrap. Research is now being directed towards the recovery of cobalt from spent catalysts in the petroleum industry. Paints and driers are dissipative applications.

### Substitution

NIckel, molybdenum, vanadium, tungsten and chromium can replace cobalt in many superalloy applications but they are generally accompanied by some loss in effectiveness. New carbon fibre or ceramic technology could reduce the need for metallic components. Magnetic material substitutes include ferrites or alloys containing less cobalt. Copper, chromium and manganese provide alternative compounds for use in paints, nickel in catalysts and molybdenum carbide, ceramics and nickel are adequate substitutes in wear-resistant applications.

## INTERNATIONAL TRADE AND WORLD PRICES

### Supply Arrangements

Cobalt is traded both on an active merchant free market and on a long term producer/consumer contract basis. Official producer prices are published but over the past two years free market prices have undercut official levels as stocks have risen. Cobalt metal is traded in a minimum of one tonne lots of broken cathode or granules, with a price in $/lb which varies according to grade. Producer prices tend to be as high as Zaire and Zambia consider the market will stand since the actual cost of production is a nebulous figure given its by-product status. In 1981-82, total cobalt stocks reached around nine months' western world output and market conditions worsened as the GSA purchased 5.2 million lbs of cobalt at a then heavily discounted price. In 1981 producers set up a Cobalt Development Institute to try and promote consumption.

### Companies

The main cobalt refiners are Metallurgie Hoboken of Belgium, (Zairean concentrates), Outokumpu Oy of Finland (integrated), Falconbridge and Inco (Canadian concentrates but refined in Norway and Wales respectively), PUK's Metaux Speciaux (Moroccan material) and SLN of France (material from New Caledonia). Nippon Mining and Sumitomo of Japan refine concentrates from Australia and the Philippines and are expanding their interests in foreign projects in an attempt to reduce vulnerability to disruptions from one source. Russian material tends to be of slightly inferior quality and exported in unpredictable quantities.

**Prices**

Cobalt has a volatile price history with a free market peak of $43/lb in January 1979 in the wake of the guerilla incursion of Zaire's Shaba province. Prices have fallen by 90 per cent from that peak and in 1982 were at their lowest point on the free market since 1975-76. Producers maintained their list price at $25/lb during 1980 but a combination of unofficial discounts and extended credit terms plus a realisation of free market forces have pushed their prices down to competitive levels.

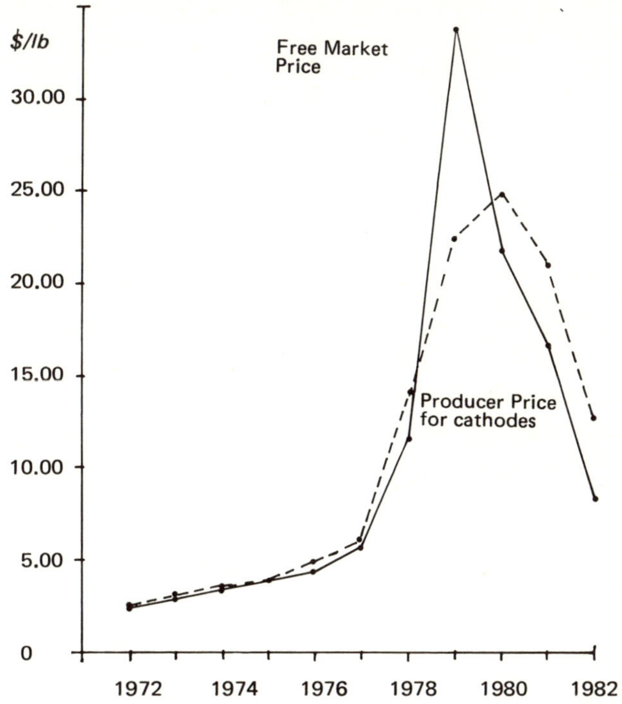


**Cobalt Metal Prices—Annual Averages and Highs/Lows for Cathodes**

# COLUMBIUM (Cb)/NIOBIUM (Nb)

| | |
|---|---|
| **FACTOR FOR LIKELIHOOD OF SUPPLY DISRUPTION** | **3.1** |
| **FACTOR FOR COST OF SUCH A DISRUPTION** | **7.2** |
| **TOTAL STRATEGIC FACTOR** | **22.3** |

*The factors for supply disruption and cost of such a disruption are assessed on a scale from 1.0 (lowest) to 10.0 (highest) risk. Multiplied together these factors give the total strategic factor rating. For components of these assessments see the table later in this section.*

Columbium has only a moderately high strategic factor although it is important in the manufacture of several specialised types of alloyed steels. It has experienced a high rate of consumption growth over the past two decades and has the potential to benefit from a resurgence in the nuclear and aerospace industries but substitutes are available in many applications. Reserves and production are concentrated in Brazil. The USA is wholly dependent on imports of columbium source materials.

### Physical Characteristics

Silver-grey metal with specific density of 8.4 and an elevated melting point of 2,415°C; corrosion resistant with useful heat and electricity transmission characteristics and a low thermal neutron capture cross-section.

## OCCURRENCE AND PRODUCTION

Columbium is chiefly extracted from the mineral pyrochlore or from mixed columbite-tantalite ores ($Cb_2O_5/Ta_2O_5$).

### World Reserves

These are estimated at 7.6 billion lbs columbium, sufficient for over 300 years of consumption at current levels. Reserves are unevenly distributed with 77 per cent in Brazil, found mainly in the form of pyrochlore. The Comecon countries possess an estimated seven per cent of world reserves, again largely as pyrochlore.

### Production Characteristics

| | |
|---|---|
| *Mining Methods:* | Mechanised open-pit/underground stoping methods for pyrochlore. Columbite mined by range of methods from pick and shovel to floating dredges |
| *Processing:* | Pyrochlore directly converted to medium grade ferrocolumbium by aluminothermic process. Columbite treated chemically to give salts, may be further purified to the metal |
| *Energy Requirements:* | Moderate |
| *By-Products:* | Tantalum from mixed ores which may occur with cassiterite deposits |
| *Environmental Factors:* | Fumes, dust and low-level radiation generated by processing, controllable by modern technology |
| *Development Lead Time:* | Dependent on mining method. |

### The Major Producers

| *Countries* | *Output 1981 (Thousand lbs Cb content)* | *Percentage of Western World Output* |
|---|---|---|
| Brazil | 32,000 | 86.9 |
| Canada | 4,200 | 11.4 |
| Nigeria | 400 | 1.1 |
| Zaire | 30 | 0.1 |
| Others | 200 | 0.5 |
| WESTERN WORLD TOTAL | 36,830 | 100.0 |

*Source: USBM, Mineral Commodity Summaries 1982.*

Brazil, Canada and Nigeria dominate world output with 90 per cent of total production between them. Both Canada and Brazil have been working well within capacity in recent years so producer prices have been dominated by cost inflation rather than by the supply/demand balance. Vertical integration has become fashionable recently with Brazil producing columbium oxide rather than exporting pyrochlore ore to the EEC and the USA for conversion. World capacity for columbium is well in excess of current needs and the costs of Brazilian production are so low as to discourage the development of alternative resources.

### Technological Developments

The development of vacuum melting in an electron beam furnace now allows 99.9 per cent pure columbium metal to be obtained.

## COMPONENTS OF RISK ASSESSMENT

| *Factor* | *Rating* | *Comment* |
|---|---|---|
| **Production Risks** | | |
| Existing capacity | 2 | Current surplus capacity and CBMM holds 300 tonnes columbium oxide stocks |
| Labour disputes | 3 | Brazil's labour record is superior to many others in South America but Canada is often a trouble spot |
| Violent conflict | 1 | Unlikely despite Brazil's well publicised financial problems. Disturbances in Nigeria/Zaire are more likely but would affect a small percentage of output only |
| Range of primary supply sources | 8 | Source concentrated in Brazil |
| Time lags for new supplies | 5 | USA and others possess vast resources which are uneconomic at current rates. Columbium-bearing tin slags could be exploited to a greater extent |
| **Transportation Risks** | | |
| Primary | 5 | Feasible considering concentration of supplies in Brazil |
| Secondary | 2 | Transported as various alloys and steel. Transport strikes relatively low |
| **Application/Use Risks** | | |
| Total economic impact | 4 | World market totals 36-37 million lbs a year contained columbium |
| Effect on key industries | 6 | High usage concentration in aerospace, oil and gas pipeline industry and chemical industry |
| Availability of substitutes | 5 | Alternatives available in all applications but at a performance or cost penalty in most cases |
| Longer term substitutability | 5 | A feasible goal |
| **Trade Risks** | | |
| Collusive price agreements | 7 | Currently occurs to a great extent |
| Embargoes | 2 | Largely irrelevant |

*These components are assessed on a scale from 1 (lowest) to 10 (highest) risk. Grouped together, they form the basis for the assessments of the factors for likelihood of supply disruption, and for cost of such disruption which appear in the first table in this section.*

## WORLD COLUMBIUM RESERVES

Total = 7,580 Million lbs

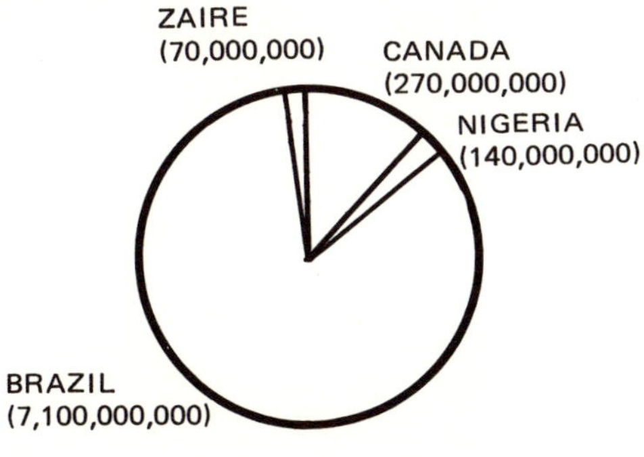


*Source: US Bureau of Mines, 1982*

## US COLUMBIUM CONSUMPTION

By Market Sector 1981

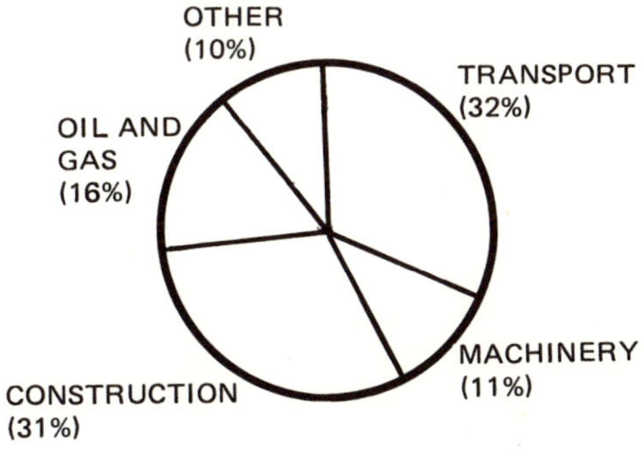


*Source: US Bureau of Mines, 1982*

## TRANSPORTATION

Columbium is generally mined in the form of columbium oxide, as tin slags (containing tantalite/columbite) or as pyrochlore concentrates. The main routes are from Brazil and Canada to the USA and EEC plus exports from Nigeria, Zaire and Thailand to Japan, the EEC and the USA. The reliability of transport links tends not to be a problem given the relatively small volumes involved.

## APPLICATIONS

### Consumption Trends

Columbium tends to be added in the form of the pentoxide or the ferro-alloy to steels and specialised non-ferrous alloys. Demand for columbium is expected to increase by six per cent a year between 1978 and 1990 with current US consumption at around 7.0-7.5 million lbs.

### The Principal Markets

Columbium is a valued additive since it enhances the hardness and corrosion resistance of alloys which can then be used in more arduous applications. High-strength low-alloy steels (HSLA steels), stainless and alloy steels account for 80-85 per cent of all columbium supplies. The HSLA steels are a major growth sector due to their application in the pipeline, construction and chemical industries since they have a higher strength/weight ratio and a greater reliability at low temperatures than other steels of a comparable cost.

Columbium is used in alloys for electrical and magnetic applications and research is continuing into the development of columbium/zirconium superconductors. Non-metallic applications include the employment of columbium ceramics as tiles for space shuttles and in drilling bits. Columbium is also used as the material for fast nuclear reactor fuel cans and in specialised chemical equipment. Demand growth in the transportation industry, encompassing both aerospace and automobile applications (where HSLA steel car bodies are under investigation), should be the most rapid since the use for oil and gas pipelines is currently low despite the notable exception of the Soviet gas pipeline project.

### Recycling

Recycling of prompt and old scrap is insignificant since it is not viable to reclaim columbium which is present in very small quantities in many alloys—0.1 per cent by weight in the case of HSLA steels.

### Substitution

Vanadium and molybdenum may be substituted in HSLA steels; tantalum and titanium in stainless and high-strength steel and superalloys; and molybdenum, tungsten, tantalum and ceramics may be used in high-temperature applications. Performance or cost penalties are likely, particularly in steels where columbium is twice as effective per unit weight as tantalum, whilst remaining less expensive than vanadium.

## INTERNATIONAL TRADE AND WORLD PRICES

### Supply Arrangements

Companhia Brasileria de Metalurgia e Mineracao of Brazil (CBMM) dominates the scene with 80 per cent of world output and prices tend to be tightly controlled by producers with little free market activity. The producers have managed to remain flexible over output and stockpiles in order to maintain a slow, steady price progression in the face of a firm demand. Columbium is traded either as the ferro-alloy with a stated columbium content of 40 per cent, 60 per cent or 70 per cent or as the mixed ore containing a minimum of 65 per cent columbite plus tantalite. Prices are quoted in $/lb columbium or columbium oxide content.

The USA and Europe are almost totally import dependent on columbium source materials and the GSA has a stockpile goal of 5.6 million lbs columbium concentrate. Japan has put columbium on its priority list for establishing secure supply sources.

### Companies

Companhia Brasileria de Metalurgia e Mineracao and its US marketing agency, Niobium Products, largely control the market. Foote Minerals and Shieldalloy of the USA are major suppliers of various grades of columbium ferroalloys. Amalgamated Tin Mines of Nigeria (Holdings) Ltd, Geomines of Zaire, Molycorp and Kennecott (co-operating in Quebec) and Pato Consolidated Gold Dredging in Canada are other major columbium ore producers.

### Prices

Columbium ore prices peaked at $11/lb contained oxide in 1980 but have declined since then. The price of various grades of columbium alloys do not necessarily move in parallel as demonstrated in 1981. The price of steel-making grade ferrocolumbium remained steady in that year but that of high-purity ferrocolumbium declined, reflecting falling demand in the aerospace industry and an oversupply of columbium oxide. Sharp cutbacks have been forced onto several Brazilian producers by poor prices.

**Columbium Ore Prices—Annual Averages**

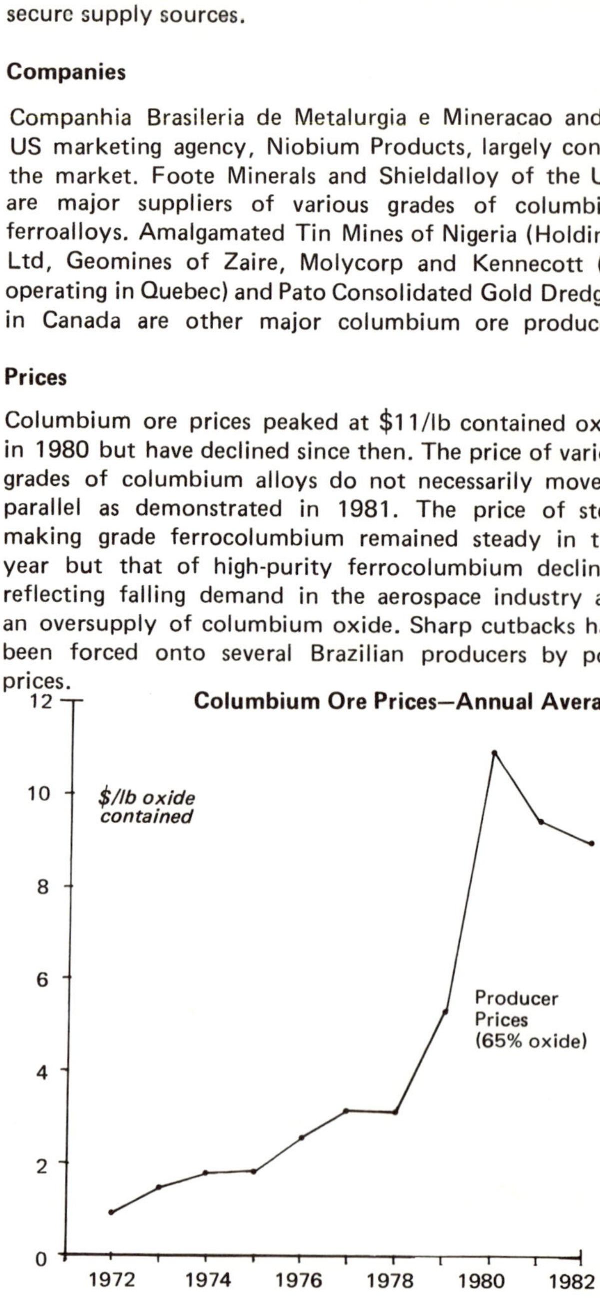

# COPPER (Cu)

| | |
|---|---|
| **FACTOR FOR LIKELIHOOD OF SUPPLY DISRUPTION** | **3.6** |
| **FACTOR FOR COST OF SUCH A DISRUPTION** | **8.0** |
| **TOTAL STRATEGIC FACTOR** | **28.8** |

*The factors for supply disruption and cost of such a disruption are assessed on a scale from 1.0 (lowest) to 10.0 (highest) risk. Multiplied together these factors give the total strategic factor rating. For components of these assessments see the table later in this section.*

Copper has a relatively high strategic factor due to its widespread importance as an industrial metal with over 50 per cent of all consumption in the electrical industry. It is widely used as a general barometer of industrial health due to its wide penetration into the construction, electrical, engineering and automobile industries. Over 50 per cent of internationally traded copper is controlled by the CIPEC nations, including Chile, Peru, Zambia and Zaire which are the more active members.

**Physical Characteristics**

Red-orange metal with a fairly high melting point (1,083$^{o}$C); high specific density (8.96); good conductor of heat and electricity; excellent mechanical properties; corrosion resistant.

## OCCURRENCE AND PRODUCTION

Occurs in over 250 minerals, but only six of these are widely exploited: chalcopyrite ($CuFeS_2$), chalcocite ($Cu_2S$), chrysocolla ($CuSiO_3.2H_2O$), malachite ($Cu_2(OH)_2CO_3$), azurite ($2CuCO_3.Cu(OH)_2$) and bornite ($Cu_5FeS_4$).

**World Reserves**

These are estimated at 505 million tonnes copper content, sufficient for over 65 years of consumption at current levels. Reserves are concentrated in distinct geological areas, principally along the western backbone of North and South America, Central Africa, South East and Central Europe, the South West Pacific and the USSR. Total land based resources estimated to total 1.6 billion tonnes copper plus another 690 million tonnes contained in deep sea nodules. The Communist bloc possesses approximately 10 per cent of world reserves.

**Production Characteristics**

*Mining Methods:* Varies according to deposit type. Open cast mining of large low grade (0.2-1.5 per cent) copper porphyritic deposits is the main method accounting for 45 per cent of the total. Underground mining is still important particularly in the Central African copper-belt

*Processing:* Copper sulphide ore concentrated then smelted to blister copper. Anode refined and then electrolytically refined to copper cathodes and cast into wirebars in preparation for final fabrication. Oxide ores chemically leached to give impure copper which is then refined

*Energy Requirements:* 7,200,000 kcal/tonne copper with 25-35 per cent of this at the smelting stage and 58 per cent used during mining and milling

*By-Products:* (i) Nickel, PGMs, silver, gold from Canada (Sudbury). (ii) Lead, zinc, silver, arsenic, cadmium from North America and Australia. (iii) Cobalt and sometimes nickel from Central Africa. (iv) Molybdenum and rhenium from Chile, parts of the USA and British Columbia. (v) Gold, silver from Papua New Guinea and Australia. Also sporadically—germanium, selenium, tellurium

*Environmental Factors:* Expensive extractors etc necessary to prevent emission of sulphur compounds during smelting

*Development Lead Time:* Up to 10 years for green field projects.

## COMPONENTS OF RISK ASSESSMENT

| *Factor* | *Rating* | *Comment* |
|---|---|---|
| **Production Risks** | | |
| Existing capacity | 2 | Current surplus in capacity but some sources predict a shortage late in the decade since 1982 prices are insufficient to promote further developments |
| Labour disputes | 7 | South America and Central Africa are labour trouble spots whilst Canadian, USA and Australian copper unions tend towards the militant |
| Violent conflict | 2 | More likely in Zambia and Zaire than elsewhere |
| Range of primary supply sources | 6 | USA is a major producer but also an importer. Major exporters number only eight and are CIPEC members |
| Time lags for new supplies | 8 | New mines have a long lead time and require extensive infrastructure. Also exporting countries tend to arrange annual supply contracts |
| **Transportation Risks** | | |
| Primary | 7 | Transportation bottlenecks for copper concentrates include the Tanzara railway and the ports of Canada and Australia which are prone to dock strikes |
| Secondary | 2 | Transportation of refined and fabricated copper tends to be subject to fewer problems |
| **Application/Use Risks** | | |
| Total economic impact | 10 | Used in all sectors of industry. Total market volume of 9.5 million tonnes a year, including recycled metal |
| Effect on key industries | 7 | Heavy emphasis on vital electrical industry plus construction, transportation and general engineering industry |
| Availability of substitutes | 9 | Available to a certain extent in many areas but often requires major technical changes to adjust to substitutes |
| Longer term substitutability | 8 | Possible in certain areas of telecommunications but likely to retain most of its markets due to its useful properties and ease of fabrication |
| **Trade Risks** | | |
| Collusive price agreements | 5 | CIPEC has so far made little impact but aid from UNCTAD could give their aims some teeth |
| Embargoes | 1 | Unlikely since producers depend heavily on sales generated |

*These components are assessed on a scale from 1 (lowest) to 10 (highest) risk. Grouped together, they form the basis for the assessments of the factors for likelihood of supply disruption, and for cost of such disruption which appear in the first table in this section.*

## WORLD COPPER RESERVES

Total = 505 Million Tonnes

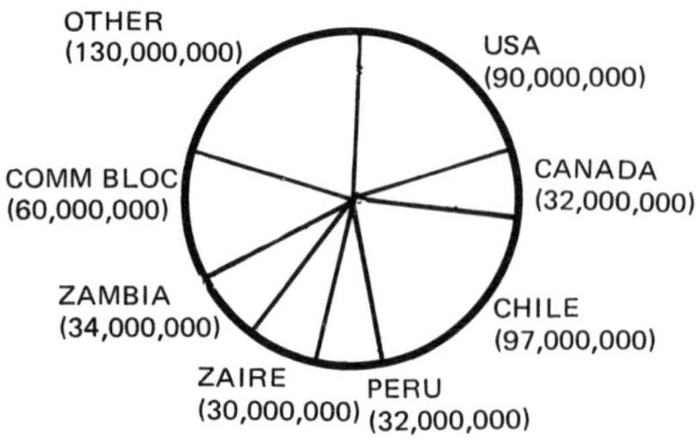


*Source: US Bureau of Mines, 1982*

## US COPPER CONSUMPTION

By Market Sector 1981

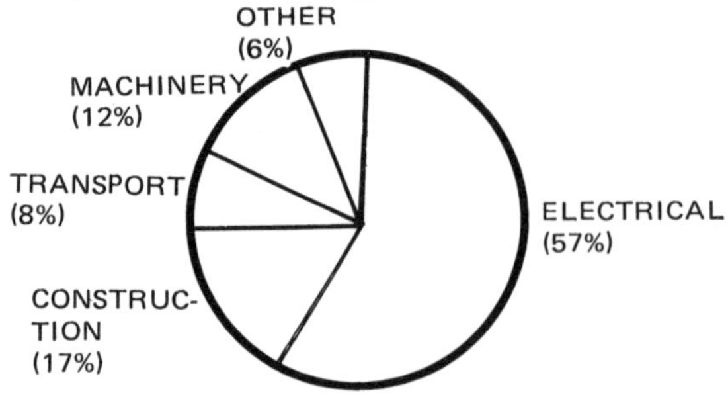


*Source: US Bureau of Mines, 1982*

### The Major Producers

World Mined Production
(producers over 200,000 tonnes a year)

| *Countries* | *Output 1981 (Thousand Tonnes)* | | *Percentage of Western World Output* | |
|---|---|---|---|---|
| EUROPE | | 297 | | 4.6 |
| AFRICA | | 1,391 | | 21.6 |
| South Africa | 209 | | 3.2 | |
| Zaire | 505 | | 7.8 | |
| Zambia | 587 | | 9.1 | |
| ASIA | | 504 | | 7.8 |
| Philippines | 302 | | 4.6 | |
| AMERICA | | 3,854 | | 59.9 |
| Canada | 693 | | 10.7 | |
| USA | 1,529 | | 23.7 | |
| Chile | 1,081 | | 16.8 | |
| Mexico | 230 | | 3.6 | |
| Peru | 328 | | 5.0 | |
| AUSTRALASIA | | 389 | | 6.0 |
| Australia | 224 | | 3.4 | |
| WESTERN WORLD TOTAL | | 6,435 | | 100.0 |

(Communist bloc production is currently estimated at 1,819,000 tonnes and production in the USSR 1,150,000 tonnes a year.) (Independently rounded figures.)

*Source: World Bureau of Metal Statistics, June 1982.*

World Refined Production
(producers over 200,000 tonnes a year)

| *Countries* | *Output 1981 (Thousand Tonnes)* | | *Percentage of Western World Output* | |
|---|---|---|---|---|
| EUROPE | | 1,354 | | 18.8 |
| Belgium | 310 | | 4.2 | |
| West Germany | 387 | | 5.3 | |
| AFRICA | | 889 | | 12.3 |
| Zambia | 573 | | 7.9 | |
| ASIA | | 1,261 | | 17.5 |
| Japan | 1,050 | | 14.5 | |
| AMERICA | | 3,517 | | 48.7 |
| Canada | 477 | | 6.6 | |
| USA | 1,959 | | 27.1 | |
| Chile | 776 | | 10.7 | |
| Peru | 211 | | 2.9 | |
| AUSTRALASIA | | 191 | | 2.6 |
| WESTERN WORLD TOTAL | | 7,212 | | 100.0 |

(Communist bloc output is estimated at 2,322,000 tonnes a year.)

*Source: World Bureau of Metal Statistics, June 1982.*

Mined output is dominated by the USA, Chile, USSR, Canada, Zaire and Zambia but copper is also mined in over 40 countries on all continents. However refined output in the Western world is concentrated in the USA, Japan (which produces 20 times as much refined copper as it mines) and Europe (which produces over four times as much refined copper as is domestically mined). In the Third World, Chile has moved furthest towards its goal of refining all its mined output but Zaire, Zambia, Peru and the Philippines export much of their production in the form of concentrates and blister copper.

Ownership of the world copper mines has changed radically over the past two decades. The industry used to be dominated by a few major international groups with headquarters in the old colonial powers. A wave of forced nationalisations in Third World countries has given their governments control over their own resources. From 1974 onwards there has been a major diversification by oil companies into holdings in mining companies with the result that the seven oil majors now control 15-20 per cent of world copper capacity. During 1982, savage copper cutbacks amounting on an annual basis to 450,000 tonnes a year were effected by Canada and the USA in the face of a drastic fall in prices and sales. Chile, Peru, the Philippines, Zambia and Zaire increased output, however, using government and IMF support to provide the needed temporary financial support.

### Technological Developments

Include one-step continuous copper making from the concentrates to the blister stage, plus improved methods for leaching copper oxide ores and mine tailings. Constant efforts are directed into improving the efficiency of the mining and milling stages which are particularly energy intensive for lean sulphidic ores.

## TRANSPORTATION

World trade in ores and concentrates totals 1.2-1.3 million tonnes a year. Unrefined copper trade totals 500-600,000 tonnes a year plus 2.5-3.0 million tonnes a year in refined copper. Transport is obviously a major link and an important consideration. Danger points involve the movements of concentrate and blister copper from Zaire and Zambia through to Tanzania or South Africa for shipment to Europe and the USA. Japan obtains the major portion of its blister and anode copper from the Philippines, Papua New Guinea, Australia and now also from the USA, due to a competitive cost structure. Distances from South America to Europe and the USA are large and sporadic labour disputes at vital ports do little to ensure regularity of shipments. Other trouble spots include the Australian and Canadian ports.

## APPLICATIONS

### Consumption Trends

Non-Communist world consumption totalled 7.09 million tonnes in 1981 showing little growth on the previous year. Non-OECD demand is expected to grow much more quickly than the 2-3 per cent a year rate predicted worldwide and is expected to reach 1.5 million tonnes a year by 1990 compared with 900,000 tonnes in 1981.

### The Principal Markets

Around 97 per cent of all copper is consumed in the metallic form, as distinct from copper sulphate etc, 70 per cent as lightly alloyed copper and 25-27 per cent as brass. Applications depend on fabricability and electrical and thermal conductivities. Copper consumption in the electrical and electronic industry is a sensitive function of GNP since the sector encompasses power cable, magnet wire, telecommunications equipment and domestic wiring systems.

Copper has largely lost control of the overhead transmission power cable market to its less dense competitor aluminium but has retained the major share of the domestic market on safety grounds. Energy demand growth profiles have been revised downward in the OECD nations leading to static levels of copper consumption in this sector but industrial and domestic electrification schemes continue in the developing countries. The growing use of optical fibre telecommunications is another threat in this field particularly for computer link-ups and telephone systems.

The construction industry uses 17 per cent of all copper consumed in the USA, as depicted in the piechart, but only nine per cent in Japan and 12-15 per cent in Europe. This sector mainly comprises plumbers' fittings and brassware, domestic service tubing, water heaters and roofing sheets. General engineering—12-15 per cent of all copper in Japan and Europe—involves valves and fittings, pumps, process plants and copper chemicals and in these metallic applications, copper tends to be heavily alloyed to give the best balance of properties. Transportation—11-12 per cent in Europe and 17 per cent of all copper in Japan—uses copper in vehicle radiators, shipbuilding, starters, generators and bearings but this is not considered a growth area.

### Recycling

Old scrap provides 25-27 per cent of consumption in the USA and Europe. About one-third of this is recovered as refined copper and the remainder in alloy form. These percentages are sensitive to prevailing price levels but it must also be noted that the average life of copper in service varies from three to five years for transportation purposes to 40 years or more in power transmission lines and cable giving a pattern to recycling trends.

### Substitution

Copper is vulnerable to substitution for many uses such as aluminium for electrical purposes, steel for chemical process plants, plastics for plumbing and germania-cored glass fibres for telecommunications. It retains its markets on a strictly cost effective basis although it must be cost per unit property, such as conductivity, that is compared rather than simply cost per tonne.

## INTERNATIONAL TRADE AND WORLD PRICES

### Supply Arrangements

Approaching 90 per cent of all copper sales are made by annual supply contracts negotiated between consumers and producers, generally in the October of the preceding year. The producers set a price for their product normally in unofficial agreement with other producers on the same continent. The remainder is sold via the two major terminal markets; the LME—25 tonne contract of 99.9 per cent higher grade or standard cathodes—and the New York Commodity Exchange—25,000 lb contract of electrolytic copper. The price structures agreed in producer/consumer contracts are often referenced to daily, weekly or monthly LME average prices so the price volatility seen on the LME has a disproportionate effect on producers' sales revenues.

CIPEC, the French acronym of the Copper Exporting Nations Association, was formed in 1974 and now comprises Chile, Peru, Zambia, Zaire, Australia, Indonesia, Papua New Guinea and Yugoslavia, controlling almost 60 per cent of internationally traded copper. However it has proved to be of little use in its main aim, that of supporting and stabilising copper prices, since the nations involved depend too heavily on foreign exchange from copper exports.

### Companies

The major mining companies include the state-owned organisations of the USSR, Chile (Codelco), Zaire (Gecamines), Zambia (Zambia Consolidated Copper Mines) plus Kennecott and Phelps Dodge of the USA, Southern Peru Copper Corporation of Peru, Mount Isa Mines of Australia and Inco of Canada. All the above produce over 150,000 tonnes a year of copper in concentrate form. The leading refiners in the Western world are Codelco, Phelps Dodge, Noranda of Canada, Zambia Consolidated Copper Mines and Gecamines.

### Prices

With USA breakeven production costs at 90 cents per lb and the Comex spot price hovering in the 60-70 cents range for much of 1982, North American producers had little chance of profitability. Third World producers were similarly affected but tended to receive support either from their governments or through some form of international aid.

**Copper Wirebars Prices—Annual Averages, Highs and Lows (London Metal Exchange)**

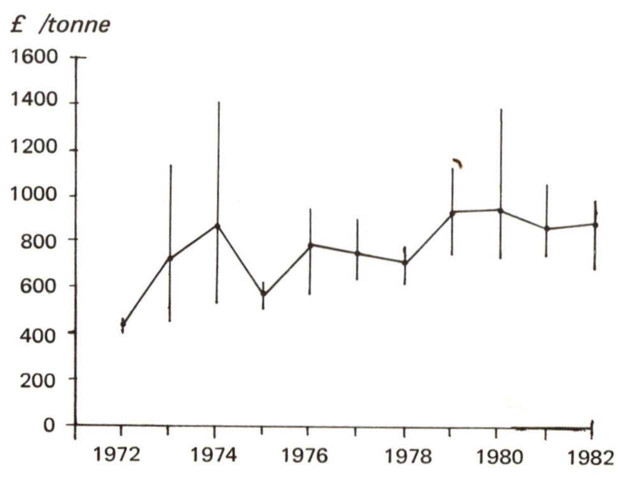

# DIAMOND (C)

| | |
|---|---|
| FACTOR FOR LIKELIHOOD OF SUPPLY DISRUPTION | 3.8 |
| FACTOR FOR COST OF SUCH A DISRUPTION | 5.0 |
| TOTAL STRATEGIC FACTOR | 19.0 |

*The factors for supply disruption and cost of such a disruption are assessed on a scale from 1.0 (lowest) to 10.0 (highest) risk. Multiplied together these factors give the total strategic factor rating. For components of these assessments see the table later in this section.*

Diamond, the crystalline form of carbon, occurs extremely rarely and has a moderate but falling strategic rating. In its industrial applications it is capable of being produced synthetically whilst as a gemstone its price is maintained at an artificially high level by a marketing cartel which faces an uncertain future.

### Physical Characteristics

Natural diamond is the hardest known substance; applications in cutting, drilling, abrasives and wire drawing; fairly easily split allowing the faceting that produces a cut or polished gemstone; commonest crystal form is an octahedron. The first successful diamond synthesis achieved in 1953 since when the commercial production of synthetic industrial stones has become established; high melting point exceeding 3,500°C and a low specific density of 2.25.

## OCCURRENCE AND PRODUCTION

Primary diamond deposits occur in what are commonly known as the 'Kimberlite pipes', dense, ultra-basic peridotite lava flows which were subjected to great heat and pressure as they pushed their way from deep in the earth's crust. Surface weathering often erodes the tops off these pipes, the diamonds being carried away by water action to form secondary deposits in river beds and estuaries.

Diamonds were first discovered and mined in India and then Brazil but the largest deposits were found in South Africa from 1870, to be followed by finds in other parts of Africa, in Zaire, Tanzania and Angola. These sources dominated production until joined in the late 1970s by Botswana and Lesotho. This still left the vast majority of non-Soviet production in Southern Africa, greatly influenced by the Central Selling Organisation until the discovery of a very large field in Western Australia. By 1983 government permission had been granted to develop what could become the world's largest mine, in the Argyle district of Western Australia.

### World Reserves

It is hardly possible to quantify world reserves in terms of their diamond content. Very many Kimberlite deposits are known, grades vary as much as qualities and depths may be extreme. The volume of secondary deposits is also very large. Known reserves will easily sustain recent outputs of about 40 million carats per year indefinitely, particularly with the growth of synthetics production. Overall grades (carats per tonne of rock) and the percentage of gemstones are critical. A broad comparison of the world's more important bodies shows:

| *Countries/Deposits* | *Carats per 100 Tonnes* | *Percentage Gem Content* |
|---|---|---|
| South Africa | | |
| Koffiefontein OFS | 11.5 | 50 |
| Premier TVL | 30.0 | 20 |
| Finsch Cape | 75.0 | N/A |
| Lesotho | | |
| Letseng le Terai | 2.8 | 80 |
| Botswana | | |
| East Botswana, Orapa | 64.0 | 10 |
| Australia | | |
| Argyle (AKI) W Australia | 700.0 (possible) | N/A (low) |
| Zaire | | |
| Massif I | N/A | 7 |

### The Major Producers

World Mine Production of Natural Diamond (Millions of Carats)

| *Area* | *1980 Total* | *Percentage of Gem* | *Percentage of Western World Diamond Output in 1980* |
|---|---|---|---|
| Botswana | 5.10 | 31 | 17.6 |
| Namibia | 1.56 | 100 | 5.4 |
| South Africa | 8.70 | 41 | 29.9 |
| Zaire | 8.60 | 7 | 29.6 |
| Other | 4.16 | 40 | 14.3 |
| TOTAL AFRICA | 28.12 | 32 | 96.8 |
| SOUTH AMERICA | 0.85 | 34 | 3.2 |
| OTHER | 0.10 | N/A | |
| WESTERN WORLD TOTAL | 29.07 | N/A | 100.0 |

## COMPONENTS OF RISK ASSESSMENT

| *Factor* | *Rating* | *Comment* |
|---|---|---|
| **Production Risks** | | |
| Existing capacity | 1 | Current surplus is growing |
| Labour disputes | 2 | Minor and sporadic |
| Violent conflict | 7 | Latent possibility throughout Africa |
| Range of primary supply sources | 4 | Largely concentrated on Africa but will soon include a large additional source in Australia |
| Time lags for new supplies | 6 | Less than for most minerals |
| **Transportation Risks** | | |
| Primary | 2 | Low as diamonds are low volume products |
| Secondary | 3 | Similarly low as small-unit volumes are shipped |
| **Application/Use Risks** | | |
| Total economic impact | 1 | World mine production of natural diamond totals only 40-42 million carats a year |
| Effect on key industries | 8 | High |
| Availability of substitutes | 7 | Growing but natural stone is still vital |
| Longer term substitutability | 4 | Already discounted |
| **Trade Risks** | | |
| Collusive price agreements | 8 | Already in place for gem diamond through the CSO and a degree of harmony exists in the industrial market |
| Embargoes | 2 | Both East and West possess large sources and are united through the CSO |

*These components are assessed on a scale from 1 (lowest) to 10 (highest) risk. Grouped together, they form the basis for the assessments of the factors for likelihood of supply disruption, and for cost of such disruption which appear in the first table in this section.*

### WORLD INDUSTRIAL DIAMOND RESERVES

Total = 620 Million Carats

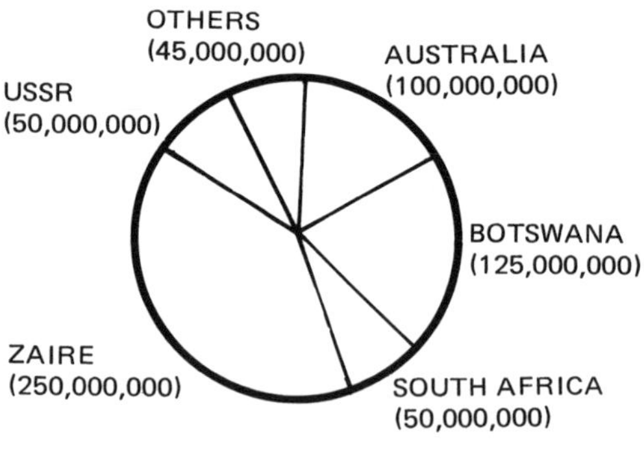


*Source: US Bureau of Mines, 1982*

### US INDUSTRIAL DIAMOND CONSUMPTION

By Market Sector 1981

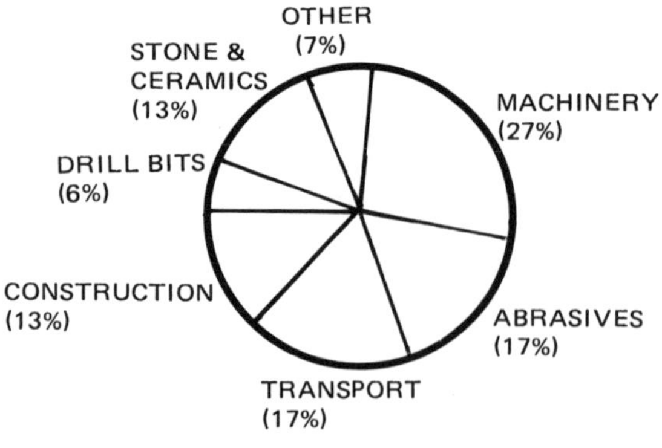


*Source: US Bureau of Mines, 1982*

If, as believed, the Australian developments are low in gem quality the above pattern will not be materially disturbed. Some of the smaller, rich gem sources, such as the Ivory Coast, Ghana, Sierra Leone, have shrunk considerably leaving South Africa the largest gem producer. Botswana, Namibia and Lesotho have measurable capacity for expansion so that orderly marketing and development are critical.

### Technological Developments

The promotion of gem diamonds for jewellery and investment accompanied by the maintenance of an orderly market and a wholly artificial price structure has been an unparalleled success. Gems provided the profits whilst sales of industrial stones, for which there were no substitutes for many years, provided a base load. The development of synthetic stones, brought to commercial reality by De Beers and General Electric of the USA has focused on the necessity to maintain the gem market profitably. The USBM estimates that synthetics output roughly doubled to 46 million carats between 1970 and 1980, or about 60 per cent more than natural industrial stones. This total could rise to 250 million carats by 1990, if the market can absorb it. Some natural stones still command a premium but the consistent quality of synthetics makes them attractive to the drilling industry, a major outlet.

## TRANSPORTATION

Finally sorted diamonds are normally flown from mine site to the marketing and polishing centres of London, New York, Antwerp, Tel Aviv, Bombay and Johannesburg. As such few threats to disruption exist. The remoteness of many production sites also renders organised disruption unlikely.

## APPLICATIONS

### Consumption Trends

The gemstone market accounts for about 30 per cent by volume, but 80 per cent by value. Since it is wholly reliant on jewellery and investment demand it is directly related to the economy. Because of the market controls imposed by the CSO a normal supply/demand/price relationship does not pertain.

Sales of industrial stones have been maintained by lowering prices sharply to match synthetic levels. This in turn has led to increasing demand which, barring a total economic collapse, should enable the two markets to coexist during the 1980s.

### The Principal Markets

Gemstones attract about 85 per cent jewellery and 15 per cent investment demand. The latter has been promoted seriously since 1975 but fails on two important points: the definition of quality and the large retail mark-up on price. Compared with other investment media such as precious metals, coins and even stamps, it suffers a marked disadvantage.

Industrial applications are associated with the unique hardness of diamond, the major outlets including: rock drilling (mining, oil exploration), abrasives, steelcutting, stone cutting and polishing, glass, automobiles. Little overall growth took place between 1979 and 1983 but the availability of synthetics and the overall lowering of price should displace some substitutes and stimulate demand.

### Recycling

Most industrial end-uses are destructive e.g. drilling and abrasion but a small volume of scrap recovery has always persisted. Synthetics have damaged the economics of scrap recovery. Gemstones are sometimes reset from one piece of jewellery to another but the intrinsic value is often in the setting and the mount as well as the stone so that the older a piece, the less likelihood of it being recycled.

## INTERNATIONAL TRADE AND WORLD PRICES

### Supply Arrangements

The Central Selling Organisation plays the pivotal role in the world gem diamond market and has a strong presence in the industrial sector.

### Companies

The CSO is controlled by De Beers, the South African based group, of whose shares almost 90 per cent are controlled by Anglo American Corporation-related companies. Gem diamonds are purchased from and marketed on behalf of CSO member countries/companies by the Diamond Trading Company (DTC) arm of the CSO and industrial stones by the Industrial Diamond Corporation (IDC). In this way some 80 per cent of the world's gems are traded and a smaller, shrinking proportion of industrial quality stones, perhaps as low as 30 per cent. When, in 1981, the Zairean government broke away from the CSO, it was indicative of a growing danger. Other countries may be tempted to break away both to sever the political connection and to avoid the application of quotas in a thin market. Producers such as Botswana, Lesotho and Namibia have a growing dependency upon diamond revenues and unless the market regains the bouyancy of the 1970s, may be tempted to secede. Ghana, the Central African Republic, Guinea, the Ivory Coast, Guyana, Venezuela and Brazil all market independently. The Australian decision in late 1982, to join the CSO was therefore of some importance. Synthetics for industrial use are marketed by the manufacturers: De Beers, General Electric, (about 80 per cent of world output); Du Pont (USA), Komatsu (Japan), USSR, China and Czechoslovakia.

### Prices

It is misleading to quote prices for gem diamonds since each stone will be individually priced on the basis of quality, colour, caratage and cut. CSO prices rose by 30 per cent in 1978 followed by increases of 12-13 per cent during the next two years. Since then levels have been static.

# GALLIUM (Ga)

| | |
|---|---|
| **FACTOR FOR LIKELIHOOD OF SUPPLY DISRUPTION** | **1.7** |
| **FACTOR FOR COST OF SUCH A DISRUPTION** | **4.5** |
| **TOTAL STRATEGIC FACTOR** | **7.7** |

*The factors for supply disruption and cost of such a disruption are assessed on a scale from 1.0 (lowest) to 10.0 (highest) risk. Multiplied together these factors give the total strategic factor rating. For components of these assessments see the table later in this section.*

Gallium has a relatively low strategic factor. It is a minor element and one which has only recently come into the spotlight of research and technology. Applications chiefly depend on the metal's electrical properties, a function of its technical status as a semi-metal. The range of uses is potentially very broad but much research is needed in the relevant areas. Resources are almost limitless since it is a byproduct of aluminium. Supplies are relatively secure and concentrated in Europe and North America.

### Physical Characteristics

Low melting point (30°C) semi-metal; has the most extended temperature range of all elements over which it is a liquid; low, but finite, electrical conductivity.

## OCCURRENCE AND PRODUCTION

Occurs in bauxite in average concentration of 50 ppm and also in zinc ore deposits in similar concentrations.

### World Reserves

World resources in bauxite estimated to exceed one million tonnes with approximately the same geographical distribution as that of bauxite. However, only small percentages of the gallium are recoverable in many deposits so an accurate assessment of viable reserves is unobtainable.

### Production Characteristics

| | |
|---|---|
| *Mining Methods:* | Those of bauxite/zinc ore parent products |
| *Processing:* | Extracted by chemical means. Final process of zone refining gives metal of 99.99 per cent purity |
| *Energy Requirement:* | Largely irrelevant due to by-product status |
| *By-Products:* | Aluminium, zinc |
| *Environmental Factors:* | Gallium metal and compounds relatively non toxic |
| *Development Lead Time:* | Dependent on zinc/bauxite mine. |

### The Major Producers

| *Countries* | *1981 Output (Tonnes)* | *Percentage of Western World Output* |
|---|---|---|
| USA | 4.5 | 18.8 |
| Canada | 4.0 | 16.7 |
| Europe (largely imported concentrates) | 15.0 | 62.5 |
| Japan | 0.5 | 2.0 |
| WESTERN WORLD TOTAL | 24.0 | 100.0 |

(Communist bloc output is estimated at 0.5 tonnes a year.)

*Source: USBM estimates and Strategic Inc 'Importance of Gallium Arsenide', 1982.*

Output data is unreliable due to the proprietary nature of the production levels and methods employed. Major producers tend to be those who mine and/or refine bauxite or zinc and include Canada, the USA, Japan and Switzerland. This last country uses imported source materials derived from bauxite.

### Technological Developments

Chemical purification and zone refining methods are undergoing constant improvements whilst methods for growing suitable crystals for electronic applications are subject to concerted research efforts.

## TRANSPORTATION

Gallium is extracted and purified at alumina refineries and zinc smelters which are near the main areas of consumption so transport problems are minimal given the variety of methods available and the short distances involved.

*GALLIUM*

## COMPONENTS OF RISK ASSESSMENT

| *Factor* | *Rating* | *Comment* |
|---|---|---|
| **Production Risks** | | |
| Existing capacity | 1 | Surplus capacity, typified by cutbacks by Alcoa and Alusuisse, partly prompted by the slump in aluminium industry |
| Labour disputes | 1 | Unlikely to affect output since the major producer is in Switzerland where labour relations are good |
| Violent conflict | 1 | Unlikely |
| Range of primary supply sources | 2 | Present in most bauxite or zinc deposits but processing facilities are available in only a few locations, mainly Switzerland, the USA, Canada |
| Time lags for new supplies | 4 | As a by-product, largely dependent on the rate of bauxite refining |
| **Transportation Risks** | | |
| Primary | 1 | Present in such small quantities in each shipment/trainload of bauxite that the loss of one shipment has little overall effect on gallium supplies |
| Secondary | 7 | Finished products containing gallium are delicate and costly, must be handled with care |
| **Application/Use Risks** | | |
| Total economic impact | 1 | Total market volume less than 30 tonnes, only 3 per cent of volume of gold mined each year |
| Effect on key industries | 8 | Applications mainly in the electronics industry with products utilised in military and aerospace equipment |
| Availability of substitutes | 6 | Substitutes are available but gallium arsenide chips have the advantage of speed over germanium and silcon |
| Longer term substitutability | 6 | Possible, depending on the technology available |
| **Trade Risks** | | |
| Collusive price agreements | 2 | Feasible considering the small number of producers but gallium sales are such a minute proportion of revenue that there is little incentive to fix prices unreasonably |
| Embargoes | 1 | Unlikely |

*These components are assessed on a scale from 1 (lowest) to 10 (highest) risk. Grouped together, they form the basis for the assessments of the factors for likelihood of supply disruption, and for cost of such disruption which appear in the first table in this section.*

## ESTIMATED WORLD GALLIUM RESERVES

Total = 110,000 Tonnes

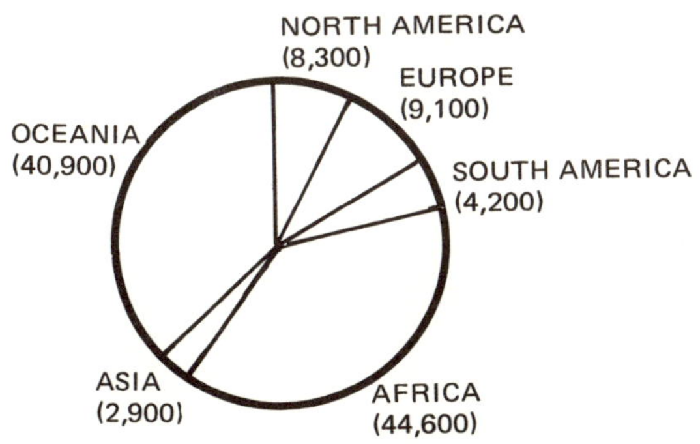


*Source: Gallium, Mineral Facts and Problems, US Bureau of Mines, 1974*

## US GALLIUM CONSUMPTION

By Market Sector 1981

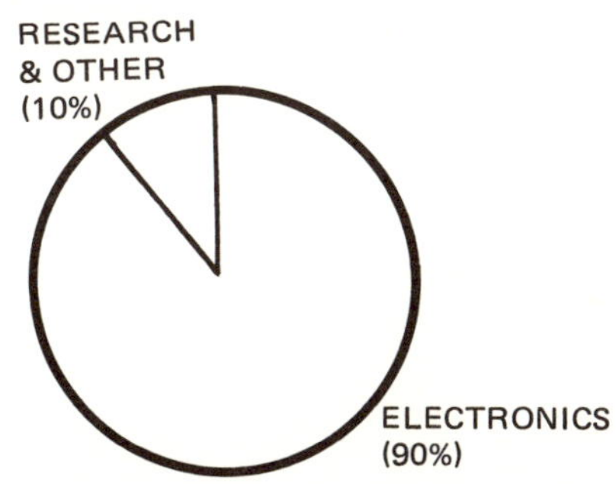


Electronics includes: LED's, bubble memories and other semi-conductor devices

*Source: US Bureau of Mines, 1982*

## APPLICATIONS

### Consumption Trends

Direct extrapolation is of little use since so much depends on semi-conductor technology, both for microchip circuitry and for solar energy conversion. The USBM consider that demand will increase at an annual rate of six per cent until 1990 from a 1978 base.

### The Principal Markets

US consumption is expected to total 7.5 tonnes in 1982. Over 90 per cent of all gallium is consumed in electronic applications spanning light emitting diodes, photomultipliers and direct imaging devices.

Research continues into other advanced applications such as solid state lasers, photovaltaic devices, high speed solid state circuitry for advanced computers and the use of gallium arsenide as signal initiators in optical fibre networks. Strategic Inc. suggest that sales of gallium arsenide micro-chip circuitry could rise from zero in 1982 to $292 million by 1990. The adoption of a solar energy scheme by NASA or the US Department of Energy could herald a similar mushrooming of the market. Military applications could grow equally strongly particularly for gallium arsenide chips which now absorb around 70 per cent of world gallium output.

### Recycling

Gallium is not normally recovered from discarded electronic products since source material costs are minute compared to those of fabrication. New scrap volumes are high and new scrap is reprocessed due to the frequent failure rate of manufactured devices.

### Substitution

Gallium is losing the battle with organic liquid crystals for the numeric display sector in watches and calculators. Silicon and germanium are alternatives in many semiconductor applications with a heavy emphasis on performance parity. Cheaper and more effective substitutes have been found for gallium gadolinium garnets in the decreasingly fashionable bubble memories.

## INTERNATIONAL TRADE AND WORLD PRICES

### Supply Arrangements

The free market is very limited in size and prices sought by producers are a sensitive function of purity and quality. Gallium of a specified purity in the range 99.9-99.9999 per cent in the form of small ingots is traded with further purifications, alloying and crystal growth done in situ by the consumer to avoid damage to the delicate gallium arsenide crystals. The USA has two refined gallium producers and, in 1981, was 40 per cent import dependent with two-thirds of imports deriving from Switzerland. The gallium market is currently depressed after the blaze of publicity directed at a limited free market in the "strategic metals boom" of 1979-80. Some analysts estimate that world production has actually fallen over the last two years.

### Companies

Alusuisse of Switzerland is the world's largest producer with an estimated capacity of 10 tonnes a year. Alcoa and Eagle-Picher in the USA, Cominco of Canada and Nippon Light Metals and Toho Zinc of Japan are other major producers. Gallium is only a tiny component of the overall metals production of these large companies so prices realised for this by-product are relatively unimportant and they can afford to cut back output to maintain a sensible supply balance.

### Prices

Producer prices have declined from the high values of the 1970s as processes to extract and refine gallium have become more efficient. A rapid price increase is unlikely in the 1982-86 period unless a new end-use suddenly becomes popular. Free market prices tend to be quoted in a nominal fashion only.

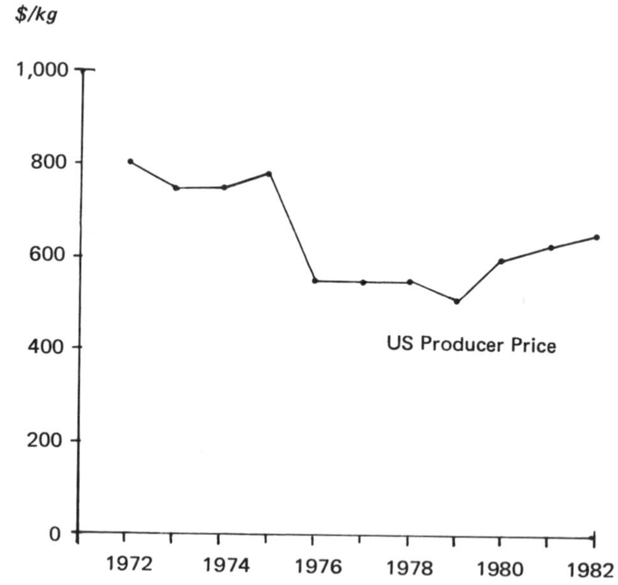


**Gallium Prices—Annual Averages**

# GERMANIUM (Ge)

| | |
|---|---|
| **FACTOR FOR LIKELIHOOD OF SUPPLY DISRUPTION** | **2.5** |
| **FACTOR FOR COST OF SUCH A DISRUPTION** | **6.2** |
| **TOTAL STRATEGIC FACTOR** | **15.5** |

*The factors for supply disruption and cost of such a disruption are assessed on a scale from 1.0 (lowest) to 10.0 (highest) risk. Multiplied together these factors give the total strategic factor rating. For components of these assessments see table later in this section.*

Germanium has a moderate strategic factor since, apart from its currently burgeoning electronic applications, it is a minor by-product of zinc and copper. Supplies are not unduly affected by political machinations but its consumption profile is biased towards military applications. It is currently a low volume metal but usage could double given a boom in fibre optics or solid state laser technology.

**Physical Characteristics**

Technically a semi-metal with an electrical conductivity which increases with temperature and is a sensitive function of impurities, termed dopants, when added deliberately.

## OCCURRENCE AND PRODUCTION

Occurs as germanium oxide in some zinc sulphide and lead/zinc/copper sulphide ores. Also occurs in carbonaceous materials but is not yet viably extracted from coal ash and flue dusts or from petroleum refinery residues.

**World Reserves**

Viable reserves are vast and the USBM estimate potential worldwide resources at several million tonnes. Reserves are widespread although not accurately delineated with sizeable reserves in the USA and Zaire.

**Production Characteristics**

| | |
|---|---|
| *Mining Methods:* | Depend on the sulphide ore body in which germanium is found |
| *Processing:* | Collected from flue dusts during copper refining and after volatilisation during zinc refining. Zone refined to required purity. Electronic applications then require the careful growth of fault-free single crystals |
| *Energy Requirements:* | Irrelevant due to by-product status |
| *By-Products:* | Copper, lead, zinc, perhaps silver, cadmium, gold, etc |
| *Environmental Factors:* | Occurs only as a trace element in ores and has little ecological impact |

*Development Lead Time:* Depends on parent metals.

**The Major Producers**

| *Countries* | *1981 Output (Tonnes)* |
|---|---|
| USA | 30 |
| Zaire | N/A* |
| Namibia | N/A* |
| Others | 80 |
| WESTERN WORLD TOTAL | 110 |

(Communist bloc output is estimated at 15 tonnes a year.)

*Plant production unknown but ore production estimated at 7.3 tonnes in Namibia and 22.5 tonnes in Zaire.

World mined production is centred in Zaire, the USSR, Namibia and the USA. Europe is a major producer of refined germanium utilising concentrates from Zaire, Namibia and less pure germanium from China. Only an approximate estimate of the USSR's output is available but the USA imports 1-2 per cent of its annual requirements from the Soviet bloc. Germanium output is ultimately dependent on current rates of base metal production and on the availability of germanium-bearing residues from earlier operations. 1981 saw a cessation in output of germanium-rich zinc concentrates in Zaire but production has resumed. New sources include increased Chinese output and exports, the new Jersey Miniere zinc facilities in Tennessee, USA, and the Canadian zinc mining industry.

**Technological Developments**

US research bodies are seeking a viable method of recovering germanium from the waste products of coal burnt in power generators which would vastly increase supplies. Research is also aimed at new methods of preparing germanium, either crystalline or amorphous, for use in electronics.

## COMPONENTS OF RISK ASSESSMENT

| *Factor* | *Rating* | *Comment* |
|---|---|---|
| **Production Risks** | | |
| Existing capacity | 5 | Not in marked over-capacity. Production to some extent limited by output of germanium-rich zinc concentrates, which occur sporadically |
| Labour disputes | 1 | Unlikely |
| Violent conflict | 3 | The importance of Zaire and Namibia may pose problems |
| Range of primary supply sources | 8 | Limited to Zaire, Namibia, USA, USSR and China |
| Time lags for new supplies | 1 | Stockpiled concentrates held in Europe and USA provide a limited buffer to temporary shortage |
| **Transportation Risks** | | |
| Primary | 2 | Only present as a small proportion of zinc or copper ore so transport risks minimised by dilution factor |
| Secondary | 1 | Only small volumes but great care must be taken due to fragility of finished crystals and other devices |
| **Application/Use Risks** | | |
| Total economic impact | 1 | Market volume totals only 110-120 tonnes a year |
| Effect on key industries | 9 | Heavy bias towards military applications for optical instrumentation plus use in growing fibre optic telecommunications industry |
| Availability of substitutes | 9 | Substitutes available but, except in semi-conductor chip technology, largely inferior on performance grounds |
| Longer term substitutability | 9 | There are few semi-conductor elements and germanium retains its markets on the strength of its unique characteristics |
| **Trade Risks** | | |
| Collusive price agreements | 2 | An active free market supplied by China and the USSR restrains the tendency of the European refiners and Eagle-Picher to dominate the market |
| Embargoes | 1 | Unlikely except as in the finished devices e.g. USA embargo of military equipment sales to Russia |

*These components are assessed on a scale from 1 (lowest) to 10 (highest) risk. Grouped together, they form the basis for the assessments of the factors for likelihood of supply disruption, and for cost of such disruption which appear in the first table in this section.*

## US GERMANIUM CONSUMPTION

By Market Sector 1981

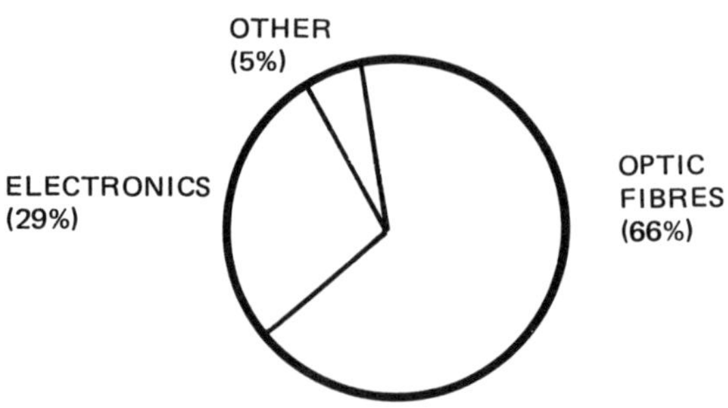


*Source: US Bureau of Mines, 1982*

## TRANSPORTATION

The most fragile primary link is for concentrates being transported from Zaire to Europe, mainly by ship. As with other by-product elements, the dilution of the required element in the concentrate is such that the loss of a single shipment has little overall effect. The USA imports 40-45 per cent of its requirements largely from Europe as the refined metal but total quantities are small and transport is not a problem.

## APPLICATIONS

### Consumption Trends

Germanium has many special applications in electronics, chemotherapy, polymer chemistry and optical instrumentation. Consumption has grown rapidly from a low base and has doubled over the last five years in the USA to reach 46 tonnes by 1982. From a 1978 base, demand for germanium is expected to increase at an annual rate of 5.5-6.0 per cent until 1990. A more rapid rate of growth is expected in Japan.

### The Principal Markets

Germanium is incorporated into solid state devices, including light emitting diodes for visual display units, lasers, etc but its general use in electronics is continuing on the slow downward trend begun in the late 1960s. Germanium is employed in infra-red detection devices for both military and civilian sectors and this application could double in volume over the next two-three years unless reflective technology, involving lower weight germanium utilised per device, overtakes refractive. Fibre optics contain 0.5-2.0 per cent germanium depending on type and this market is expected to increase in total value by an order of magnitude to a global $1 billion by 1985. These applications are all at the forefront of technology and current materials and methods could possibly be superceded leaving germanium without a secure base. But full scale development of optical fibre telecommunications systems, using germanium oxide-cored fibres as a transmission medium, is continuing with notable success by the Japanese. Germanium oxide is also used as a polyester fibre condensation catalyst although this is a relatively price sensitive application.

### Recycling

Over 50 per cent of the metal used in the manufacture of most electronic and infra-red sensing devices is routinely recycled as new scrap due to inevitable production faults. Old scrap volumes are minimal and the increased use of fibre optics containing under two per cent germanium will add to this.

### Substitution

Cheaper silicon has been substituted increasingly for germanium in many electronic applications, whilst indium, gallium and arsenic compounds may be used on grounds of performance and speed. Germanium remains more reliable in some high-frequency and high power requirements. Zinc selenide is a poor substitute in infra-red guidance systems. The use of germanium oxide in fibre optics appears secure although the percentage used in the glass may be subject to reduction.

## INTERNATIONAL TRADE AND WORLD PRICES

### Supply Arrangements

Germanium is traded via producer/consumer contracts and on a free market which has recently gained in volume and volatility. It is traded either as germanium dioxide of stated purity or as germanium metal with a stated electrical conductivity, (a sensitive function of purity) either as ingots or as single crystals. Prices are quoted in US dollars per kilo in minimum 20 kilo lots. The brand of origin is important since in such a technically precise market, small quality differences are vital. Prices tend not to be correlated to zinc but are dependent on the level of supplies offered by Russia and China.

### Companies

Eagle-Picher is the only US primary germanium producer, handling around 18-20 tonnes a year, although Kawecki Berylco, a division of Cabot Corporation maintained a plant using imported source materials. The Bunker Hill plant closure in the USA in December 1981 has materially reduced germanium-rich zinc concentrate supplies. Other major producers include:

—Penarroya (France) 33 tonnes a year
—Metallurgie Hoboken-Overpelt (Belgium) 30 tonnes a year
—Soc. Mineraria e Metallurgica di Pertusola (Italy) 9 tonnes a year
—Preussag AG (West Germany) 10 tonnes a year
—Bleiburger Berwerks Union AG (Austria) N/A

The refined output of the state-owned companies of Russia and China is considered to total between 16 and 17 tonnes a year.

### Prices

Germanium's free market has seen great activity and price volatility over the 1980-82 period with a peak of $1,250/kilo for the metal and $850/kilo for the oxide in mid-1981. The influx of speculators and a rush of enthusiasm for germanium's applications in electronics and optic fibres caused the rise, but the peak was not sustained and prices fell 40 per cent in the ensuing 18 months. Producer prices showed a more gentle, sustained increase.

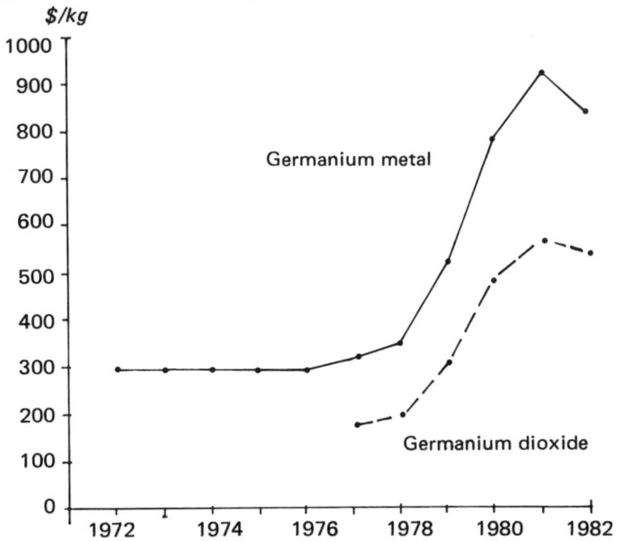


**Germanium Prices—Annual Averages for zone refined metal and electrolytic grade dioxide (Free market)**

# GOLD (Au)

| | |
|---|---|
| **FACTOR FOR LIKELIHOOD OF SUPPLY DISRUPTION** | **4.0** |
| **FACTOR FOR COST OF SUCH A DISRUPTION** | **6.8** |
| **TOTAL STRATEGIC FACTOR** | **26.4** |

*The factors for supply disruption and cost of such a disruption are assessed on a scale from 1.0 (lowest) to 10.0 (highest) risk. Multiplied together these factors give the total strategic factor rating. For components of these assessments see table later in this section.*

Gold has a relatively high strategic factor due to its unique status among all commodities as a long term store of value and as a sensitive indicator of geo-political tensions. Its use as an industrial metal is totally overshadowed by its importance as a monetary instrument. Gold continues to be hoarded by official agencies and by investors in many areas of the world. World reserves are relatively large, with 50 per cent concentrated in South Africa. Above-ground stocks total around two billion ounces representing 55 years' mined production at current levels. A genuine gold shortage is unlikely although a sudden speculative rush can create the appearance of insufficiency.

### Physical Characteristics

Soft, yellow, corrosion resistant metal with fairly high melting point (1,064$^{o}$C) and a high specific density (19.3); excellent conductor of heat and electricity.

## OCCURRENCE AND PRODUCTION

Gold is found in finely disseminated metallic form in quartzose and other hard rocks or in alluvial deposits formed from the debris produced by the weathering of such rocks.

### World Reserves

Gold reserves are estimated at 1.2 billion troy ounces, sufficient for 30 years of mining activity at current levels. World resources are estimated at two billion ounces of which 15-20 per cent are by-product resources. South Africa possesses about half of all world resources.

### Production Characteristics

| | |
|---|---|
| *Mining Methods:* | Alluvial deposits mainly mined by large scale hydraulic methods. Disseminated reef deposits, usually containing under 10 grams of gold per tonne of ore, mined in deep underground operations |
| *Processing:* | Extracted by simple crushing and amalgamation when iron pyrites not present. When pyrites present a high fraction of gold passes into the tailings and is then extracted by a chemical cyanidation process. |
| *Processing (contd):* | The extraction of gold as a base metal by-product requires methods specific to the particular mineral |
| *Energy Requirements:* | Moderately high dependent on grade |
| *By-Products:* | Copper, antimony, lead, tellurium, selenium, nickel, molybdenum, zinc; uranium; depending on ore type |
| *Environmental Factors:* | Aesthetic problems of disposing of tailings and disrupting the ecology by exploiting alluvial deposits |
| *Development Lead Time:* | Up to ten years for underground mines, three to five years for large scale mining of alluvial deposits. Up to three months for one-man gold panning operations from shallow, rich alluvial deposits. |

### The Major Producers

South Africa is the world's leading gold producer and the country's investment and production policies are of paramount importance. Output is concentrated under the control of three major mining houses and their paternalistic view of the market means that they balance the average grade of ore mined with price levels, with the purpose of obtaining a steady income. South African output is expected to remain in the 650-700 tonnes a year region over the next five years but then fall off steadily to about 350 tonnes a year at the end of the century. But this is a very price sensitive forecast. Soviet gold production is estimated to be in the 280-350 tonnes

## COMPONENTS OF RISK ASSESSMENT

| *Factor* | *Rating* | *Comment* |
|---|---|---|
| **Production Risks** | | |
| Existing capacity | 1 | Currently adequate to meet requirements whilst adjustment of grades mined allows some control of output to suit demand |
| Labour disputes | 1 | Labour problems in South Africa, the leading producer, occur occasionally but are quickly resolved |
| Violent conflict | 10 | Concentration of output in Russia, China and South Africa make gold very sensitive to any major conflict whilst prices would be drastically affected by such events |
| Range of primary supply sources | 10 | South Africa and USSR (exports) provide 75 per cent of total mined world gold supplies. Reserves are concentrated in South Africa |
| Time lags for new supplies | 10 | Up to ten years for deep underground mine, up to three years for major alluvial deposit exploitation but only a few months for small panning operations in Brazil and elsewhere |
| **Transportation Risks** | | |
| Primary | 1 | Transportation tends not to be a problem since refining is carried out as near to the mine as possible |
| Secondary | 1 | As a high value commodity all precautions are taken and, barring a major robbery, problems are minimised |
| **Application/Use Risks** | | |
| Total economic impact | 1 | Gold bullion supply to the non-Communist private sector totals 900-1,100 tonnes a year |
| Effect on key industries | 8 | Extensively used as an investment medium and store of value in troubled areas of the world. Industrial uses in electronics etc, are in the high technology sectors |
| Availability of substitutes | 6 | Substitutes are available for industrial uses although these are often inferior in performance. Economies of use favoured for jewellery in times of high prices, but its investment role remains intact |
| Longer term substitutability | 6 | Gold's status is unlikely to be replaced although substitutes could be developed and used for all practical applications |
| **Trade Risks** | | |
| Collusive price agreements | 8 | A well developed free market gives the semblance of an unregulated price, but a large degree of cooperation over the timing and quantity of gold sales by South Africa and the USSR is suspected. Central bank intervention has also been noted at certain price levels in the past |
| Embargoes | 3 | Possible, and these would be likely to be in conjunction with East/West trade |

***These components are assessed on a scale from 1 (lowest) to 10 (highest) risk. Grouped together, they form the basis for the assessments of the factors for likelihood of supply disruption, and for cost of such disruption which appear in the first table in this section.***

## WORLD GOLD RESERVES

Total = 1,200 Million Troy Ounces

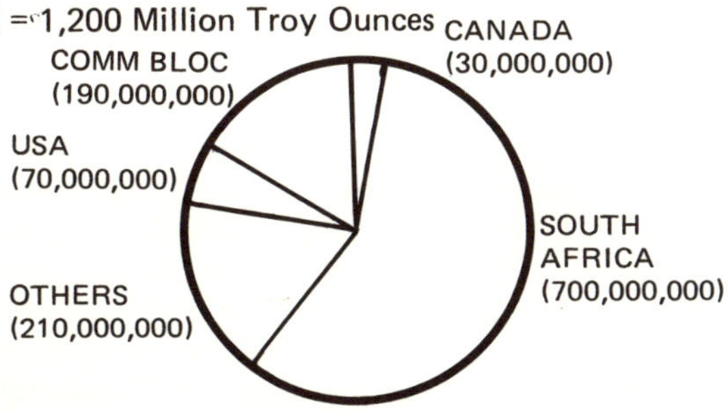


*Source: US Bureau of Mines, 1982*

## DEVELOPED COUNTRIES FABRICATED GOLD CONSUMPTION

By Market Sector 1981

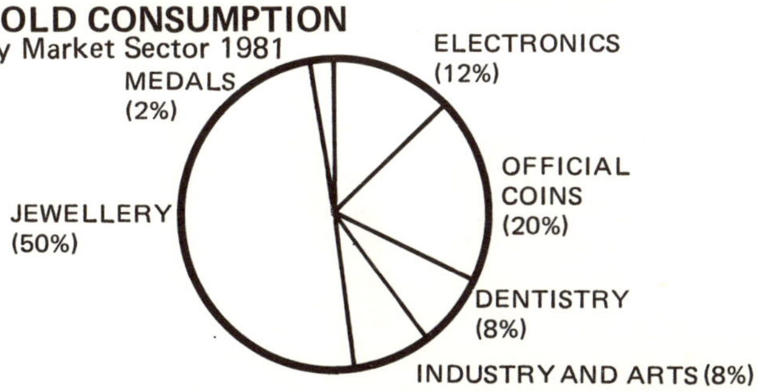


*Source: Gold 1982, Consolidated Gold Fields PLC*

| Countries | Output 1981 (Tonnes) | | Percentage of Western World Output | |
|---|---|---|---|---|
| AFRICA | | 698.0 | | 72.6 |
| South Africa | 657.6 | | 68.4 | |
| LATIN AMERICA | | 96.1 | | 10.0 |
| Brazil | 35.0 | | 3.6 | |
| ASIA | | 34.5 | | 3.6 |
| Philippines | 24.9 | | 2.5 | |
| N AMERICA | | 90.1 | | 9.4 |
| Canada | 49.5 | | 5.1 | |
| USA | 40.6 | | 4.2 | |
| OCEANIA | | 34.5 | | 3.6 |
| EUROPE | | 8.5 | | 0.8 |
| WESTERN WORLD TOTAL | | 961.7 | | 100.0 |

(USSR output estimated at 260 tonnes.)

*Source: Consolidated Gold Fields, PLC, Gold 1982.*

a year range and is expected to increase. Elsewhere the 'second line' producers such as Brazil, Canada, Australia, USA, Zimbabwe and Ghana are extending their exploration efforts to offset the depletion of established mines. Brazil and, to some extent, Australia have a large number of small operations which can switch production on and off very easily depending on prevailing prices. Much current development is a legacy of the historic price peaks reached in 1979-80. China is a relative newcomer to the field, but gold mining has been singled out as a priority for development. Reserves are thought to be large, offering good prospects for future growth in output. Gold is mined in over 30 countries on all continents of the world. South Africa's favourable position relative to other gold producers is enhanced when weighted average break-even costs to produce gold are considered. Consolidated Gold Fields estimated a production cost in 1980 of $176/oz for South Africa (equating to perhaps $210 in 1982) and $224/oz average for the rest of the non-Communist world ($270 in 1982).

### Technological Developments

The South African gold mining houses are at the forefront of developing deep mining techniques with improved ventilation and overall efficiency. Methods of heap-leaching gold ores using cyanide solutions remain under constant review and the USBM is on the point of perfecting a way of treating cyanide process wastes.

## TRANSPORTATION

Gold forms such a minute constituent of the ores or sands in which it is found that processing up to at least the final refining stage is accomplished near to the mine site. Refined gold is transported by air or train under heavy surveillance. Its high unit value makes it an attractive target for terrorists and thieves whilst producing countries have the constant spectre of smuggling with which to contend. The main transport routes are from South Africa, the Philippines and Oceania to the fabricating centres of Western Europe, Japan and the USA.

## APPLICATIONS

### Consumption Trends

The consumption pattern for gold is to some extent an artificial one because of its role as a medium of exchange and its heavy reliance on the luxury (jewellery, objets d'art and medallions) sectors whose health depends largely on the level of freely disposable income, the prevailing gold price, and the performance of other investment vehicles. In almost all types of gold purchases the decorative and investment elements tend to be inextricably mixed with jewellery at one end of the spectrum and bullion holdings at the other. The USBM estimate fabricated gold demand to increase by 2.2 per cent a year during the 1980s.

### The Principal Markets

The pie chart on page 79 shows the end-use markets for fabricated gold in the developed world. The developing countries used approaching three-quarters of their gold consumption for jewellery in 1981, although this sector fluctuates widely with price and the readiness to dishoard when prices appear high to the individual from the Middle and Far East. Most areas noticed a move towards lower caratage, lighter pieces during 1980-81 but preliminary 1982 figures have highlighted an increase in jewellery fabrication.

The use of gold in electronics is concentrated in the USA, Japan and Western Europe and remained static at 85-86 tonnes a year worldwide during 1980-81. The continuing reduction in the thickness of plated surfaces, the trend towards miniaturisation and the development of non-gold alloy substitutes have offset the growth in the electronics industry as a whole. Dental alloys consume 61-62 tonnes a year gold, a declining sector, largely because the rising costs of both dental treatment and the gold itself have forced patients and their national or private state insurance schemes to look for cheaper alternatives.

Other industrial and electronic uses include gold in various plating applications, rolled gold and chemicals. Official coins used 227 tonnes of gold worldwide in 1981 and are an attractive investment medium with a known weight and purity of gold content. The Krugerrand remains dominant in this area. Medals, medallions and imitation coins formed a declining sector over the past five years, partly due to assorted governmental controls and to increased interest in official coins and jewellery. Identified bar holdings totalled 280 tonnes in 1981, 250 tonnes of this in Asia.

### Recycling

Old gold scrap supplies around 60 per cent of total gold consumed in the USA. This area includes recycled jewellery, electrical contacts from old computers, discarded dental goods, old coins and other items. The bulk of identifiable dishoarding occurs as jewellery in the Middle and Far East and scrap quantities are a sensitive function of prices. There is a developed network of secondary collectors and refiners which cope with the continuous flow of industrial scrap and the sporadic miscellany of old scrap.

### Substitution

Silver and lower caratage gold jewellery has come into vogue whilst base or platinum group metals are increasingly being used in electrical and electronic applications. Palladium/silver and nickel/chrome alloys are gaining in popularity for dental use. A tendency to economise by switching from rolled to plated gold has been seen in the industrial and decorative areas.

## INTERNATIONAL TRADE AND WORLD PRICES

### Supply Arrangements

Gold bullion supply to the non-Communist private sector includes Western world mine production plus erratic exports by the Communist bloc plus net official sales by such bodies as the IMF and US Treasury Department. Supplies totalled 806 tonnes in 1980 and 985 tonnes in 1981. The South African gold mines sell their refined output via the South African Reserve Bank mainly in the Zurich bullion market and tend to spurn the futures markets. Supplies from the USSR and other producers are directed via Zurich or London. The London gold bullion market is controlled by five brokers who officially "fix" the gold price twice a day, reflecting world trading conditions. Leading Swiss banks control the Zurich market and the two main centres interact closely. Gold futures markets exist in Chicago, New York, London, the Far East, Canada and Australia, of which the most important is the New York Commodity Exchange (COMEX). The usual contract size is 100 troy ounces of 0.995 purity gold with a price quoted in $/troy oz. The futures markets register huge volumes traded and are mainly used for speculative purposes with under one per cent of contracts taken to delivery. Gold coins, particularly Krugerrands, are traded on a physical or futures basis. Despite the official demonetisation of gold, faith in its intrinsic value has been retained such that international gold movements are of great political interest. The pattern has been complicated by the conversion of surplus petrodollars into huge Middle East deposits. Nonetheless, the continuous question mark remains the level of Communist sale to the West whilst vastly increased gold imports by the Japanese during 1981 and 1982 have boosted total imports into the Far East.

### Companies

The four mining houses, Consolidated Gold Fields, Anglo-American, Gencor and Barlow Rand, of South Africa effectively control that country's output. Other producers include the state-owned operations of the USSR and China, RTZ's Bougainville Copper of Papua New Guinea, Newmount Pty Ltd of Australia and Homestake Mining Co of the USA. Major gold refiners include Johnson Matthey and Engelhard in the UK, the South Africa Chamber of Mines, Metallurgie Hoboken Overpelt in Belgium and Degussa of West Germany.

### Prices

The gold price is a sensitive function of political and economic factors particularly on the futures markets where fluctuations mirror the underlying bullion price trends but are more exaggerated and volatile. Gold prices reached a peak of $850 in January 1980 in the wake of the Russian invasion of Afghanistan but then weakened over the next two years as the recession deepened and interest rates rose to historic highs.

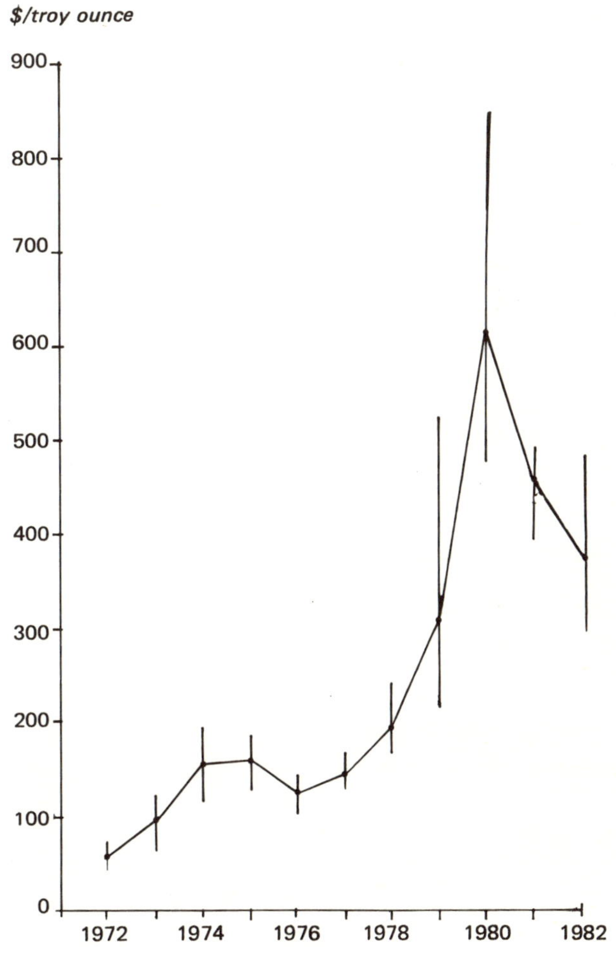


**Gold Prices—Annual Averages, Highs and Lows (London bullion afternoon fixes)**

# INDIUM (In)

| | |
|---|---|
| **FACTOR FOR LIKELIHOOD OF SUPPLY DISRUPTION** | **1.7** |
| **FACTOR FOR COST OF SUCH A DISRUPTION** | **6.2** |
| **TOTAL STRATEGIC FACTOR** | **10.5** |

*The factors for supply disruption and cost of such a disruption are assessed on a scale from 1.0 (lowest) to 10.0 (highest) risk. Multiplied together these factors give the total strategic factor rating. For components of these assessments see table later in this section.*

Indium has a moderate strategic factor. It is an expensive metal with a small but broadly based market and is extracted as a by-product of lead and zinc. Reserves are large but the availability of good grade ores can be a problem leading to sporadic supply/demand imbalances and a volatile price. Production is dominated by relatively stable sources in the USA, Canada and Japan.

### Physical Characteristics

Fairly high density (7.3) semi-metal with low melting point (156$^{o}$C); useful electrical properties; no mammalian toxicity.

## OCCURRENCE AND PRODUCTION

Indium occurs mainly in zinc and sulphide ores and in some tin, lead, tungsten, manganese and copper ores in quantities that vary from a trace upwards to about 2.8 per cent.

### World Reserves

These are estimated at 50 million oz (22,700 tonnes) indium content, sufficient for over 35 years of consumption at current levels. Commercial recovery comes mainly from lead and zinc smelters but the indium content varies widely between different zinc ores.

### Production Characteristics

| | |
|---|---|
| *Mining Methods:* | Depending on lead/zinc ore body |
| *Processing:* | Extracted from flue dusts at lead and zinc smelters and from the cadmium-bearing residues during the purification of zinc sulphate. Electrolysed to 99.9999 per cent pure metal |
| *Energy Requirements:* | Largely irrelevant due to by-product status and necessity of removing impurities from flue gases before emission |
| *By-Products:* | Lead, zinc, cadmium, silver |
| *Environmental Factors:* | None unique to indium extraction |
| *Development Lead Time:* | Depends on zinc mine plus 3-6 months for collection of sufficient flue dusts to warrant indium extraction. |

### The Major Producers

| *Countries* | *Output 1981 (Thousand Troy Oz)* | *Percentage of Western World Output* |
|---|---|---|
| USA* | 320 | 41.5 |
| Canada | 150 | 19.5 |
| Japan | 150 | 19.5 |
| Peru | 150 | 19.5 |
| Others | N/A | N/A |
| WESTERN WORLD TOTAL | 770 | 100.0 |

(USSR production is currently estimated at 200,000 troy oz and production in other Communist bloc countries 100,000 troy oz a year.)

*Figure withheld by USBM, estimated from Roskill's Metal Databook, 1982.

*Source: USBM, Mineral Commodity Summaries, 1982.*

These tend to be synonymous with major zinc mining and refining countries and include the USA, Canada, Peru and Japan. The Communist bloc is estimated to produce 300,000 troy oz a year indium. Refined output is also supplied by Belgium and West Germany, largely from imported zinc concentrates. The closure of particular zinc mines can sharply tighten the availability of source materials. This is emphasised in the USA due to a continuing shift away from western state zinc deposits, comparatively rich in indium, to leaner Mississippi ores.

### Technological Developments

Research is principally directed towards developing ways of manufacturing pure indium metal, alloys and semiconductors more cheaply.

## COMPONENTS OF RISK ASSESSMENT

| *Factor* | *Rating* | *Comment* |
|---|---|---|
| **Production Risks** | | |
| Existing capacity | 3 | Currently in over-capacity coinciding with glut in zinc market but often the markets are not in synchrony and upper limit of indium output is dictated by zinc production |
| Labour disputes | 1 | No history of major disruption |
| Violent conflict | 1 | Unlikely to affect indium output |
| Range of primary supply sources | 7 | Resources are large and widespread but there is a shortage of zinc ore bodies with high indium grades, concentrated in the USA, Canada, Peru, Japan, USSR and China |
| Time lags for new supplies | 7 | Depends on discovery of suitable ores plus up to ten years to develop zinc ore body and three to six months to accumulate sufficient flue dusts to treat |
| **Transportation Risks** | | |
| Primary | 1 | Transported as minor component of zinc concentrates |
| Secondary | 1 | Indium products, except perhaps as sophisticated solid state devices, unlikely to be subject to disruption |
| **Application/Use Risks** | | |
| Total economic impact | 1 | Relatively minor volume of around 55 tonnes a year consumed |
| Effect on key industries | 9 | Broadly based market in civilian and military infra-red sensing systems, solders and alloys, electronic displays, nuclear reactors. Important possible future use in photovoltaic cells |
| Availability of substitutes | 7 | Silicon in transistors, gallium as an expensive alternative in alloys, hafnium in nuclear reactor control rods, liquid crystals in displays |
| Longer term substitutability | 6 | Possible if necessary but with cost and quality penalties |
| **Trade Risks** | | |
| Collusive price agreements | 3 | Fairly likely considering the mere handful of refined metal producers, but so far in 1981-82 not had much success |
| Embargoes | 1 | Unlikely except as component of military equipment |

*These components are assessed on a scale from 1 (lowest) to 10 (highest) risk. Grouped together, they form the basis for the assessments of the factors for likelihood of supply disruption, and for cost of such disruption which appear in the first table in this section.*

### WORLD INDIUM RESERVES

Total = 50,000,000 Troy Ounces

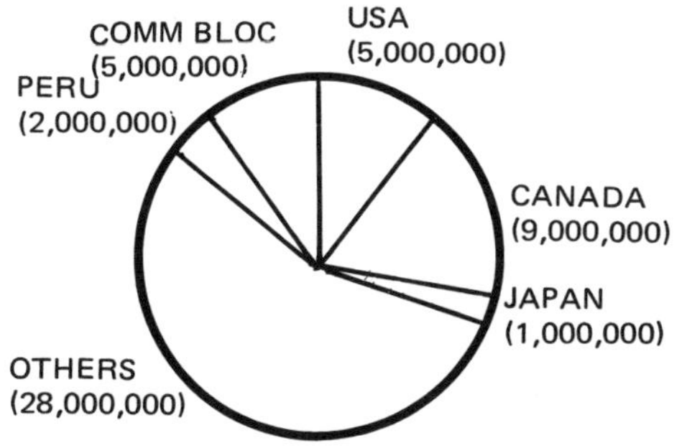


*Source: US Bureau of Mines, 1982*

### US INDIUM CONSUMPTION

By Market Sector 1981

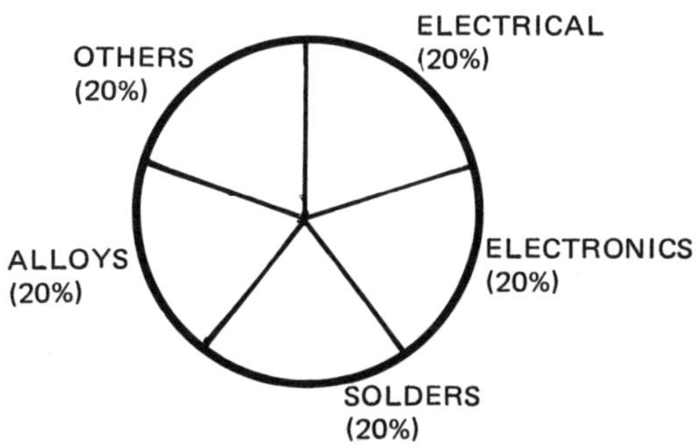


*Source: US Bureau of Mines, 1982*

## TRANSPORTATION

The USA is an importer of indium metal, 380,000 oz in 1981, mainly from Peru, Belgium and Canada. Japan is also a major producer but consumes most of its own output. The major trade links are from Belgium, West Germany, South America and Canada to the USA and several European countries.

## APPLICATIONS

### Consumption Trends

Indium has seen a moderately high growth rate since the advent of widespread solid state devices. Annual world demand is currently around 53-55 tonnes with a growth rate of three per cent a year predicted by the USBM. This is almost three times as fast as that predicted for its parent metal, zinc.

### The Principal Markets

A pie chart showing the end-use markets in the USA is given on page 83. The solder and alloy sector is relatively stable and here indium is used for plating on electronic components, low melting point alloys, coatings on electrical contacts, low friction bearings and in dental alloys. The electronic applications have the potential for major growth since indium is an integral part of infra-red detectors for military surveillance or heat seeking missile systems and also for civilian fire detection and scanning devices. Indium is employed in electronic displays in instruments and also as a transparent, conductive glass coating for liquid crystal displays. Nuclear control rods are often made of indium but this area is flagging as the need for the rapid development of nuclear power subsides.

Research is continuing into the use of indium in solar power cells and the US Solar Energy Research Institute optimistically predicted an 800 per cent increase in indium demand by 1990 if a new photovoltaic cell containing a copper/indium/selenium film becomes marketable. Increased outlets for indium's applications in other discrete electronic devices are also being investigated.

### Recycling

Recycled supplies are minor being mostly from alloy scrap materials since other applications are either dissipative or employ only minute quantities of indium per unit application.

### Substitution

Silicon has largely replaced germanium-indium alloys in transistors. Gallium is a possible, but expensive, substitute in many alloys. Boron carbide and hafnium can replace indium in nuclear reactor control rods. Indium has no serious competitors for use in infra-red detectors. Liquid crystal displays have largely overtaken electronics in small portable devices such as watches.

## INTERNATIONAL TRADE AND WORLD PRICES

### Supply Arrangements

Indium is traded both on a producer and free market basis and has had a volatile price history. Prices are quoted in $/troy oz for 1,000 troy oz lots in the USA and in $/kilo for 100 kilo lots in Europe and Japan. It is important to note that different grades of the metal are required for various applications with metal of 99.9999 per cent purity for electronic applications. Different producers have taken the lead in setting the tone for indium prices over the past 20 years and it is a specialised market due to the high unit cost and small market volume. The free market is supplied by material from the USSR and China plus surplus material (after long term consumer contracts) from Western producers.

### Companies

Refined indium metal output is dominated by Metallurgie Hoboken-Overpelt of Belgium, Nippon Mining of Japan, Indium Corporation of America, USA, Preussag of West Germany, Cominco of Canada and the state-owned operations of the USSR and China.

### Prices

In January 1982 it was felt that the minimum price for profitable extraction of indium was $150-185/kilo and below that companies would simply not process residues. In 1981 Nippon Mining stopped output but it is likely that others did not follow suit and as a result prices halved during 1981-82.

**Indium Prices—Annual Averages**

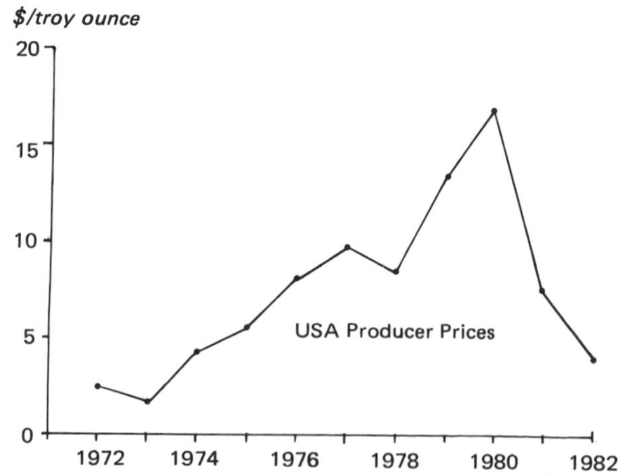

# IRON (Fe)

| | |
|---|---|
| **FACTOR FOR LIKELIHOOD OF SUPPLY DISRUPTION** | **2.2** |
| **FACTOR FOR COST OF SUCH A DISRUPTION** | **4.3** |
| **TOTAL STRATEGIC FACTOR** | **9.5** |

*The factors for supply disruption and cost of such a disruption are assessed on a scale from 1.0 (lowest) to 10.0 (highest) risk. Multiplied together these factors give the total strategic factor rating. For components of these assessments see table later in this section.*

Iron has a low strategic factor. It is mined on all continents of the world, supplies are large and the metal appears in many cases to be taken for granted. Steel, the best-known iron alloy, is vital in all aspects of modern technology, and has the largest volume market of any metal or alloy. Iron lacks strategic quality since its availability is the foundation on which the industrialised world is built.

### Physical Characteristics

Specific density of 7.88; convenient melting point (1,531°C) with good weldability, malleability and strength; principal advantage is its ability to form a wide range of alloys, including those with other metals and such non-metals as carbon, oxygen and nitrogen.

## OCCURRENCE AND PRODUCTION

Comprises five per cent of the earth's crust of which a small part has been concentrated into rich deposits. The main ores are magnetite ($Fe_3O_4$), haematite ($Fe_2O_3$), quartz with accessary iron silicates and iron carbonates.

### World Reserves

These are estimated at 108 billion tonnes recoverable iron sufficient for over 315 years of consumption at current levels. The largest reserves are in Precambrian sedimentary formations locally enriched to high grade deposits (50-68 per cent Fe) found in the USA, Canada, South America, Africa, India, Australia, the Far East and the USSR. An alternative source is the oolitic ironstones containing 20-40 per cent iron plus troublesome impurities, still mined in the USA and Western Europe.

### Production Characteristics

*Mining Methods:* Usually open cast, large scale mines

*Processing:* Ore is beneficiated, sintered and pelletised prior to reduction to iron by smelting in a blast furnace, by direct reduction or by partial reduction then converted to cast irons, stainless or other steels

*By-Products:* Manganese, phosphate, copper, silver, gold, vanadium, titanium, nickel and sulphur

*Environmental Factors:* Their importance is magnified by the volume of the iron ore industry and include disposal of solid wastes, elimination of dust and reclamation of process water and tailings sites

*Development Lead Time:* Around ten years with extensive infrastructure needed to service mine sites.

### The Major Producers

| *Countries* | *Output 1981 (Million Tons)* | *Percentage of Western World Output* |
|---|---|---|
| USA | 74 | 14.0 |
| Australia | 90 | 17.1 |
| Brazil | 98 | 18.6 |
| Canada | 47 | 8.9 |
| India | 41 | 7.8 |
| South Africa | 26 | 4.9 |
| Sweden | 25 | 4.8 |
| Others | 126 | 23.9 |
| WESTERN WORLD TOTAL | 527 | 100.0 |

(Communist bloc production is currently estimated at 326,000,000 tonnes, production in the USSR 241,000,000 tonnes and China 70,000,000 tonnes a year.)

*Source: US Bureau of Mines, 1982.*

The USSR, Brazil, Australia and the USA dominate world production of iron ore with 59 per cent of world output between them. World pelletising capacity is now in the 250-270 million tonnes a year range with the principal

## COMPONENTS OF RISK ASSESSMENT

| *Factor* | *Rating* | *Comment* |
|---|---|---|
| **Production Risks** | | |
| Existing capacity | 1 | Current over-capacity in all stages from iron ore to iron and steel to scrap |
| Labour disputes | 3 | Canada, Australia and the USA have relatively poor mining industry labour relations |
| Violent conflict | 1 | Unlikely to affect the bulk of iron ore production |
| Range of primary supply sources | 1 | Iron ore widely available on all continents although 59 per cent of Western output concentrated in the USA, Australia, Canada and Brazil |
| Time lags for new supplies | 1 | Lead time for a new mine is up to ten years but current overcapacity and plentiful scrap supplies ensure that alternative supplies are usually available |
| **Transportation Risks** | | |
| Primary | 4 | Liable to dock strikes affecting exports from Australia and Canada, rail strikes in Canada and elsewhere. Long ocean transport routes for large volumes of ore |
| Secondary | 4 | Liable to dock strikes and to rail strikes in the UK, the USA etc, disrupting transport patterns |
| **Application/Use Risks** | | |
| Total economic impact | 10 | Vital to all branches of industry from construction to transportation |
| Effect on key industries | 3 | High volume carbon and cast steels and irons are low cost, convenient materials but the specialised and stainless steels have sophisticated technological applications |
| Availability of substitutes | 10 | Little likelihood of being able to replace iron and steel in short order |
| Longer term substitutability | 1 | Iron and steel could in the long-term be replaced but at high cost |
| **Trade Risks** | | |
| Collusive price agreements | 3 | The Association of Iron Ore Exporting Nations is currently embryonic but could become influential. Steels are subject to excessive price and trade regulations |
| Embargoes | 1 | Unlikely except for particular high technology steels needed in very low volumes |

*These components are assessed on a scale from 1 (lowest) to 10 (highest) risk. Grouped together, they form the basis for the assessments of the factors for likelihood of supply disruption, and for cost of such disruption which appear in the first table in this section.*

## WORLD IRON ORE RESERVES

Total = 108 Billion Tons Recoverable Iron

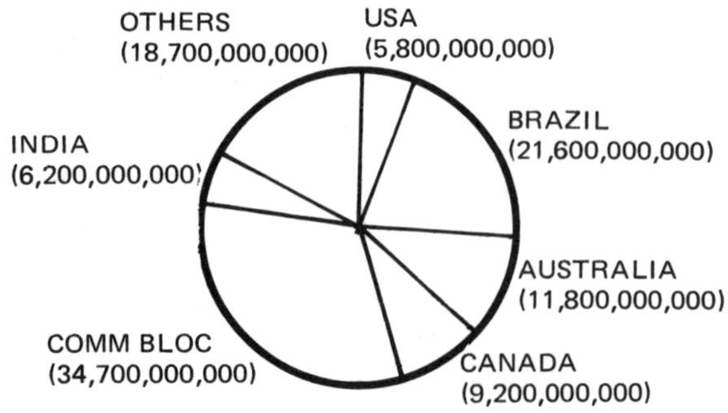


*Source: US Bureau of Mines, 1982*

## US IRON AND STEEL CONSUMPTION

By Market Sector 1981

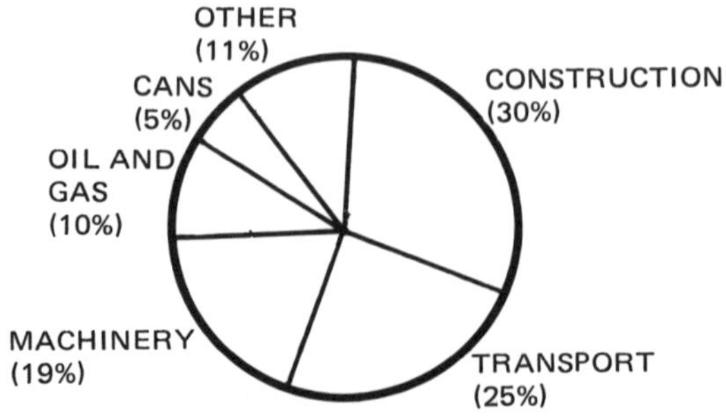


*Source: US Bureau of Mines, 1982*

increases supporting the concentration of output in the USA, Brazil and the USSR. Pig iron production is dominated by the USA, EEC, Japan and the USSR with other free market economies producing only 17 per cent of the world total. Raw steel output reflects the pattern of pig iron output with the USA at 16 per cent of world output, the EEC with 18 per cent and Japan with 14 per cent. The production of stainless and specialised steel alloys is a relatively high technology industry and has tended to be the province of the developed world with relatively little competition except from Korea, Taiwan and a few other Third World countries. The mining of iron ore and its conversion to pig iron and steel is a matter of national economic importance to many countries and of national pride to those least able to afford the environmental problems the industry produces or the financial support it requires. The future expansion of crude steel production is limited in the USA and EEC where overcapacity has forced closures, government subsidies and a rise in protectionism. Japan is fighting hard to keep crude steel output steady and all three are concentrating on the high-value steels. Many less-developed countries have large steel expansion projects and can exploit relatively cheap labour and indigenous fuel supplies to produce lower cost steel.

### Technological Developments

Pelletising to reduce blast furnace losses is a relatively new innovation. Blast furnace and direct reduction processes are subject to constant efforts to reduce costs with computerised control of process variables being a major step forward. Totally new steel alloys are thin on the ground with emphasis placed on meeting the realistic aim of gradually improving existing alloys by minor compositional and production technique changes.

## TRANSPORTATION

This forms a major component of this high volume sector of the mining industry. Iron ore is always beneficiated and usually pelletised or sintered prior to transportation to increase the grade and reduce volumes. Transport is usually by railway or huge adapted tankers and great efforts are being directed towards improving the efficiency of port facilities and of conveyor systems, particularly by Australia, Canada and Brazil—the major exporters. The transport of iron slurries by pipeline is increasing, particularly in South America. Ocean transport costs tend to account for 30-40 per cent of cif prices and are particularly sensitive to oil price increases.

## APPLICATIONS

### Consumption Trends

The USBM predicts that demand for iron ore and for steel will increase by 1.6 per cent a year during the 1980s. Demand for stainless steel is expected to grow more rapidly at three per cent a year. The next decade is likely to prove difficult for world steel and iron ore producers, given the trend towards lighter, high quality steels and the reduction in global economy growth predictions. A steel/GDP growth ratio of 0.4-0.6 for the industrialised countries and 1.2-1.9 for the developing countries is now the norm, suggesting that steel producers will benefit less than those in other industries from economic recovery.

### The Principal Markets

Over 99 per cent of all iron ore is used to produce iron and steel and the diversity of the end uses of these ferro-alloys is shown in the piechart. Cast irons containing two to four per cent carbon plus other additives are used for agricultural and engineering machinery components. Plain carbon steels with up to 1.7 per cent carbon comprise a high-volume, low cost sector and the steels combine fabricability with strength in combinations depending on the carbon content. The higher carbon content steels are used for springs and cutting tools. The properties of steel can be developed by alloying with a range of metals, including chromium, nickel, tungsten, manganese, vanadium and others, to increase toughness, strength, corrosion resistance and melting point. Iron alloys provide relatively cheap and effective soft magnetic materials for electricity generators.

From 1970-1980 worldwide steel production increased by 20 per cent but demand has contracted sharply during the recession. Demand growth is likely to be highest for high-quality low-alloy steels in the pipeline and automobile markets and for stainless steels in the chemical industry.

### Recycling

Iron and steel scrap collection and recycling is a major industry with apparent consumption of ferrous scrap in the USA put at 82.4 million tonnes in 1981, 68 per cent of 1981 apparent consumption of steel and cast iron. Europe and Japan produce proportionately less scrap since their machinery, automobiles, etc have a higher life expectancy. Steel scrap is directly used to make steel indistinguishable from that made solely from virgin pig iron, provided that the scrap level is strictly controlled according to the type of process. Around one per cent of scrap is used for ferro alloys and the chemical industry. In contrast to primary production, the scrap industry consists of a pyramid structure based on the efforts of individual collectors armed with automobile shredders etc. The USA is a net exporter of scrap to Western Europe and Japan.

### Substitution

Direct reduced iron is a potential substitute for the iron and steel scrap content (30-40 per cent) of steel. Steel substitutes include aluminium for tinplate in the canning industry, aluminium and plastics in the motor vehicle industry and concrete, aluminium and wood in construction. Steel tends to retain its markets on the grounds of convenience, cost and versatility but is superseded by aluminium in strength/unit weight comparisons and by titanium for corrosion resistance. Plastics and glass are additional non-metallic alternatives to steel in packaging, and both tend to be extremely competitive on price.

## INTERNATIONAL TRADE AND WORLD PRICES

### Supply Arrangements

Iron and steel are virtually never traded on a free market but are generally sold directly from producer to consumer, often within an integrated concern. The bargaining position of iron ore producers has started to improve by the formation of the Association of Iron Ore Exporting Countries but the power of this group is small since both Brazil and Canada have decided not to join. More than half the world's iron ore trade takes place under long-term delivery contracts that specify both prices and quantities and 30 per cent occurs through spot or short-term—under one year—contracts. The balance, which is declining, represents sales from "captive" mines, wholly owned or controlled by steel companies. Prices are quoted cif per tonne of ore, depending on the origin, iron and impurity contents, whilst there is a vast plethora of quotations for various steel and iron types. Steel is similarly traded on a long-term contract basis with producers often situated near their consumers, traditionally close to coal and limestone supplies and a port. The iron and steel scrap industry is an important side-issue whose size has encouraged many medium-sized scrap dealers, sorters and distributors to flourish leading to an active and competitive free market.

### Companies

The major companies are: the state-owned operations of China and the USSR; Cia. Vale do Rio Doce, Brazil; Hamersley Iron Pty Ltd, Australia; Mt. Newman Mining Co Pty Ltd, Australia; Reserve Mining Co, USA; and South African Iron & Steel Industrial Corp Ltd. Major Western integrated iron and steel concerns include: US Steel; National Steel of the USA; British Steel Corp; Thyssen and Krupp of West Germany. Steel companies tend to be state-owned or aided to a great extent from the nationalised British Steel to the US companies helped by protectionist tariffs on imports.

### Prices

Iron ore producer prices have remained on a moderately upward trend and a large component of the final cost of the ore to the blast furnace operator depends on freight rates. Heavy competition in the shipping sector caused ocean freight rates to decline in 1981, despite rising oil prices, thus muting the effect of increased ore prices. Ferrous alloy prices tend to be determined within each country and, for example, the USA has a trigger price mechanism to prevent the dumping of cheaper foreign steels and alloys.

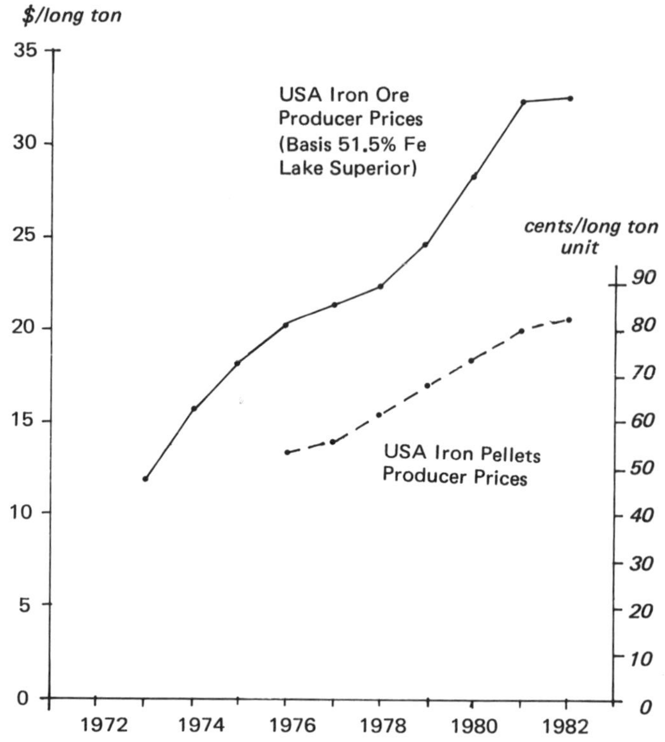


**Iron Ore Prices—Annual Averages**

# LEAD (Pb)

| | |
|---|---|
| FACTOR FOR LIKELIHOOD OF SUPPLY DISRUPTION | 2.0 |
| FACTOR FOR COST OF SUCH A DISRUPTION | 4.5 |
| TOTAL STRATEGIC FACTOR | 9.0 |

*The factors for supply disruption and cost of such a disruption are assessed on a scale from 1.0 (lowest) to 10.0 (highest) risk. Multiplied together these factors give the total strategic factor rating. For components of these assessments see table later in this section.*

Lead is an industrial metal with a fairly low strategic rating. Over the last 40 years it has assumed a potentially vulnerable 65 per cent dependency on the automobile industry with the remainder in construction, solders and other relatively minor uses. It has an unusual market structure which tends to reduce supply risks since the Third World accounts for under 20 per cent of both refined lead production and consumption whilst about 40 per cent of annual refined output derives from secondary materials. Reserves and mine production of lead are widespread.

**Physical Characteristics**

High specific density (11.4); low melting point (327°C); soft, malleable, corrosion resistant metal. Its density grants it relative immunity to X-ray and gamma radiation. It is also a cumulative mammalian poison.

## OCCURRENCE AND PRODUCTION

Lead is present in a wide range of minerals, the most important of which are galena (PbS), cerussite ($PbCO_3$), and anglesite ($PbSO_4$). Mainly occurs in mixed ores with zinc.

**World Reserves**

These are estimated at 165 million tonnes lead content, sufficient for over 45 years of consumption at current levels. Reserves are concentrated in the Americas and in Australia although commercial deposits have been identified in over 45 countries and on all continents except Antarctica. The Communist bloc possesses about 16 per cent of reserves.

**Production Characteristics**

*Mining Methods:* Mainly underground mines except for a few large open cast operations in Canada

*Processing:* Ore concentrated, generally by flotation methods, then sintered or roasted and finally smelted, drossed and refined to 99.9 per cent pure lead. Different refining processes (the Betts method in Canada and Peru, the Imperial Smelting method in Europe) to cope with different mixed ore types

*Energy Requirements:* 35-40 million Btu/tonne refined lead, fairly low

*By-Products:* Zinc, silver, gold, cadmium, bismuth, antimony, arsenic, copper, tellurium, fluorine, sulphuric acid. Ratio of lead to zinc contained in the ore is of major importance

*Environmental Factors:* Extensive environmental problems which are likely to make refining in the less developed countries a refuge of last resort. The US Environmental Protection Agency rightly continues to tighten restriction on ambient air lead content and Western Europe is following suit. Nonetheless environmentalism is a costly burden on primary and secondary producers

*Development Lead Time:* Up to ten years for underground operations.

Mined output is dominated by North and South America which jointly produce 45 per cent of the Western total. Australia is also a significant force responsible for 16 per cent of the total. Refined output is concentrated in Europe, which refines three times as much lead as it mines and the USA which refines twice as much as is domestically mined.

Lead is one of the few metals which Africa does not produce in abundance although South Africa has expanded lead output at the Aggeneys to around the 100,000 tonnes a year level.

## COMPONENTS OF RISK ASSESSMENT

| *Factor* | *Rating* | *Comment* |
|---|---|---|
| **Production Risks** | | |
| Existing capacity | 1 | Currently in over-capacity although output does, to some extent, depend on world demand for zinc, its major co-product |
| Labour disputes | 5 | Canada, the USA, Peru and Mexico (the major producers) are all prone to labour problems |
| Violent conflict | 1 | Unlikely to affect lead output |
| Range of primary supply sources | 2 | Viable ore deposits located in over 45 countries but Canada, the USA and Australia jointly control 48 per cent of Western mined output |
| Time lags for new supplies | 4 | Mainly underground mines, up to 10 years lead time, with some shorter lead times for open cast mines in Canada. Secondary materials provide another major, but sometimes erratic, supply source |
| **Transportation Risks** | | |
| Primary | 2 | Main possibilities include dock strikes at Canadian and Australian ports but disruptions not aimed directly at lead shipments |
| Secondary | 2 | No major problems recorded |
| **Application/Use Risks** | | |
| Total economic impact | 7 | Large market volume of 3.7-4.0 million tonnes a year including secondary material |
| Effect on key industries | 4 | Heavily dependent on the automobile industry as batteries, solders and petrol additives. Military use as radiation screening |
| Availability of substitutes | 5 | No viable substitute yet for lead in lead/acid car batteries but PVC is alternative in construction and cable sheathing, and variety of chemicals can be used instead of lead oxide in pigments etc |
| Longer term substitutability | 5 | Lead is a convenient low cost metal in its current applications and it would not be easy to totally replace it |
| **Trade Risks** | | |
| Collusive price agreements | 2 | Producers can, to some extent, manipulate the LME prices on which their supply contracts are based. Anti-trust laws in the USA promote competitiveness |
| Embargoes | 1 | Unlikely |

***These components are assessed on a scale from 1 (lowest) to 10 (highest) risk. Grouped together, they form the basis for the assessments of the factors for likelihood of supply disruption, and for cost of such disruption which appear in the first table in this section.***

## WORLD LEAD RESERVES

Total = 165 Million Tonnes Lead Content

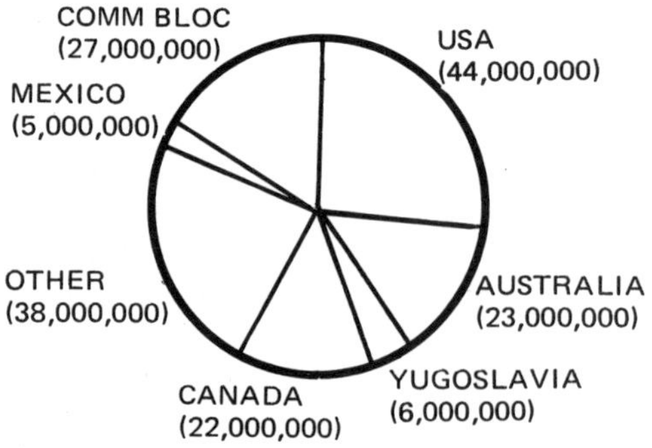


*Source: US Bureau of Mines, 1982*

## US LEAD CONSUMPTION

By Market Sector 1981

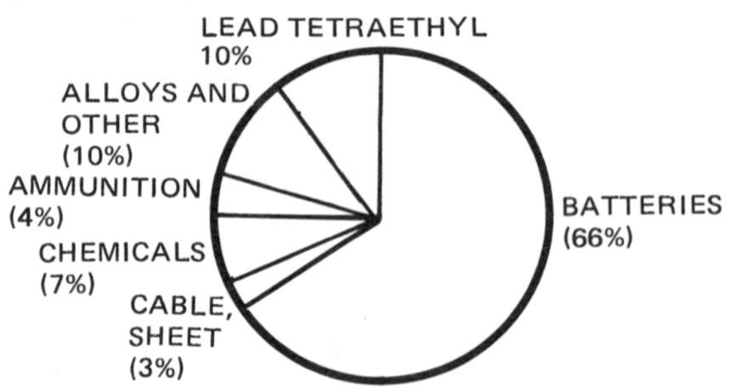


*Source: International Lead & Zinc Study Group, 1982*

**The Major Producers**

World Mined Production
(Producers over 100,000 tonnes a year)

| *Countries* | *Output 1981 (Thousand Tonnes)* | | *Percentage of Western World Output* | |
|---|---|---|---|---|
| EUROPE | | 445 | | 18.1 |
| Yugoslavia | 119 | | 4.8 | |
| AFRICA | | 284 | | 11.6 |
| Morocco | 112 | | 4.5 | |
| ASIA | | 120 | | 4.9 |
| AMERICA | | 1,208 | | 49.3 |
| Canada | 332 | | 13.5 | |
| USA | 452 | | 18.4 | |
| Mexico | 142 | | 5.7 | |
| Peru | 187 | | 7.6 | |
| AUSTRALIA | | 395 | | 16.1 |
| WESTERN WORLD TOTAL | | 2,452 | | 100.0 |

(Communist bloc output is estimated at 1,057,000 tonnes a year.)

World Refined Production
(Producers over 200,000 tonnes a year)

| *Countries* | *Output 1981 (Thousand Tonnes)* | | *Percentage of Western World Output* | |
|---|---|---|---|---|
| EUROPE | | 479 | | 16.2 |
| France | 228 | | 5.7 | |
| W Germany | 348 | | 8.7 | |
| UK | 333 | | 8.4 | |
| AFRICA | | 148 | | 5.0 |
| ASIA | | 404 | | 13.7 |
| Japan | 317 | | 8.0 | |
| AMERICA | | 1,673 | | 56.6 |
| Canada | 238 | | 6.0 | |
| USA | 1,065 | | 26.9 | |
| AUSTRALASIA | | 251 | | 8.5 |
| Australia | 239 | | 6.0 | |
| WESTERN WORLD TOTAL | | 2,955 | | 100.0 |

(Communist bloc output is estimated at 1,335,000 tonnes a year.)

*Source: World Bureau of Metal Statistics, June 1982.*

Lead mining operations vary vastly in size with some operations as small as a few tonnes per day ore, such as in Mexico. Others, however, such as the Missouri Mines south of St Louis and the Mount Isa mine in Queensland produce as much as 10,000 tonnes a day ore. Lead has been exploited since 3,000 BC. It has escaped the fate of being controlled by a few recently formed multinationals. There is a good degree of competitiveness within the industry due to the large number of companies involved in production.

**Technological Developments**

Fairly stable production technology with the main efforts concentrated on automated control systems in treating plants to cope with ore composition variations and fulfilling environmental legislation.

## TRANSPORTATION

This forms a fairly major link in the chain from lead concentrate to primary refined lead. Secondary lead tends to be collected and treated within the country of origin, and with the scrap industry being so fragmented, transport over long distances is unusual. The USA varies from being a minor exporter to a net importer and is effectively neutral in the trade pattern. Canada, Mexico, Peru and Australia are major ore and refined metal exporters, with destinations primarily to Europe and Japan. The Communist bloc is traditionally a net importer of refined lead. The major trade routes are from Canada, Mexico, Peru and Morocco to Western Europe and from Australia to Japan, all by ship. Disruptions in deliveries tend to be a consequence of port strikes rather than a direct attempt to halt lead shipments.

## APPLICATIONS

**Consumption Trends**

Lead is uncomfortably vulnerable to the health of the automobile industry, for use in batteries, solder and petrol additives, and thus to vagaries of freely disposable customer income levels and to interest rates. Consumption in the West was relatively stable in the late 1970s at around 4.1 million tonnes a year but fell seven per cent in 1980, three per cent in 1981 and a further 3.5 per cent annualised in the first half of 1982. The USBM is probably optimistic in its expectation of US demand growth of 2.5 per cent a year until 1990. In the developing countries, growth in demand is also relatively small.

**The Principal Markets**

Demand is based primarily on the use of lead in car batteries, as is emphasised by the pie-chart showing the end-use markets in the USA. This sector has been hard hit by the trend towards smaller, lighter cars with maintenance-free batteries and those containing less lead. The traditional 13-14 kilo battery has given way to models 30 per cent lighter and new developments such as GNB Batteries' 'battery-and-a-half' Cathanode are due to affect the market over the next five years. Japan uses 53 per cent of its lead in batteries, a lower proportion due to its bias towards smaller cars. The UK, a declining car producer uses only 29 per cent of its lead in batteries. The use of lead as the tetra-ethyl in petrol is still a major use in the USA (nine per cent) and the UK (21 per cent) but is minimal in Western Europe and Japan. Tighter legislation in the USA and the UK is likely to bring an end to this market over the next decade. Lead is used in the construction industry as cable and sheet but health risks have limited application in water piping systems. The increase in the number of apartment dwellers has reduced the use of lead in construction per inhabitant but lead cable sheathing has regained popularity for use on buried power cables in severe conditions. Lead oxides are used as paints, pigments and as additions to glass and crystal but these sectors show little growth, particularly since the lead content of paints is subject to regulation. The lead-tin system is the basis of many bearing metals, solders and printing type alloys. Growth areas include the use of lead in sound deadening and radiation shielding plus the long-term industrial battery market.

### Recycling

Lead recycled from old scrap comprised 44 per cent of US consumption in 1981 with similar percentages in the rest of the Western world. The majority of this derived from old car batteries, most collected by small operators. Secondary lead production has suffered more from the recession than primary production or total consumption with many US secondary plants at only 50 per cent capacity by mid 1982. The secondary market has been squeezed by high recycling costs, low lead prices, an increase in the road life of automobiles (up from three to over six years in the USA) and the trend towards lighter, longer-life batteries.

## INTERNATIONAL TRADE AND WORLD PRICES

### Supply Arrangements

Lead producers contract to sell their lead using either of two main methods. In the USA, primary and secondary producers each establish an individual list price for their products in cents/lb in keeping with the cartel laws or use the average price which appears in the US publication, Metals Week. The rest of the world sells its lead using the LME cash weekly or monthly average price as a basis on which to calculate contract pricings. The LME contract is for 99.97 per cent refined pig lead in 25 tonne lots with a price quoted in £/tonne and a premium may be charged by producers for delivery of higher purity grades. It is obviously to the producers' advantage to maintain LME prices at their highest possible level and there is an informal lead producers' association, with the ostensible exception of the US producers due to strict anti-trust laws, which aims to support the LME rather than use production quotas. There is fierce competition and narrow profit margins in the lead market where a large number of producers are aiming to sell metal to a limited number of major consumers and a variety of minor consumers. There is also a large market in secondary lead.

### Companies

The major lead-in-concentrate producers are the state-owned operations of the USSR and China, Cominco of Canada, St Joe Minerals of the USA, Australian Mining and Smelting Ltd and Mt Isa Mines, both of Australia. Major smelting companies include the state-owned operations of the USSR and China, Broken Hill Associates Smelters Pty of Australia, Soc. Miniere et Metallurgique de Penarroya, France, St Joe Minerals, Cominco and Preussag of West Germany. Asarco is the acknowledged price leader in the US lead market although it is not the largest producer in the country.

### Prices

In common with most industrial metals of significant economic importance, lead saw high prices in 1974 and again in 1979-80. 1982 saw a weaker free market and lower producer prices with a high proportion of lead output sold at a loss.

**Lead Prices—Annual Averages, Highs and Lows**

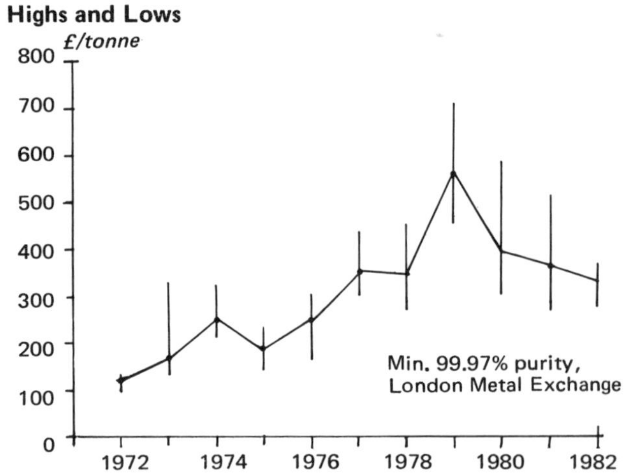

# LITHIUM (Li)

| | |
|---|---|
| FACTOR FOR LIKELIHOOD OF SUPPLY DISRUPTION | 1.3 |
| FACTOR FOR COST OF SUCH A DISRUPTION | 2.7 |
| TOTAL STRATEGIC FACTOR | 3.5 |

*The factors for supply disruption and cost of such a disruption are assessed on a scale from 1.0 (lowest) to 10.0 (highest) risk. Multiplied together these factors give the total strategic factor rating. For components of these assessments see table later in this section.*

Lithium has a low strategic factor mainly because the USA dominates Western world supply and its violent reactiveness prevents extensive use being made of its low density in weight-sensitive applications. It is mainly used as lithium compounds in the production of primary aluminium, ceramics, lubricants and in synthetic rubber production. Reserves are large and located in mainly stable countries. Demand for lithium could progress well, largely on the back of a surge in demand for aluminium alloys, but a supply shortage is unlikely.

**Physical Characteristics**

Lowest specific density of all metals (0.53); very reactive, soft, ductile metal; low melting point 170°C.

## OCCURRENCE AND PRODUCTION

Major ores are spodumene, petalite and lepidolite whilst lithium chloride from brine is another important source.

**World Reserves**

These are estimated at 2.22 million tonnes contained lithium, sufficient for over 250 years of consumption at present levels. Reserves are concentrated in Chile, the USA, Canada and Zaire with the Communist bloc possessing eight per cent of world reserves. World resources top eight million tonnes.

**Production Characteristics**

| | |
|---|---|
| *Mining Methods:* | Open cast mines or extracted from certain brines |
| *Processing:* | Lithium compounds recovered from the ores by chemical means and by a leaching and flotation process from the brines. Reduced to metal by electrolysis |
| *Energy Requirements:* | Moderate to produce the carbonate, hydroxide and other compounds, high to produce the metal |
| *By-Products:* | None of major importance |
| *Environmental Factors:* | Few problems |
| *Development Lead Time:* | Six to eight years for new projects. |

**The Major Producers**

| *Countries* | *Output 1981 (Short Tons contained Li)* | *Percentage of Western World Output* |
|---|---|---|
| USA | 5,300 | 90.0 |
| Argentina | 10 | 0.2 |
| Brazil | 60 | 1.0 |
| Portugal | 19 | 0.3 |
| African Continent | 500 | 8.5 |
| WESTERN WORLD TOTAL | 5,889 | 100.0 |

(Production in China is currently estimated at 400 short tons and production in the USSR 1,300 short tons a year.)

*Source: USBM Mineral Commodity Summaries, 1982, plus estimates of US output.*

The USA dominates the market supplying its own needs and exporting around 2,200 tonnes a year of contained lithium. US exports total more than the rest of the free world's production although a 6,350 tonnes a year lithium carbonate operation is due to come onstream in Chile in 1984. Southern Africa is the second biggest producer. The USSR and China are major lithium producers with a joint annual total of 1,540 tonnes. Due to the small number of companies extracting lithium, the market tends to be well integrated.

**Technological Developments**

US interest is focused on the recovery of lithium from the highly saline geobrines in Imperial Valley, California.

## TRANSPORTATION

The USA is a net exporter and major transport links are from North Carolina and Nevada to other parts of the

## COMPONENTS OF RISK ASSESSMENT

| *Factor* | *Rating* | *Comment* |
|---|---|---|
| **Production Risks** | | |
| Existing capacity | 1 | In oversupply due to weakness of aluminium industry |
| Labour disputes | 2 | The two major US producers have a good record of labour relations, but Brazil, Zimbabwe and Namibia do not |
| Violent conflict | 1 | Unlikely to affect mining of lithium ores |
| Range of primary supply sources | 2 | Reserves are concentrated in the USA, Chile, Canada and Zaire. The USA is self-sufficient in lithium and exports 60-70 per cent of requirements for the rest of the Western world but it is a safe source |
| Time lags for new supplies | 1 | Lead time of up to six years for open cast mine |
| **Transportation Risks** | | |
| Primary | 1 | Ore tends to be treated in plants near the mine site, few problems |
| Secondary | 1 | Transported as commercial purity lithium carbonate or other compounds. Not a sensible target for disruption |
| **Application/Use Risks** | | |
| Total economic impact | 2 | Total market of around 6,500-7,000 tons lithium metal per year |
| Effect on key industries | 4 | Important in production of aluminium and has growth possibilities in new light weight alloys. Use in ceramics, lubricants and synthetic rubbers is relatively stable |
| Availability of substitutes | 4 | Available in most cases but lithium retains its markets because of cost and convenience. Extensive process changes needed to convert from lithium compounds to alternatives |
| Longer term substitutability | 3 | Possible, if necessary |
| **Trade Risks** | | |
| Collusive price agreements | 2 | Feasible due to small number of producers with Foote and Lithium Corporation of the USA dominating the market. However the price sensitivity of several applications precludes the setting of unreasonable prices |
| Embargoes | 1 | Not a high profile metal so unlikely to be subject to this type of action |

*These components are assessed on a scale from 1 (lowest) to 10 (highest) risk. Grouped together, they form the basis for the assessments of the factors for likelihood of supply disruption, and for cost of such disruption which appear in the first table in this section.*

## WORLD LITHIUM RESERVES

Total = 2,450,000 Tons Contained Lithium

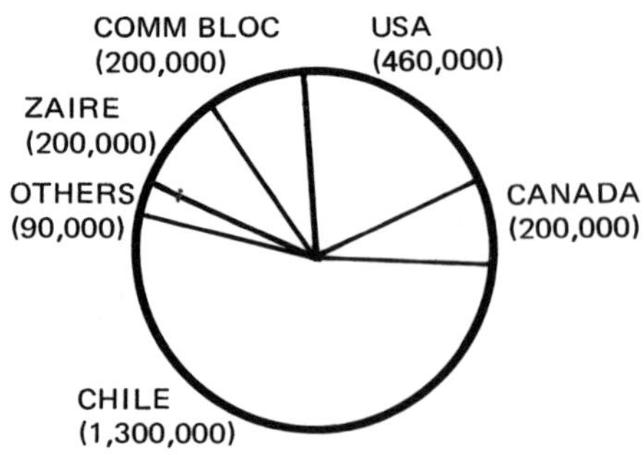


*Source: US Bureau of Mines, 1982*

## US LITHIUM CONSUMPTION

By Market Sector 1981

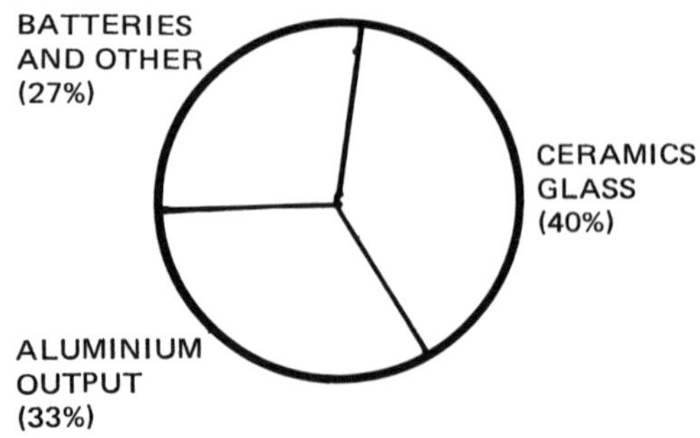


*Source: US Bureau of Mines, 1982*

USA, Japan and Europe. Few problems have been encountered.

## APPLICATIONS

### Consumption Trends

Western world consumption of lithium carbonate equivalent is currently an estimated 23,000 tonnes a year, equating to around 12,000 tonnes lithium contained. The USBM estimates US consumption of lithium products at 2,800 tonnes in 1982 with a demand growth rate of 5.7 per cent a year until 1990.

### The Principal Markets

Demand is based on the use of lithium to improve the efficiency of the aluminium smelting process and on its use as a component of certain porcelains, glasses and glass-ceramics. Lithium-stearate based soaps now provide about 55 per cent of multi-purpose greases used in the industrialised world. Lithium metal is also used to make catalysts for synthetic rubber production, in sophisticated long life batteries and as a metallurgical additive in steel and copper alloy production.

Lithium's use in aluminium production has the greatest growth potential but this depends on the future of the aluminium industry. However research has reduced the use of lithium in aluminium potlines from 5-7 lb carbonate per ton of aluminium to 3 lb. The advent of electric cars using a lithium based battery is a possibility and would certainly boost demand but this is unlikely before the end of the century. Another possible application is for the use of liquid lithium as a medium in nuclear fission power generation, but this remains a market opening for the next century.

### Recycling

There are effectively no recycling opportunities since all uses except those in ceramics and glasses are dissipative. It is totally unrealistic to reclaim minerals from glass although there are glass bottle/container recycling schemes.

### Substitution

Substitutes are available in all applications. Sodium and potassium compounds are viable in ceramic and glass manufacture, calcium and aluminium soaps are substitutes in greases and zinc, magnesium and mercury as battery anodes. Increasing the use of cryolite, calcium fluoride and aluminium fluoride can exclude lithium carbonate from use in aluminium potlines.

## INTERNATIONAL TRADE AND WORLD PRICES

### Supply Arrangements

Lithium is traded as the metal and as certain compounds. Lithium carbonate and lithium hydroxide monohydrate are both traded as pellet or powder with prices quoted in $/lb. Lithium metal is too reactive to be frequently traded. Prices have tended to be fairly stable over the last 15 years since free market activity is small and confined to the intermittent marketing of Russian material. There are only a handful of Western producers, and rumours circulate of an informal cartel to maintain prices on a sensibly inclined upward trend. The US share of world lithium production is expected to decrease in the long term and political pressures may affect the development of projects, which currently have a major US interest, in South America and in Southern Africa. The piechart opposite shows end-uses for lithium in the USA. Japan differs by using only around nine per cent of lithium in aluminium production but proportionately more in lubricants and synthetic rubber production.

### Companies

Two US companies; Foote Mineral Co (89 per cent owned by Newmont) and Lithium Corporation, control the market with around three quarters world mined output. The state-owned operations of the USSR and China, Companhia Estanifera do Brazil and Metallgesellschaft of West Germany which processes ore from the USA and USSR, Bikita Minerals Limited of Zimbabwe and SWA Lithium of Namibia are other major lithium ore and metal producers.

### Prices

Producer prices are effectively controlled by the US companies, Foote and Lithium Corp, and there is virtually no free market activity. Price increases are limited by the cost consciousness of aluminium producers, a major end-use sector; and producers' lithium prices tend to reflect merely the cost of production plus the expected return on capital.

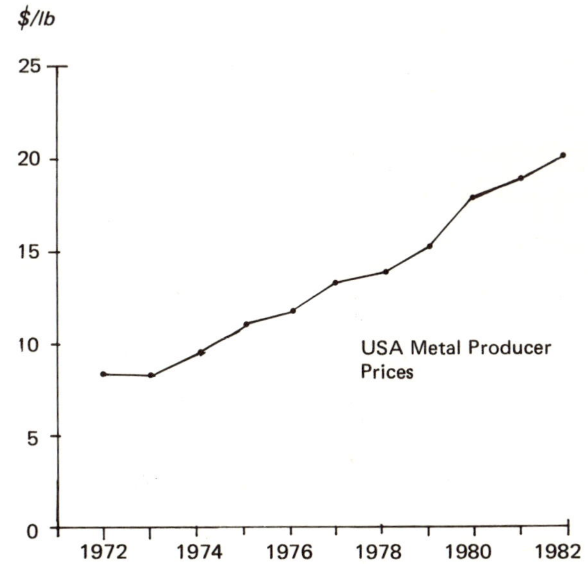


**Lithium Prices—Annual Averages (99.9% ingot)**

# MAGNESIUM (Mg)

| | |
|---|---|
| FACTOR FOR LIKELIHOOD OF SUPPLY DISRUPTION | 1.3 |
| FACTOR FOR COST OF SUCH A DISRUPTION | 4.5 |
| TOTAL STRATEGIC FACTOR | 5.9 |

*The factors for supply disruption and cost of such a disruption are assessed on a scale from 1.0 (lowest) to 10.0 (highest) risk. Multiplied together these factors give the total strategic factor rating. For components of these assessments see table later in this section.*

Magnesium is an industrial metal of second line importance with a fairly low strategic rating. Reserves are effectively infinite and globally widespread although requiring heavy capital investment and energy input for extraction. Its applications depend heavily on the aluminium industry both as a substitute for and an additive to aluminium alloys. It has been involved mainly with the faster growing sectors of aluminium usage and hence has shown an enviable growth rate during the 1970s, aided by its high strength/unit density ratio. The USA produces around half of the world's output and is considered a safe source.

**Physical Characteristics**

Low density (1.74); grey metal with a fairly low melting point of 660°C; readily oxidises on contact with air.

## OCCURRENCE AND PRODUCTION

Occurs as magnesite ($MgO_2$) in dolomitic rock, and as olivine and in natural brine solutions; each cubic mile of seawater contains six tons of magnesium.

**World Reserves**

Reserves of magnesite are estimated at 2.7 billion tonnes and reserves of magnesium compounds in well and lake brines and seawater are vast and hence possessed by all countries. The Communist bloc has about a quarter of magnesite reserves.

**Production Characteristics**

*Mining Methods:* Dolomite mined in open quarries or brines collected and concentrated

*Processing:* Dolomite ore calcined to give dolime (40 per cent magnesia) then reduced by a variety of silicothermic processes to magnesium. Brine dehydrated to powdered 100 per cent $MgCl_2$ (magnesium chloride) then electrolysed to magnesium metal

*Energy Requirements:* 18 kWh/lb magnesium metal via the electrolytic method (compared to 15 kWh/lb in aluminium production); high energy needs for both routes

*By-Products:* Sulphur from electrolytic process

*Environmental Factors:* Expensive flue-gas desulphurisation programs are necessary in the silicothermic process

*Development Lead Time:* Under six years for open quarry dolomite mine but the processing methods in each route require sophisticated plants needing up to 10 years to come onstream.

**The Major Producers**

| *Country* | *Output 1981 (Thousand short tons Mg content)* | *Percentage of Western World Output* |
|---|---|---|
| USA | 165 | 65.7 |
| Canada | 10 | 4.0 |
| France | 9 | 3.6 |
| Italy | 9 | 3.6 |
| Japan | 12 | 4.8 |
| Norway | 44 | 17.5 |
| Yugoslavia | 2 | 0.8 |
| WESTERN WORLD TOTAL | 251 | 100.0 |

(Production in China is currently estimated at 8 short tons and production in the USSR 83 short tons a year.)

*Source: USBM Mineral Commodity Summaries 1982*

Magnesium output is dominated by the USA, Norway, Japan and Canada. The USSR and China are also major producers with 26.5 per cent of world output. Electrolytic magnesium reduction plants are always sited on a

## COMPONENTS OF RISK ASSESSMENT

| *Factor* | *Rating* | *Comment* |
|---|---|---|
| **Production Risks** | | |
| Existing capacity | 1 | Producers currently operating well under capacity due to glutted aluminium market |
| Labour disputes | 2 | The magnesium industry has never been as strike prone as the copper, coal and other mining sectors since the industry developed later with fewer restrictive practices |
| Violent conflict | 1 | Unlikely to occur or to affect output if it does |
| Range of primary supply sources | 3 | Refined metal derives mainly from the USA, Canada and Norway but reserves are globally widespread |
| Time lags for new supplies | 3 | Reserves easily available but electrolytic/silicothermic reduction plants have a long lead time |
| **Transportation Risks** | | |
| Primary | 1 | Not usually liable to disruptions, since transport distances tend to be short |
| Secondary | 1 | Low risk-profile |
| **Application/Use Risks** | | |
| Total economic impact | 4 | Total market volume of 300,000 tonnes a year magnesium content |
| Effect on key industries | 7 | Used in aluminium aerospace, packaging and automobile alloys. Also important as magnesium alloy castings, a major growth area |
| Availability of substitutes | 6 | Available in most areas, except in aluminium alloys. Aluminium and zinc are substitutes for magnesium alloy components |
| Longer term substitutability | 4 | Possible if necessary but extensive engineering changes necessary to adapt to aluminium alloys without magnesium |
| **Trade Risks** | | |
| Collusive price agreements | 3 | Occurs unofficially already due to small number of major producers, of which four operate in the USA. However prices tend to be sensibly aligned to production costs |
| Embargoes | 1 | Unlikely |

*These components are assessed on a scale from 1 (lowest) to 10 (highest) risk. Grouped together, they form the basis for the assessments of the factors for likelihood of supply disruption, and for cost of such disruption which appear in the first table in this section.*

## US MAGNESIUM CONSUMPTION

By Market Sector 1981

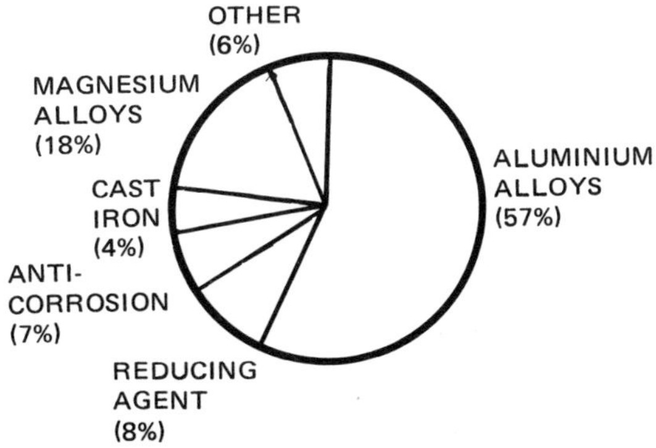


*Source: US Bureau of Mines, 1982*

sea or lake shore with ready access to energy, usually HEP. This explains the prominence of Norway and Canada in the ranking. Silicothermic reduction plants are sited near the dolomite quarries so as to reduce transport costs. Capacity is expected to continue growing as producers complete planned expansions in the USA, Brazil and Norway. Magnesium production is slowly declining in Asia and Oceania largely due to energy availability and pricing problems. The market is currently oversupplied due to weakness in the aluminium sector and the producers are showing admirable restraint in cutting back output.

### Technological Developments

Dow Chemical has pioneered an innovative energy saving electrolytic process by which seawater is reacted with a dolime-water mix and the resulting compounds electrolytically reduced. The high energy requirements mean that producers are continually searching for new process improvements.

## TRANSPORTATION

Transport is not a high-profile link in the production chain of this mineral since brines and seawater are refined to metal near the water shore whilst dolomite is reduced to metal within a short travelling distance of the quarry. Dolomite for refractories is transported in its original state but as a low-cost, high-volume product, disruptions are mainly as a consequence of general rail strikes etc and not directed specifically at magnesium/dolomite stoppages.

## APPLICATIONS

### Consumption Trends

Magnesium has experienced a growth trend comparable to that of aluminium over the past two decades and its future is dependent on its cost in relation to aluminium. It has a density of two-thirds that of aluminium and wide-scale substitution becomes viable at a price ratio of 1:1.5. Over past years the ratio has hovered between 1:1.6 and 1:1.7, too high for substitution but allowing magnesium to be used in aluminium alloys. US demand for magnesium metal is in the 110-115,000 tonnes a year range and the USBM expects magnesium demand to grow by five per cent a year over the next decade.

### The Principal Markets

A pie chart showing the end-use markets in the USA is given on page 97. Western Europe follows much the same pattern but Japan uses approximately 70 per cent of its magnesium as additives to aluminium alloys. The addition of magnesium increases the strength and ductility of the alloy. Can production consumes slightly under half of all magnesium used in the sector—with up to 1.3 per cent manganese in the body of the can and five per cent magnesium in the can ends. Aluminium alloys with higher magnesium contents are used in sophisticated corrosion resistant and aerospace applications. There is attractive growth potential in the use of magnesium alloy castings, replacing zinc or aluminium components, but the barriers are those of cost and the capital investment necessary to make the switch by automobile and aircraft producers. The typical car now contains 0.5lb magnesium but Ford, which had expected to use 5lb per car by 1985, has now revised its estimate to 1lb per car. Japanese car manufacturers use no magnesium in their cars due to price and corrosion problems. Nonetheless the magnesium diecastings sector is expected to grow by 6.8 per cent to reach 62,000 tonnes in 1986.

Other stable uses include magnesium as a reducing agent, as an additive to cast irons and in dissipative chemical uses. A minor but growing field is for magnesium as a sacrificial anode to prevent steel structures from corrosion.

### Recycling

The USA consumed around 18,000 tonnes of magnesium scrap in 1981, one-fifth of total consumption. Certain uses are dissipative, but magnesium is recycled from castings and also as a contained additive in aluminium alloys. It cannot be extracted from aluminium alloys and this is a major problem in can recycling, where the presence of the can ends pushes up the overall magnesium content above the maximum for can body stock alloy.

### Substitution

Aluminium and zinc may be substituted for magnesium in cast products depending on the current ratios of price/unit weight and price/unit density. Magnesium does have corrosion and fabrication problems but has the lowest density. Calcium carbide competes with magnesium for the steel desulphurisation sector.

## INTERNATIONAL TRADE AND WORLD PRICES

### Supply Arrangements

Magnesium is mainly traded on a producer basis with only a limited free market. Production is centred in the industrialised countries so there is no vast international volume of trade. The specification usually traded is 99.8-99.9 per cent ingots in five tonne lots. Domestic producers tend to supply their own market at an individually fixed producer price although the Dow Peglock fob Gulf producer price is often used as a bench mark.

### Companies

The major magnesium metal producers are Dow Chemical, Amax, and Northwest Alloys (an Alcoa subsidiary), all of the USA, the USSR state-owned operations, Norsk Hydro of Norway and Magnesio e Leghe di Magnesio of Italy. With the current glut in the aluminium market and depression in the automobile industry, all the USA magnesium producers are working at under 70 per cent capacity.

### Prices

Price is magnesium's greatest obstacle to increased consumption in the automobile industry. Producer prices have moved steadily upwards over the last 20 years, except for a sharp peak following the 1973-74 oil crisis, since the producers have maintained a sensibly firm grip on the supply/demand balance.

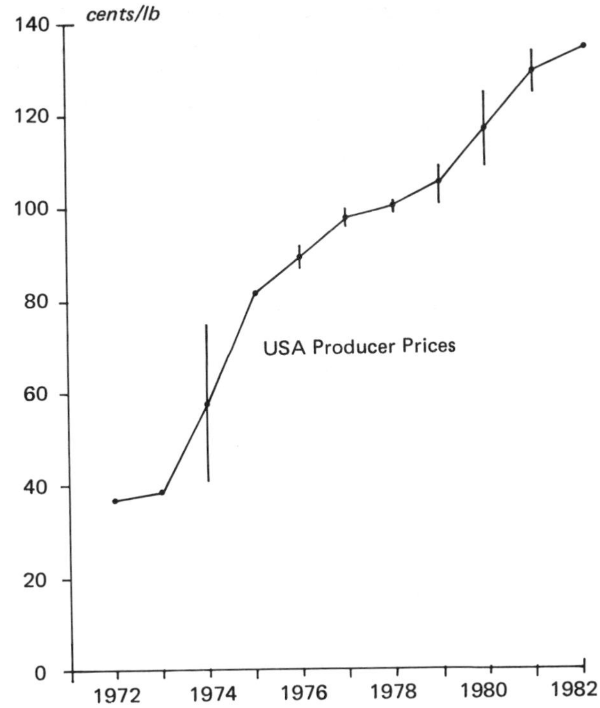


**Magnesium Prices—Annual Averages, Highs and Lows**

# MANGANESE (Mn)

| | |
|---|---|
| **FACTOR FOR LIKELIHOOD OF SUPPLY DISRUPTION** | **4.7** |
| **FACTOR FOR COST OF SUCH A DISRUPTION** | **7.8** |
| **TOTAL STRATEGIC FACTOR** | **36.7** |

*The factors for supply disruption and cost of such a disruption are assessed on a scale from 1.0 (lowest) to 10.0 (highest) risk. Multiplied together these factors give the total strategic factor rating. For components of these assessments see table later in this section.*

Manganese is an important industrial metal with a relatively high strategic factor because of its necessity in the commercial production of nearly all steels. Land based resources are large but over 80 per cent are found in the USSR and South Africa whilst extensive deep sea resources cannot yet be viably exploited. Recent expansion in ferro-manganese output, a master alloy for steels, has been concentrated in locations outside Europe and North America, adding to their vulnerability to interruptions in supplies from South Africa. There are no substitutes for manganese as a general, low cost, useful steel additive.

### Physical Characteristics

Specific density similar to that of iron (7.43); high melting point (1,245°C); hard strong metal; an essential trace constituent of man and animals.

## OCCURRENCE AND PRODUCTION

Manganese ore, containing various oxides, is defined as containing 35 per cent or more manganese. Ferruginous manganese ore and manganiferous iron ore are the names given to inferior grades.

### World Reserves

Land based reserves are estimated at 5.4 billion tons, sufficient for over 210 years of consumption at current levels. There are also extensive deep sea resources in the form of manganese oxide nodules, containing up to 30 per cent manganese, over large areas of the ocean floor particularly in the equatorial Pacific Ocean. The USSR and South Africa jointly possess 84 per cent of all manganese reserves.

### Production Characteristics

*Mining Methods:* Mechanised open cast mining

*Processing:* The ore is beneficiated and then converted to ferro alloys and silicon alloys in blast or electric furnaces. Alternatively, manganese dioxide is made electrolytically or chemically. The metal is produced electrolytically

*Energy Requirements:* Relatively high. Important in the competitive ferro-alloy industry

*By-Products:* Iron as co-product. Gold, silver, zinc, copper and lead as by-products

*Environmental Factors:* No major problems but disposal of tailings and mine wastes can be unsightly

*Development Lead Time:* Up to eight years for green field opencast mines.

### The Major Producers

| *Countries* | *Output 1981 (Thousand Tonnes)* | *Percentage of Western World Output* |
|---|---|---|
| Australia | 2,000 | 13.9 |
| Brazil | 2,100 | 14.6 |
| Gabon | 1,650 | 11.5 |
| India | 1,650 | 11.5 |
| South Africa | 6,000 | 41.7 |
| Others | 1,000 | 6.8 |
| WESTERN WORLD TOTAL | 14,400 | 100.0 |

(Communist bloc production is currently estimated at 11,900,000 tonnes, production in the USSR 10,350,000 tonnes and China 1,550,000 tonnes a year.)

Mined output is dominated by the USSR, South Africa, Australia and Brazil. The USA, Western Europe and Japan are almost totally deficient in reserves and have no domestic production of manganese ore. Recent increases in energy costs have encouraged ferro-manganese production near to where the ore is mined, both to cut down on carriage costs and to utilise indigenous coal or other cheap electricity supplies. This mirrors the rationalisations in the ferro-alloy industries of the USA and Western Europe. The USA is almost totally dependent on manganese ore and ferro-alloy imports whilst recent ownership

## COMPONENTS OF RISK ASSESSMENT

| *Factor* | *Rating* | *Comment* |
|---|---|---|
| **Production Risks** | | |
| Existing capacity | 2 | Global decline in the steel industry has ensured a generous supply/demand balance |
| Labour disputes | 1 | Few lasting strikes in USSR and South Africa, the major producers |
| Violent conflict | 7 | Such risks in the politically volatile areas of Africa and the USSR cannot be ignored |
| Range of primary supply sources | 8 | 84 per cent of mined supply derives from the USSR and South Africa and ferro-alloy production is being concentrated near mines but deep sea resources are extensive |
| Time lags for new supplies | 6 | Up to eight years for greenfield open cast mines. Deep sea resources unlikely to be commercially exploited before the end of the century |
| **Transportation Risks** | | |
| Primary | 8 | Ore processed as near to mine as possible to reduce transports costs and exploit cheap, indigenous energy supplies but transport routes are still extended |
| Secondary | 8 | Shipping routes extend from South Africa, the Western world's chief supplier, round the Horn of Africa to the USA and W Europe. Also long routes to Japan |
| **Application/Use Risks** | | |
| Total economic impact | 8 | Large volume market of 26-26.5 million tonnes a year manganese (gross weight) |
| Effect on key industries | 7 | Steels containing ferro-alloys employed in almost all major industries including construction and transportation |
| Availability of substitutes | 9 | None in the major applications, either on cost or performance comparisons |
| Longer term substitutability | 9 | Suitable alternative to manganese in steels and other alloys unlikely to be found |
| **Trade Risks** | | |
| Collusive price agreements | 5 | Feasible considering the concentration of output in a few hands but it would be counter productive to maintain unrealistically high prices |
| Embargoes | 5 | The political status of South Africa suggests this could be a bargaining point in trade discussions with the USA and Europe |

*These components are assessed on a scale from 1 (lowest) to 10 (highest) risk. Grouped together, they form the basis for the assessments of the factors for likelihood of supply disruption, and for cost of such disruption which appear in the first table in this section.*

### WORLD MANGANESE RESERVES

Total = 5.4 Billion Tons Manganese Gross Weight

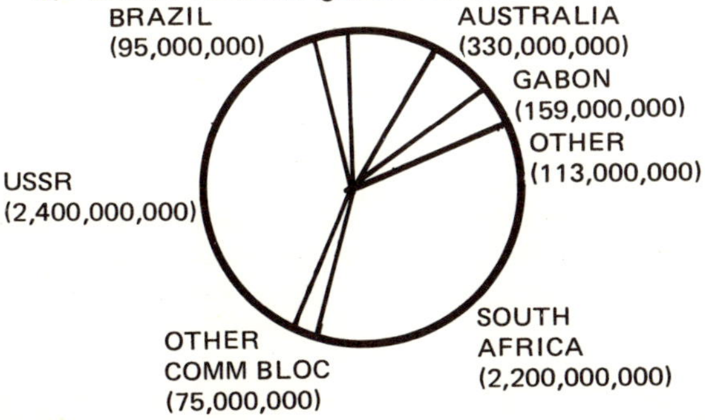


*Source: US Bureau of Mines, 1982*

### US MANGANESE CONSUMPTION

By Market Sector 1981

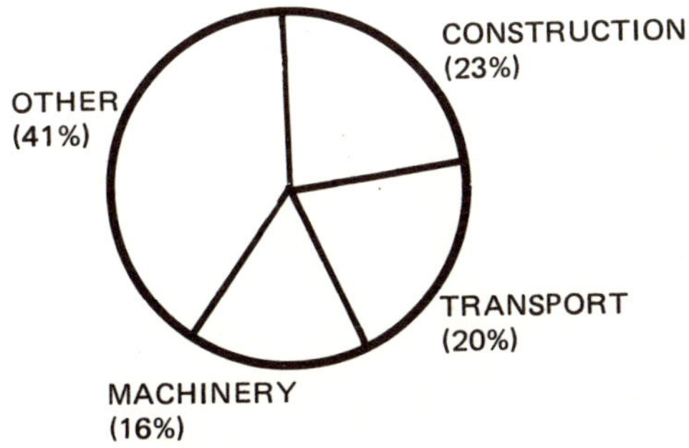


*Source: US Bureau of Mines, 1982*

changes in domestic electric-furnace ferro-alloy plants has put US ferro-alloy production predominantly under foreign control. China is a major producer and total Communist bloc output represents 45 per cent of the world total.

The USA has recently revised its GSA goal for manganese source materials which now stand at approximately three years' consumption. The USSR is an erratic ferro-manganese exporter and has recently halted exports to the West while Poland has begun to import around 30-33,000 tonnes a year. Russia does however continue to export silico-manganese at a rate of 5-10,000 tonnes a year.

### Technological Developments

Blast furnace production of ferro-manganese is being steadily replaced by the use of electric furnaces allowing tighter control of the process variables. Improved alloy control during steel making tends to be slowly reducing manganese use per tonne of steel.

## TRANSPORTATION

This is a major link in the chain from ore to steel products. Manganese is a large volume market with a low unit value and transportation costs consequently form a sizeable percentage of production costs. The location of ferro-alloy plants near to the mine keeps rail and shipping costs down but, the shipping of ferro-alloys necessitates long routes to destinations in Western Europe, the USA and Japan.

## APPLICATIONS

### Consumption Trends

90 per cent of all manganese is used in the production of ferro-alloys which increase the strength, toughness, hardness and hardenability of steels. The USBM expects US demand for manganese to grow by 1.6 per cent a year over the next decade with 1982 consumption at 1.05-1.10 million tonnes contained manganese. Nonetheless the ratio of manganese ore to steel production is steadily falling as steel alloying and treating methods improve and the demand for crude, bulk steel stagnates.

### The Principal Markets

The pie-chart shows the wide penetration of steels, containing manganese, into all branches of US industry. General, all-purpose steels contain 1.5 per cent manganese, primarily as a desulphurising agent but specialised steels can contain up to 14 per cent. A rule of thumb guide is that 5-6kg manganese are required per tonne of iron for steel making and 31kg for each tonne of cast iron. Regional consumption patterns largely follow those of general steels: away from the ailing steel industries of the USA and Western Europe and moving towards Australia, Mexico, South Africa and India.

Small amounts of manganese (0.1-1.0 per cent) are added to aluminium and other light alloy systems to give strength, hardness and stiffness. Manganese dioxide finds a use as the depolariser in common dry cell batteries for radios and other portable equipment. Other manganese compounds are employed as glass and ceramic colouring agents, paint driers, fungicides and certain catalysts. The miscellaneous section includes a sizeable proportion of process losses but also the use of manganese in specialised alloys such as manganese bronzes for ship propellors and other corrosion resistant applications.

World consumption of manganese is expected to grow slightly more rapidly than US demand but it is important to note the form in which the manganese is used. Different markets exist for the various specifications of ferro-alloy whilst the relatively high price of electrolytic manganese metal limits its use to the production of special stainless steels, aluminium alloys and certain chemicals.

### Recycling

Insignificant except as an intrinsic component of steel scrap for which there is a large and well organised secondary metal marketing and processing structure.

### Substitution

There is no adequate substitute for manganese in its major applications, either on grounds of cost or of performance. Titanium and zirconium are effective, but very expensive, substitutes for the desulphurising aspect of manganese usage in steel but do not provide the additional advantages of toughness and hardenability.

## INTERNATIONAL TRADE AND WORLD PRICES

### Supply Arrangements

Manganese is almost wholly priced and sold using long term producer/consumer contracts with a limited free market. The main specifications traded include: manganese ore containing 48-50 per cent manganese and a maximum of 0.1 per cent phosphorus; and various grades of ferro-manganese generally containing 78 per cent manganese and a specified carbon content. Electrolytic manganese metal of 99.5 per cent purity is the grade usually traded on the free market. Manganese ore producers have traditionally maintained a firm grip on the ore price, leaving the dwindling number of non-integrated ferro-alloy producers at a distinct disadvantage. Ferro-manganese is a bulk commodity and, as such, profit margins are pared by the competitive pricing of producers, who ship the material directly to the steel works who are their sole consumers.

### Companies

The major manganese ore producers are the state-owned organisations of the USSR and China, South African Manganese Amcor Ltd and Associated Manganese Mines of South Africa Ltd, Cie Miniere de l'Ogooue of Gabon; and the Broken Hill Pty Co Ltd of Australia. Ferromanganese output is dominated by Acieries de Paris et d'Outreau of France, South African Manganese Amcor Ltd; August Thyssen-Hutte-AG of West Germany and Sauda Smeltewerk A/S of Norway. Nippon Kokan KK, Japan Metals and Chemicals Co Ltd and Nippon Denko KK, all of Japan, lead the world in silico-manganese production capacity. Manganese metal production capacity is concentrated in the hands of Delta Manganese (Pty) Ltd and Electrolytic Metal Corp Ltd, both of South Africa, plus Chemetals Corp of the USA and PUK of France.

**Prices**

Prices and market conditions have remained relatively stable over the past few years reflecting a decline in manganese additions to each tonne of steel plus the malaise in the bulk steel making industry. Demand for manganese ore fluctuates with the health of the steel industry and thus prices have been influenced by this basic business cycle. Prices in this producer controlled market have remained on a moderate upward trend without startling peaks or troughs. Increases in ore prices tend to be paralleled by ferro-alloy increases.

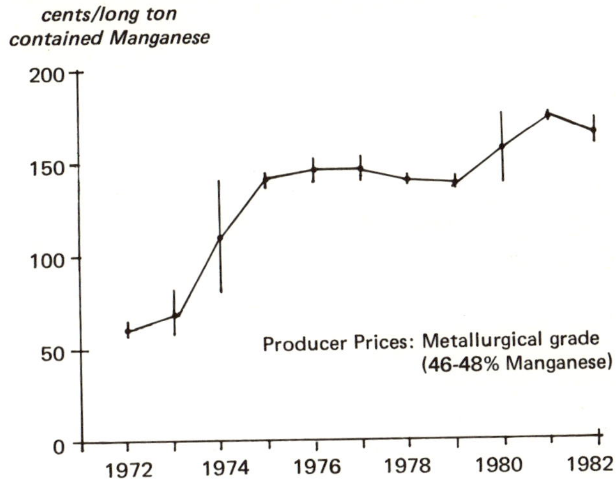


**Manganese Ore Prices—Annual Averages, Highs and Lows**

# MERCURY (Hg)

| | |
|---|---|
| FACTOR FOR LIKELIHOOD OF SUPPLY DISRUPTION | 2.5 |
| FACTOR FOR COST OF SUCH A DISRUPTION | 3.5 |
| TOTAL STRATEGIC FACTOR | 8.8 |

*The factors for supply disruption and cost of such a disruption are assessed on a scale from 1.0 (lowest) to 10.0 (highest) risk. Multiplied together these factors give the total strategic factor rating. For components of these assessments see table in this section.*

Mercury is a minor metal with a low strategic factor. Its main uses lie in the fields of electrical equipment, chemical catalysts and batteries, but research has been directed towards the replacement of mercury whenever this is possible given its toxicity. Mercury reserves and output are dominated by Spain and the USSR. Resources are large and widespread but increasing environmental control costs impose an upper limit on production in the Western world.

### Physical Characteristics

High specific density (13.5); the only metal which is liquid at room temperature with a low melting point (-38.9°C); high toxicity with a long biological half life, which causes brain damage and mutagenicity in humans.

## OCCURRENCE AND PRODUCTION

Commercially extracted from its sulphidic ore, cinnabar.

### World Reserves

These are estimated at 153,000 tonnes, sufficient for almost 25 years of consumption at current levels. Western world reserves are concentrated in Spain, the USA, Algeria and Mexico whilst the Communist bloc possesses about a quarter of the total.

### Production Characteristics

| | |
|---|---|
| *Mining Methods:* | Mainly small one-man operations exploiting surface deposits. Small amount of mercury recovered as by-product of gold refining |
| *Processing:* | Ore crushed and roasted for the mercury to sublime out and be collected |
| *Energy Requirements:* | Low due to simple process of refining |
| *By-Products:* | Occasionally gold |
| *Environmental Factors:* | Dangerous toxicity levels known to affect mined output levels, particularly in the USA |
| *Development Lead Time:* | Very short due to simple extraction methods, causing basic supply fluctuations. |

### The Major Producers

| *Countries* | *Output 1981 (Tonnes)* | *Percentage of Western World Output* |
|---|---|---|
| USA | 965 | 26.9 |
| Algeria | 1,035 | 28.8 |
| Mexico | 52 | 1.4 |
| Spain | 1,150 | 32.0 |
| Others | 390 | 10.9 |
| WESTERN WORLD TOTAL | 3,592 | 100.0 |

(Communist bloc production is currently estimated at 2,985 tonnes and production in the USSR 2,137 tonnes.)

*Source: USBM, Mineral Commodity Summaries 1982*

The major producers are Spain, Algeria, the USA, Mexico and the Communist bloc. Mercury mine production in the USA is on a decreasing trend with numerous mines closed because of increasing environmental pollution control costs. Mexican mining operations continue to be run down whilst Canadian mines, shut down during the low prices of 1975, have remained closed. Italy has recently rejoined the ranks of mercury producers with a new mine of planned production 5,000-6,000 flasks a year (each flask of 76lb mercury content). China and Russia are large mercury producers of which most is exported to the West. The USA was 39 per cent dependent on mercury imports in 1981. Earlier purchases have left the GSA with 191,391 flasks of mercury in its stockpile, of which 50,000 flasks have been authorised for disposal.

### Technological Developments

Research is currently being directed towards recovery of mercury as a by-product from other sulphide deposits and from plant and mine wastes to benefit the environment as well as adding to world supplies.

*MERCURY*

## COMPONENTS OF RISK ASSESSMENT

| *Factor* | *Rating* | *Comment* |
|---|---|---|
| **Production Risks** | | |
| Existing capacity | 1 | Currently in over-capacity with mines in Canada shut down and operations in Mexico curtailed |
| Labour disputes | 2 | Unlikely to affect output particularly from small one-man operations |
| Violent conflict | 1 | The USA and Spain regarded as relatively stable supply sources |
| Range of primary supply sources | 3 | Reserves fairly widespread but concentrated in Spain and Algeria. Communist bloc provides 45 per cent of total mined world output |
| Time lags for new supplies | 2 | Mainly outcrop mines on small scale, short lead time of under a year |
| **Transportation Risks** | | |
| Primary | 2 | The transport of mercury as cinnabar or metal does not have a high risk profile |
| Secondary | 2 | Few problems except for toxicity |
| **Application/Use Risks** | | |
| Total economic impact | 2 | World consumption totals 7,500-8,000 tonnes a year and is declining in many countries with a US demand growth of under one per cent a year |
| Effect on key industries | 3 | Use concentrated in batteries, industrial and control instruments, other electrical apparatus and fungicides but few high technology or military applications |
| Availability of substitutes | 6 | Few satisfactory substitutes in electrical apparatus and control instruments, diaphragm cells are an alternative in chlor-alkali industry and organotins and other chemicals can be used in paints and fungicides |
| Longer term substitutability | 5 | Research concentrated on reducing mercury consumption due to toxicity but these efforts have not yet been wholly successful |
| **Trade Risks** | | |
| Collusive price agreements | 6 | The main producers currently maintain a loose cartel to try to support prices |
| Embargoes | 3 | Russia periodically halts exports to the West and China is also an erratic exporter |

*These components are assessed on a scale from 1 (lowest) to 10 (highest) risk. Grouped together, they form the basis for the assessments of the factors for likelihood of supply disruption, and for cost of such disruption which appear in the first table in this section.*

### WORLD MERCURY RESERVES

Total = 4,440,000 Flasks (each 76 lbs)

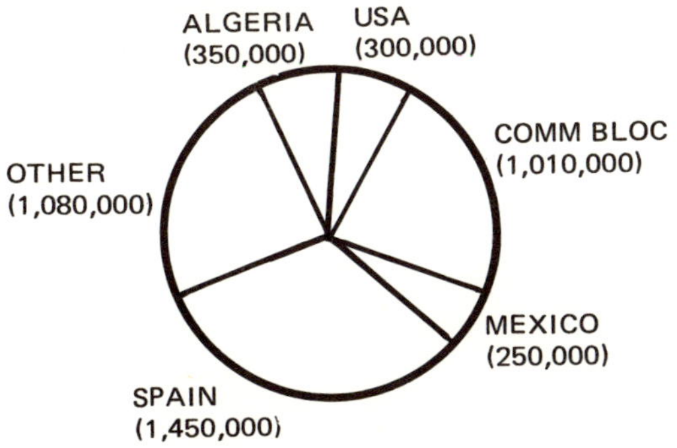


*Source: US Bureau of Mines, 1982*

### US MERCURY CONSUMPTION

By Market Sector 1981

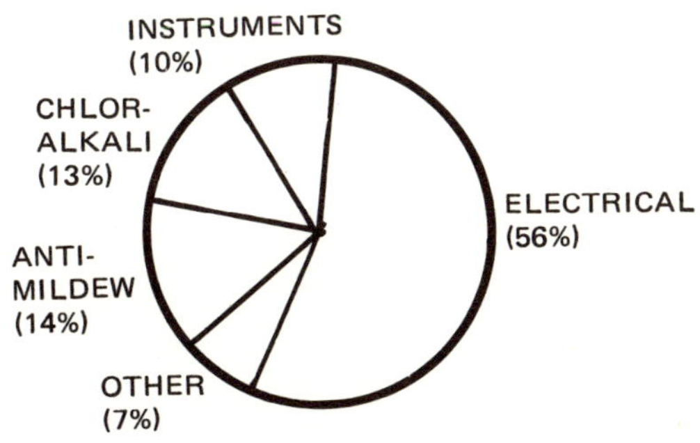


*Source: US Bureau of Mines, 1982*

## TRANSPORTATION

Not a major link in the usage cycle of mercury from ore to refined metal. Major exporters include Spain, Italy and Algeria with transport routes to the USA, Western Europe and Japan.

## APPLICATIONS

### Consumption Trends

The USBM expects demand for mercury to grow by less than one per cent a year over the next decade. Evironmental pollution control factors will continue to dampen growth in consumption.

### The Principal Markets

A pie chart showing the end-use markets in the USA is given on page 105. Japan uses 44 per cent of its mercury in inorganic chemicals and only 3.6 per cent in electrical equipment. Total Japanese consumption of mercury however, is only 13 per cent that of the USA. In Western Europe, around 20 per cent of all mercury is used in electrical equipment with 24 per cent used in the chlor-alkali industry.

Mercury is used in the electrolytic preparation of chlorine and caustic soda but this application is declining in the USA and Western Europe where no new mercury cell chlor-alkali plants are planned and existing plants have been made more efficient; consuming less mercury and recovering more. This sector has dwindled almost to zero in Japan. The electrical apparatus sector includes mercury batteries as its main component. Mercury is also applied as an all-purpose pesticide and fungicide and is still used in some pharmaceutical and dental preparations. It is also employed in arc, sun and cadmium mercury lamps and as a catalyst.

### Recycling

About 10 per cent of all mercury is recycled, mainly from batteries, dental amalgams and industrial scrap. Chlor-alkali plants internally recycle mercury and only need occasional topping-up.

### Substitution

There are few satisfactory substitutes for mercury in electrical apparatus and industrial and control instruments. Possible alternatives include nickel-cadmium battery systems, diaphragm cells in the chlor-alkali industry, organotin compounds in paints and solid state devices for industrial control instruments.

## INTERNATIONAL TRADE AND WORLD PRICES

### Supply Arrangements

Active trading of mercury on the free market is sporadic and trading is also carried out on a producer price basis. The usual specification is 99.99 per cent pure metal with a price quoted in \$/76lb flask. The producer price is generally fixed by complicity between Spanish and Italian producers and tacitly accepted by China. The long term future for mercury is poor. The producers have found it necessary to impose a loose cartel but they are not able to force this onto the small operators, particularly in Spain and Algeria, who sell their output below the fixed price.

### Companies

Leading mine producers of mercury include the state-owned operations of the USSR and China; Cia Minas de Almaden of Spain; Placer Amex of USA; Sonarem of Algeria; and Amescua y Asociados, Comercial de Mercurio SA and Almex SA of Mexico.

### Prices

Mercury has had an erratic price history tracing back to ancient times. Although it is not vital to a modern war effort, free market prices have always rallied during major conflicts.

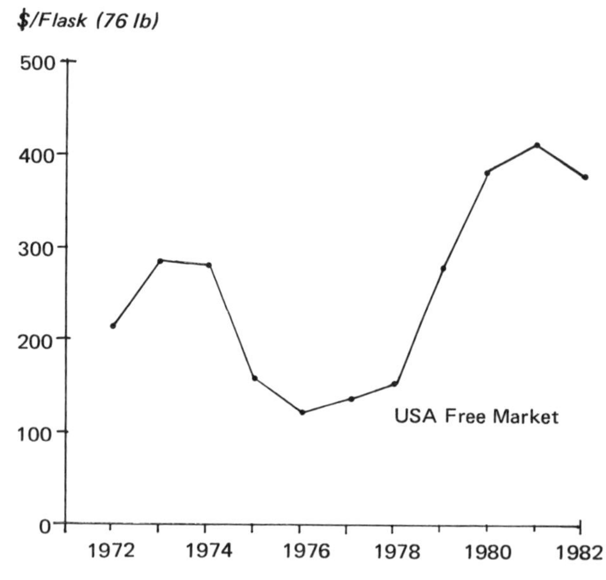


**Mercury Prices—Annual Averages**

# MOLYBDENUM (Mo)

| | |
|---|---|
| FACTOR FOR LIKELIHOOD OF SUPPLY DISRUPTION | 2.2 |
| FACTOR FOR COST OF SUCH A DISRUPTION | 7.3 |
| TOTAL STRATEGIC FACTOR | 16.1 |

*The factors for supply disruption and cost of such a disruption are assessed on a scale from 1.0 (lowest) to 10.0 (highest) risk. Multiplied together these factors give the total strategic factor rating. For components of these assessments see table in this section.*

Molybdenum has a moderately high strategic factor due to the optimum combination of its physical properties which encourage its extensive use in the manufacture of alloy steels and other materials for defence applications. Reserves are heavily concentrated in the USA and Chile. Over half annual mined production derives from the USA although large capital investment in the 1970s has helped develop deposits in other countries. The metal is currently in gross over-supply but it has shown an excellent record of growth in consumption since 1946 mainly in energy-related and high technology applications. Substitutes are, in most sectors, inferior on cost and performance grounds.

### Physical Characteristics

Tough, durable, hard metal with platinum-like appearance; high melting point (2,620°C) and high specific gravity (10.2); extremely low co-efficient of thermal expansion and corrosion resistance at moderate temperatures.

## OCCURRENCE AND PRODUCTION

Molybdenum is present in several ores, but the only one commercially mined is the sulphide $MoS_2$ (molybdenite).

### World Reserves

These are estimated at 21.7 billion lbs (9.8 million tonnes) contained molybdenum, sufficient for 90 years consumption at current levels. Reserves are associated with the broad sweep of porphyry copper reserves that ring the Pacific Ocean and span the middle of the Russian landmass. Both are areas which have been subject to volcanic disruptions in earlier times.

### Production Characteristics

| | |
|---|---|
| *Mining Methods:* | Underground and open pit methods, depending on ore body type, usually on a large scale |
| *Processing:* | Ore is beneficiated and the concentrate roasted to technical grade molybdic oxide, the starting point for ferromolybdenum, chemicals and molybdenum metal powder |
| *Energy Requirements:* | Fairly high due to complex beneficiation techniques required since ore grades are relatively low |
| *By-Products:* | 50 per cent mined as copper co-product. Other by-products include tin, tungsten, rhenium, and occasionally uranium or silver and gold |
| *Environmental Factors:* | Few problems except aesthetic disposal of mine tailings |
| *Development Lead Time:* | Up to ten years with much infrastructure needed to support the mines which are often located in remote areas. |

### The Major Producers

| *Countries* | *Output 1981 (Thousand Tonnes)* | *Percentage of Western World Output* |
|---|---|---|
| USA | 65.8 | 72.9 |
| Canada | 10.9 | 12.1 |
| Chile | 11.8 | 13.1 |
| Peru | 0.9 | 1.0 |
| Others | 0.8 | 0.9 |
| WESTERN WORLD TOTAL | 90.2 | 100.0 |

(Communist bloc production is currently estimated at 11.3 thousand tonnes).

There are some 50 primary and secondary (with copper as co-product) molybdenum mines currently operable, but the top five have the capacity to produce about 70 per cent of the total. The USA produced 65 per cent of world output in 1981 with Chile and Canada in second and third places. The Communist bloc produces nine per cent of total output. Significant additions to world production were scheduled to come onstream from 1981-85 primarily in the USA and Canada but also in Mexico, Peru and elsewhere, broadening the production base.

## COMPONENTS OF RISK ASSESSMENT

| *Factor* | *Rating* | *Comment* |
|---|---|---|
| **Production Risks** | | |
| Existing capacity | 1 | Market currently glutted with 1981 year-end surplus of almost six months' world production. Also several large projects on the horizon |
| Labour disputes | 6 | The copper and molybdenum mining industries of Canada, the USA, and South America are particularly prone to extended strikes |
| Violent conflict | 1 | Unlikely to affect molybdenum output since the main producers are relatively stable |
| Range of primary supply sources | 5 | Reserves are large but are concentrated in the USA, Canada and Chile. In 1981 the USA produced 65 per cent of world mined output |
| Time lags for new supplies | 8 | Up to ten years for new mine, with a high degree of infrastructure to support operations which are often in remote locations. There are several major projects in North America near the development stage |
| **Transportation Risks** | | |
| Primary | 1 | Major transport routes from North and South America to Europe, Japan and Russia. Few risks encountered |
| Secondary | 1 | Transport of steels and other alloys containing molybdenum have rarely been disrupted |
| **Application/Use Risks** | | |
| Total economic impact | 5 | Medium sized market of around 100,000 tonnes a year |
| Effect on key industries | 9 | Use is concentrated in steels for energy generation, oil and gas pipelines and weight sensitive (energy-conserving) applications plus super alloys for military and aerospace industries |
| Availability of substitutes | 9 | Main substitute as alloying element is tungsten but molybdenum is usually preferred due to lower cost and superior performance |
| Longer term substitutability | 8 | Molybdenum's optimum combination of physical properties has encouraged industry to develop new materials containing it and it is a long term growth metal |
| **Trade Risks** | | |
| Collusive price agreements | 3 | Does happen due to small number of major producers but effectiveness of this informal 'cartel' is limited by US antitrust laws, the free market and the current glut |
| Embargoes | 1 | Unlikely, except as a component of HSLA steels for gas pipelines |

*These components are assessed on a scale from 1 (lowest) to 10 (highest) risk. Grouped together, they form the basis for the assessments of the factors for likelihood of supply disruption, and for cost of such disruption which appear in the first table in this section.*

## WORLD MOLYBDENUM RESERVES

Total = 21.7 Billion lbs Molybdenum Content

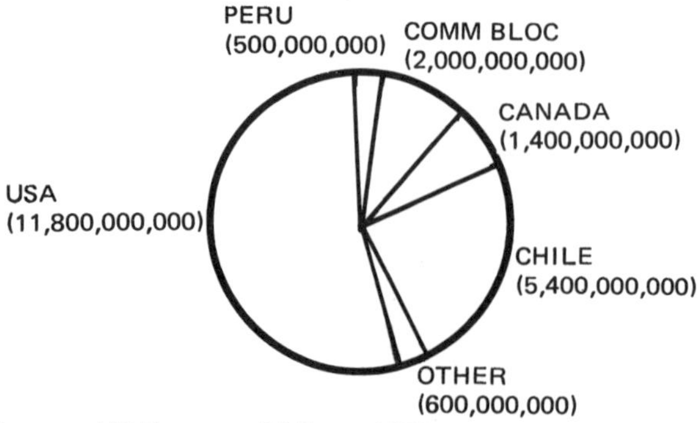


*Source: US Bureau of Mines, 1982*

## US MOLYBDENUM CONSUMPTION

By Market Sector 1981

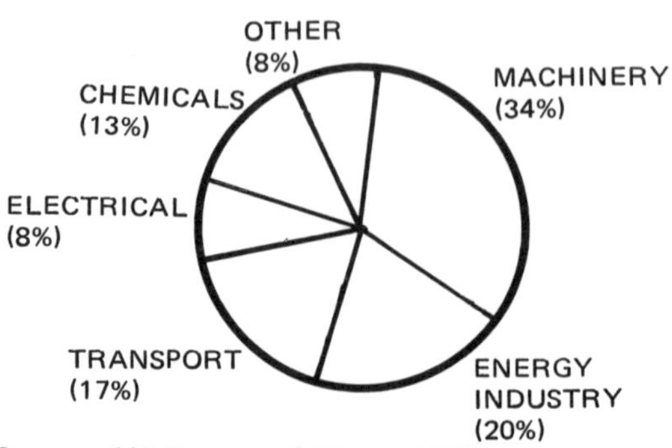


*Source: US Bureau of Mines, 1982*

Overall recovery of molybdenite ($MoS_2$) at primary ore concentrating plants ranges between 75 per cent and 90 per cent but production of molybdenite concentrate from copper ores is more difficult since molybdenite often constitutes less than 0.1 per cent of the ore. This balance consolidates the US position as leading producer with the major share of output deriving from primary molybdenite deposits.

### Technological Developments

These are mainly directed towards the improvement of methods by which technical grade oxide is converted to purer products. Ferro-molybdenum is now almost totally produced in electric furnaces while multi-layer casting techniques have reduced handling costs for the ingots. Newly developed metallurgical techniques are used to form components and alloys from molybdenum powder.

## TRANSPORTATION

The USA, Canada, Chile and the Communist bloc are net exporters, so the main trade routes are from North and South America to Europe and Japan. The main problems include the incidence of strikes disrupting rail transport within America, particularly the USA, and occasionally Chile, plus the length of the shipping routes involved. Molybdenum has only been utilised in volume within the past 40 years and has had virtually no history of transport disruptions.

## APPLICATIONS

### Consumption Trends

Since 1946 molybdenum production and consumption has increased from 10,000 to 100,000 tonnes a year with a cumulative annual growth rate of 4.5 per cent over the period. Molybdenum has suffered badly during the recession with consumption down by approaching one-third in the Western world during the first half of 1982. The USBM expects demand to increase when the recession lifts by 4.2 per cent per annum until 1990.

### The Principal Markets

Demand is based on the use of molybdenum to improve the properties of alloyed steels. A pie chart showing the end-use markets in the USA is given opposite, emphasising that metallurgical applications consume 90 per cent of all molybdenum. Steel and iron alloys contain 0.1-10 per cent molybdenum depending on the end-use. The recently developed high-strength low-alloy (HSLA) steels contain 0.2-0.3 per cent Mo and are a major growth area for use in oil and gas pipelines and in automobile and other transportation sectors. Stainless steels containing molybdenum are used in the chemical and electrical power generation industries. Molybdenum is an important component of high temperature superalloys which now make up 46 per cent and 58 per cent respectively of the weights of commercial and military aircraft. They are also extensively used in gas turbines, power plants etc. Cast irons containing molybdenum provide high strength components for rail and other applications.

Two new types of steel containing molybdenum are the ferritic grades for use in corrosive conditions and the dual phase steels whose high strength/unit weight is utilised to make automobile parts. Other minor uses for molybdenum are in a variety of electrical and electronic components and as part of wear resistant coatings.

Molybdenum compounds are used as lubricants, pigments and high profile catalysts for several petroleum refining processes.

### Recycling

Some secondary molybdenum in the form of metal or superalloys is recovered but only in small volumes. Limited amounts of molybdenum is reclaimed from spent catalysts.

### Substitution

Tungsten is molybdenum's traditional nearest rival but molybdenum has retained its market due to cost, price stability and fabricability advantages. There is little substitution for molybdenum in its major alloying applications and the trend has rather been to develop new materials utilising molybdenum.

## INTERNATIONAL TRADE AND WORLD PRICES

### Supply Arrangements

The USA is a net exporter of molybdenum and its producers effectively control the producer prices on the basis of which a large proportion of material is contracted for sale to consumers. The importance of Eastern bloc imports should not be under-estimated since the Communist world produces around 28 million lbs a year molybdenum, but consumes nearer 45 million lbs a year. The USSR is only 50-60 per cent self-sufficient and the other Comecon countries produce no more than five per cent of their needs. China is potentially a much larger producer than at present but the main bulk of its exports go to the Western world.

Molybdenum is traded in a variety of forms including the metal powder, technical grade molybdic oxide, ferro-molybdenum containing 65-70 per cent molybdenum and molybdenite concentrate. Prices are quoted in $/lb or kilo molybdenum content. The free market is sporadically active in times of a supply shortage and the specifications usually traded are the technical grade oxide and molybdenite concentrate.

**Companies**

The market is dominated by Amax of the USA, which produces around 40 per cent of total world mined output. Other major producers are Codelco of Chile; Duval Corporation of the USA; Placer Development of Canada; and the state-run organisations of the USSR and China. During the glut of 1981-82, US companies cut their output radically with Amax operating at an annualised level of under 45 per cent capacity during 1982. Amax and the other major producers have also postponed or cancelled several major projects in North America in order to try and control over-supply and maintain their relatively firm grip on the market.

**Prices**

Molybdenum prices showed a steady uptrend in the 15 years from 1964-65 onwards but the recent glut has forced producers to lower their prices dramatically in order to maintain some semblance of competitiveness with a free market price which has fallen by over a half within twelve months.

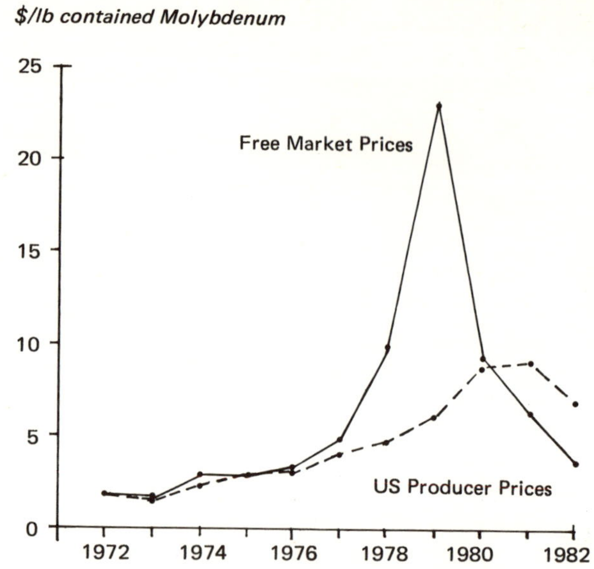


**Molybdenum Oxide Prices—Annual Averages**

# NICKEL (Ni)

| | |
|---|---|
| FACTOR FOR LIKELIHOOD OF SUPPLY DISRUPTION | 3.3 |
| FACTOR FOR COST OF SUCH A DISRUPTION | 5.7 |
| TOTAL STRATEGIC FACTOR | 18.8 |

*The factors for supply disruption and cost of such a disruption are assessed on a scale from 1.0 (lowest) to 10.0 (highest) risk. Multiplied together these factors give the total strategic factor rating. For components of these assessments see table later in this section.*

Nickel has a moderately high strategic factor due to its vital importance in the iron and steel industry since it gives strength and corrosion resistance to a wide range of alloys. The principal end uses are in the chemical, petroleum, electrical and aircraft industries. Canada, Australia, New Caledonia and Cuba possess the major share of reserves whilst Canada produces 30 per cent of world mined output. Deep sea resources are massive but unlikely to be exploited within this century. Consumption growth is expected to level at three per cent a year in the 1983-90 period, a rate similar to that of the stainless steel industry on which it depends so heavily.

### Physical Characteristics

Silvery ductile metal with high melting point (1,453$^{o}$C); a specific density of 8.9; good corrosion resistance and useful magnetic properties.

## OCCURRENCE AND PRODUCTION

Two main types of economic ores; nickel-copper sulphides, containing 2-4 per cent nickel, and lateritic deposits rich in iron, containing one per cent nickel. Nickel also occurs in other minerals.

### World Reserves

Land based reserves are estimated at 54 million tonnes contained nickel of which four-fifths is found in laterites and the balance in sulphides. This is equivalent to over 80 years of consumption at current levels but there are also extensive deep sea resources in the form of manganese modules, containing one per cent nickel, found particularly on the floor of the Pacific Ocean. Sulphidic ore reserves are concentrated in Canada, Australia, Southern Africa and Finland, with lateritic ores in the equatorial areas of New Caledonia, the Dominican Republic, the Philippines and Cuba. The Communist bloc, excluding Cuba, possesses 13 per cent of total world reserves.

### Production Characteristics

| | |
|---|---|
| *Mining Methods:* | Sulphide ores mined by conventional underground methods. Laterites mined from open pits. Sulphidic ores provide 55-65 per cent of world mined output |
| *Processing:* | Sulphidic ores beneficiated and roasted to nickel oxide which is reduced to impure metal. Lower grade concentrates smelted first to give an impure nickel/copper matte. This is electrolytically refined to cathodes or purified via the Mond carbon monoxide process. Lateritic ores must be dried first and are more difficult to treat since they are lower grade |
| *Energy Requirements:* | Very high energy requirements which average 230-240 million Btu/tonne metal |
| *By-Products:* | Copper, cobalt, the platinum group metals, silver, gold, selenium and tellurium, all mainly from sulphidic ores |
| *Environmental Factors:* | The control of sulphur dioxide emission during sulphide ore treatment requires expensive anti-pollution equipment |
| *Development Lead Time:* | Up to ten years for greenfield underground projects, up to six years for open cast mines. |

### The Major Producers

Canada totally dominated mined output until the late 1960s, but a protracted strike in 1969 encouraged the development of nickel deposits in other countries. Now Canada produces only 25-30 per cent of world output

SOUTHAMPTON LIBRARY UNIVERSITY

## COMPONENTS OF RISK ASSESSMENT

| *Factor* | *Rating* | *Comment* |
|---|---|---|
| **Production Risks** | | |
| Existing capacity | 1 | Vast oversupply of metal, more than double normal stock requirements during 1982. Much capacity, including recently installed plants, has been indefinitely shut |
| Labour disputes | 6 | Canada plagued by strikes of varying magnitude almost every three years, to coincide with labour contract negotiations |
| Violent conflict | 2 | Major producers located in Australia, New Caledonia, and Canada without history of serious conflicts |
| Range of primary supply sources | 1 | Currently no problem since, although Canada provides 30 per cent of Western mined output, stocks are large, and new lateritic projects have recently come on stream plus the possibility of exploiting deep sea nodules |
| Time lags for new supplies | 7 | 7-10 years needed to develop new lateritic or sulphidic ore bodies. Deep sea nodule recovery unlikely to be viable this century |
| **Transportation Risks** | | |
| Primary | 8 | Main problems in shipping material from strike-prone Canadian and Australian docks, plus rail strikes in the USA and rail disruptions in Southern Africa |
| Secondary | 2 | Few risks in transporting nickel briquettes, cathodes and ferro-nickel |
| **Application/Use Risks** | | |
| Total economic impact | 6 | World market of 750,000-800,000 tonnes a year |
| Effect on key industries | 8 | Use is concentrated in the chemical, petroleum, electrical and aircraft industries, generally in the more expensive and sophisticated alloys |
| Availability of substitutes | 6 | Substitutes tend to be inferior on grounds of cost and performance. Nickel-free speciality steels have been developed, and titanium and plastics in severe corrosion service applications |
| Longer term substitutability | 6 | Extensive changes in several major industries would be needed to allow the full replacement of nickel |
| **Trade Risks** | | |
| Collusive price agreements | 5 | Inco, Falconbridge and several other producers have had notable past successes in price control but the discounting price wars of 1977 and of 1981-82 have undermined their strength |
| Embargoes | 1 | Unlikely. The Communist bloc is a net exporter of nickel (produced as a platinum group metal by-product) and needs the foreign exchange generated |

*These components are assessed on a scale from 1 (lowest) to 10 (highest) risk. Grouped together, they form the basis for the assessments of the factors for likelihood of supply disruption, and for cost of such disruption which appear in the first table in this section.*

## WORLD NICKEL RESERVES

Total = 59.8 Million Tons Nickel Content

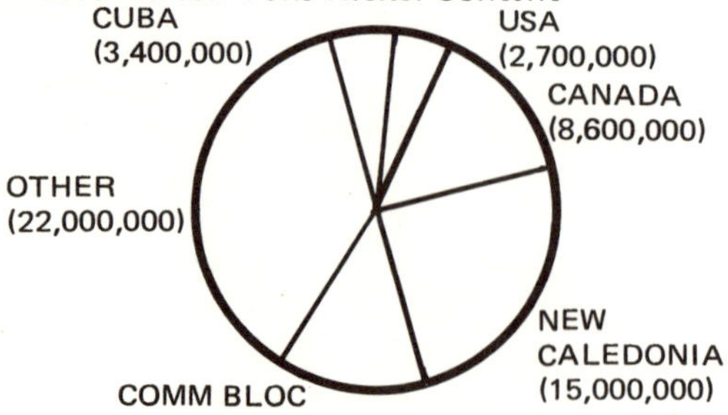


*Source: US Bureau of Mines, 1982*

## US NICKEL CONSUMPTION

By Market Sector 1981

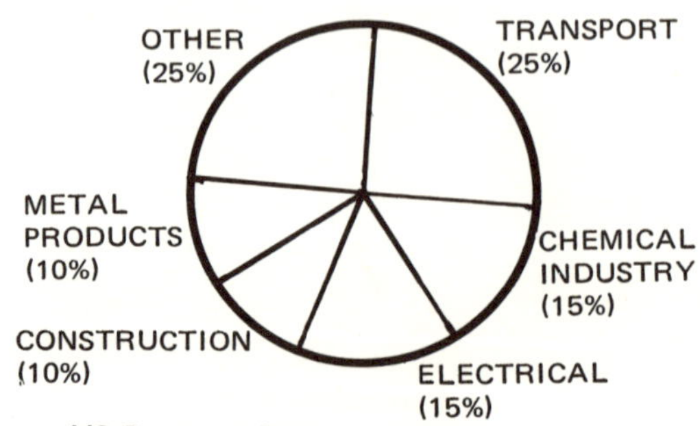


*Source: US Bureau of Mines, 1982*

**The Major Producers**

| *Countries* | *Output 1981 (Thousand Tonnes)* | | *Percentage of Western World Output* | |
|---|---|---|---|---|
| EUROPE | | 20.8 | | 4.2 |
| AFRICA | | 60.3 | | 12.3 |
| South Africa | 26.4 | | 5.3 | |
| ASIA | | 70.0 | | 14.3 |
| Indonesia | 45.5 | | 9.3 | |
| Philippines | 29.2 | | 6.0 | |
| AMERICA | | 190.0 | | 38.8 |
| Canada | 155.0 | | 31.6 | |
| AUSTRALASIA | | 149.0 | | 30.4 |
| Australia | 74.0 | | 15.1 | |
| New Caledonia | 74.0 | | 15.1 | |
| WESTERN WORLD TOTAL | | 490.1 | | 100.0 |

(Communist bloc production is currently estimated at 209 thousand tonnes a year.)

*Source: World Bureau of Metal Statistics, June 1982.*

from its Sudbury Basin deposits in Ontario and those in Manitoba. Other major producers are New Caledonia, Australia, Indonesia and the Philippine Republic. The supply shortages of the late 1960s triggered rapid expansion in output and led to chronic overcapacity in the 1970s. Increased energy and capital costs have eroded cash flow expectations from recently developed nickel projects. Energy costs have had an adverse effect on laterite producers in particular. Lateritic ores require about three times as much energy to process as the sulphidic variety. Another problem is the lack of transport and administrative infrastructure in the tropical regions where laterites are found. With energy now comprising two-thirds of the overall costs of running an oil-fired lateritic ore refinery the percentage of nickel extracted from sulphide ores has increased. The USA has a low level of domestic nickel mining due to the poor grades of reserves. In 1981 it was 72 per cent dependent on imports from Canada, Botswana and Australia. The sulphide concentrates and nickel-rich mattes from the main producers are often exported for further refining, largely in the UK, Norway and Japan. Japan has no domestic nickel mining industry but refines about one-fifth of Western world nickel concentrates to give pure metal and ferro-alloys. Similarly Europe refines five times as much nickel concentrate as is mined locally.

### Technological Developments

Research is directed towards reducing the energy required to extract nickel from lateritic ores with emphasis on the development of chemical leaching techniques. The high cost of pollution control equipment is an added incentive to improving processing methods for sulphidic ores.

## TRANSPORTATION

The main trade routes are for nickel sulphide concentrate or matte. This is transported from Canada to the USA, UK and Finland for treatment. Supplies also come from New Caledonia, the Dominican Republic and other ore producers to Europe and Japan. The Communist bloc is a net exporter of 25,000-40,000 tonnes a year nickel, fluctuating in response to foreign currency requirements. Transport is mainly by ship and by rail on long routes. The major disruptions are due to dock disputes, rail strikes, etc.

## APPLICATIONS

### Consumption Trends

Nickel usage grew spectacularly from 10,000 tonnes a year at the beginning of the century to 700,000 tonnes a year in 1980. The 5-6 per cent a year growth rate from 1946 to 1977 has given way to a period of moderate or even negative growth since 1977. But between 1984 and 1990 Amax Nickel expect an annual rate of 3.7 per cent in consumption growth.

### The Principal Markets

Demand for nickel is 90 per cent based on its use in the metallic form for the manufacture of steel and other alloys. Europe, the USA and Japan account for over 90 per cent of Western consumption and a pie chart showing the end-use markets in the USA is given opposite. Stainless steels for the chemical and construction industries account for nearly a half of all nickel, with the nickel content varying between 1.25 and 37 per cent. Nickel is also used in several types of high technology steels for tools, the petroleum pipeline and motor industries.

Superalloys for use in aerospace and power generation contain 50-55 per cent nickel which provides corrosion resistance and strength at high temperatures. The chemical and marine industries utilise a very corrosion resistant series of nickel/copper alloys.

Nickel chemicals are useful catalysts in oil cracking. They are also used in batteries and fuel cells and nickel carbides provide hard-facing materials. Nickel is a component of permanent magnet alloys and the chief constituent of many high resistance alloys, both types used in electrical machinery.

Nickel consumption is expected to change significantly in the next ten years with the chemical and petroleum industry taking a much larger share, and the automotive sector declining in importance.

### Recycling

Recovery of secondary nickel in the USA from old and prompt industrial scrap accounted for 23 per cent of demand. The chemical and catalytic sectors are dissipative and in the steel sector, nickel tends to be reclaimed as an alloy component rather than being extracted and treated separately.

### Substitution

Substitutes tend to cost more or not perform as well. Alternatives include aluminium coated steel and plastics in the construction and transportation industries, nickel-free speciality steels in the power generating, petrochemical and petroleum industries; titanium and plastics in severe corrosion service applications and platinum, cobalt and copper in some catalytic uses.

## INTERNATIONAL TRADE AND WORLD PRICES

### Supply Arrangements

Over 95 per cent of all nickel supplies are sold on a producer/consumer basis and, historically, nickel was controlled by a limited number of powerful groups led by Inco of Canada, who set the authoritative producer price. This dominance has been challenged by a proliferation of smaller producers, weakening demand in 1977-78 and again in 1981-82 (which provoked damaging price wars between the major producers), plus the advent of an LME nickel contract in 1979. Producer prices are quoted in c/lb for cathodes, granules, briquettes and pellets, and the LME contract is for six tonnes 99.8 per cent nickel in specified forms with a price in £/tonne. There is also an informal free market based in Rotterdam where marginal volumes are traded since many major producers do not feel comfortable with the LME. The Communist bloc producers channel their material through the free market and tend to price their products very competitively.

Nickel has a high elasticity of demand. About two-thirds are used mainly in capital goods and purchases are delayed during times of recession. This led to a massive surplus in 1981-82 and the main Western producers responded positively with sharp cutbacks.

### Companies

Inco Ltd of Canada is the acknowledged market leader, with subsidiaries in Indonesia and elsewhere, and the state-run organisations of the USSR follow just behind. Second rank producers include Botswana RST Ltd of Botswana; SLN/Imetal SA in New Caledonia; Cuba Niquel; Western Mining Corporation of Australia; and Falconbridge Nickel Mines of Canada. The leading electrolytic nickel producers are Inco; Falconbridge Nikkelverk A/S of Norway; the state-run organisations of the USSR and Sumitomo Metal Mining Co Ltd of Japan. Ferro-nickel output is concentrated in the hands of SLN; Pacific Metals, Japan; and Nippon Kogyo of Japan.

### Prices

Nickel demand is extremely sensitive to the business cycle, reflected in free market prices, whilst production costs are becoming increasingly dependent on oil prices, reflected in official producer prices. The USBM estimates that every 10 per cent increase in the price of fuel oil puts seven cents on the production costs of nickel and it is obvious that when, as in 1982, free market prices vastly undercut producer prices, the producer price was unofficially discounted to the break-even point or beyond.

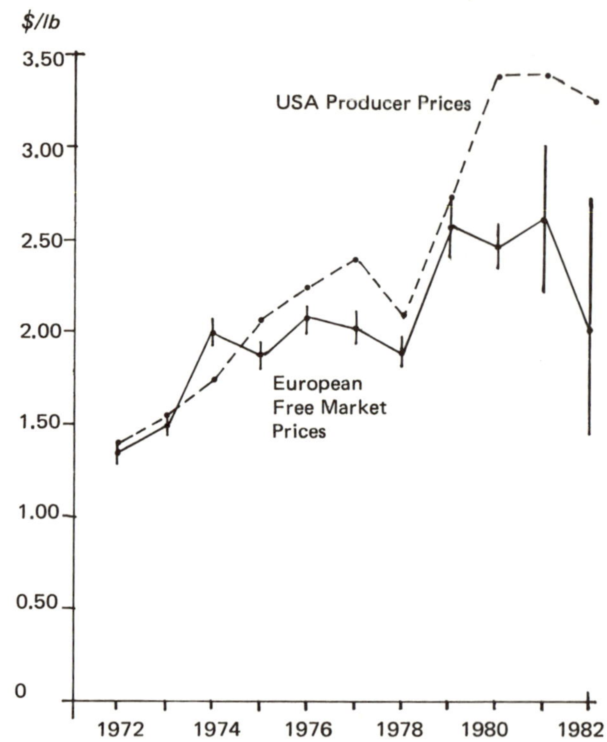


**Nickel Prices—Annual Averages and High/Lows for European Free Market Prices, Annual Average USA producer prices**

# PLATINUM GROUP METALS (PGM)

| | |
|---|---|
| FACTOR FOR LIKELIHOOD OF SUPPLY DISRUPTION | 4.8 |
| FACTOR FOR COST OF SUCH A DISRUPTION | 6.0 |
| TOTAL STRATEGIC FACTOR | 28.8 |

*The factors for supply disruption and cost of such a disruption are assessed on a scale from 1.0 (lowest) to 10.0 (highest) risk. Multiplied together these factors give the total strategic factor rating. For components of these assessments see the table later in this section.*

The platinum group metals can be considered together because they normally occur in association and have properties in common. They have a high strategic factor due to their rarity and unparalleled physical and chemical properties which ensure their use in the oil and automobile industries as well as for aesthetic purposes. South Africa and Russia produce 95 per cent of newly mined PGM supply and a degree of supply and price collaboration between the two is assumed. There are no viable substitutes in the majority of applications for the most important metals in the group; platinum and palladium.

**Physical Characteristics**

All have high melting points (platinum 1,774$^{o}$C), relatively high specific densities (platinum 21.5), extraordinary catalytic ability, great corrosion resistance and high thermal and electrical conductivities.

## OCCURRENCE AND PRODUCTION

They are found in disseminated elemental form in various hard rocks.

**World Reserves**

These are estimated at 1,195 million troy ounces PGMs, sufficient for over 175 years of mining at current levels. Reserves are 81 per cent concentrated in three extensive flat bedded reef deposits in South Africa. Communist countries, essentially the USSR, possess 17 per cent of world reserves.

**Production Characteristics**

| | |
|---|---|
| *Mining Methods:* | Mainly in deep underground mines |
| *Processing:* | By-product platinum obtained during processing of nickel or copper. South African ore beneficiated and chemically treated to give slimes containing 25-75 per cent PGMs. Concentrates refined to separate out the six metals |
| *Energy Requirements:* | Fairly high due to energy intensive crushing and milling processes to concentrate the ore |
| *By-Products:* | (a)Nickel: USSR, Canada, Philippines<br>(b)Copper: USA |
| *Environmental Factors:* | In Canada and USA, those associated with nickel and copper mining. In South Africa, the problems associated with deep underground mines have been ignored |
| *Development Lead Time:* | Up to 10 years. |

**The Major Producers**

South Africa possesses the only major ore bodies to be mined solely for their PGM content which is of the order of 10g/tonne ore, in the proportions 60 per cent platinum, 25 per cent palladium, 4 per cent rhodium, and 11 per cent ruthenium, iridium and osmium. The USSR produces a similar amount of PGMs to South Africa but in the proportions of 56 per cent palladium, 30 per cent platinum, 10 per cent rhodium, 4 per cent ruthenium, iridium and osmium. The USSR has the reserves and capacity to increase production by at least 10 per cent with the minimum of effort but its PGM output is a co-product of nickel mining.

In Canada, the PGMs are a by-prduct of nickel, providing up to 20 per cent of mine revenue. Other small producers include the USA, New Caledonia, Columbia, Australia and the Philippines. PGM refining in the western world is often carried out in a different country to that in which the ore is mined. Traditionally, South Africa sent all its concentrates to Johnson Matthey Ltd, in the UK, but now three new domestic refining plants have been built. Canada's concentrates are transported to the UK and Norway for treatment. The USSR refines its own output at the world's largest plant in Krasnoyarsk.

## COMPONENTS OF RISK ASSESSMENT

| *Factor* | *Rating* | *Comment* |
|---|---|---|
| **Production Risks** | | |
| Existing capacity | 1 | The major South African producers are operating at only 80-85 per cent capacity and possess mounting stocks |
| Labour disputes | 1 | South Africa and the USSR are the main production areas |
| Violent conflict | 6 | Usage increases in times of war whilst PGM mines would be prime candidates for being bombed and disrupted |
| Range of primary supply sources | 8 | 95 per cent of annual mined PGM supplies come from the USSR and South Africa |
| Time lags for new supplies | 8 | Up to 10 years to develop new underground mines, often in difficult conditions |
| **Transportation Risks** | | |
| Primary | 7 | Political problems and threat of thefts during transportation of platinum concentrate and refined metal from South Africa and USSR to consuming centres |
| Secondary | 1 | Few problems in transporting fabricated platinum in catalysts, jewellery, etc |
| **Application/Use Risks** | | |
| Total economic impact | 2 | Fairly minor in volume terms with world PGM annual output at 6.7-6.9 million troy ounces |
| Effect on key industries | 7 | Platinum and palladium, particularly, useful in electronic/electrical industries in high performance applications. Platinum used for oil cracking catalysts and with palladium and rhodium in car exhaust emission catalysts |
| Availability of substitutes | 5 | Available in most areas but generally poorer performance |
| Longer term substitutability | 6 | To some extent the PGMs can be substituted for each other and gold/silver are possible substitutes which could be developed on a long-term basis |
| **Trade Risks** | | |
| Collusive price agreements | 9 | South Africa and Russia known to collaborate closely on pricing and supply structures and have the financial muscle to support the market |
| Embargoes | 8 | Could be a major problem in the event of political realignments and the USA, EEC and Japan have separately expressed concern at the possibility of PGM supply disruptions. |

*These components are assessed on a scale from 1 (lowest) to 10 (highest) risk. Grouped together, they form the basis for the assessments of the factors for likelihood of supply disruption, and for cost of such disruption which appear in the first table in this section.*

### WORLD PLATINUM GROUP METAL RESERVES

Total over 1,195 Million Troy Ounces, not all reserves delineated

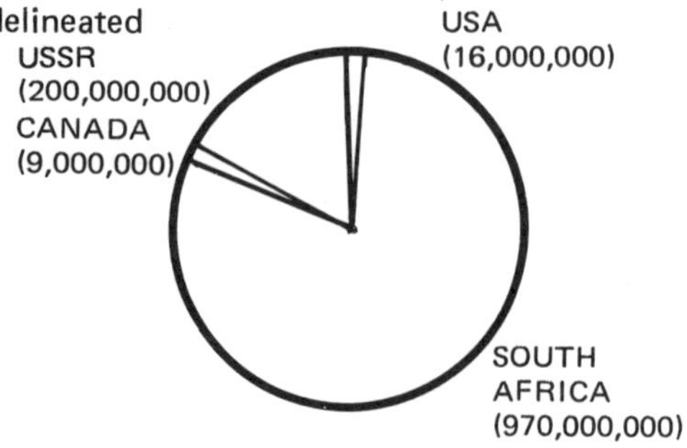


*Source: US Bureau of Mines, 1982*

### US PLATINUM GROUP METALS CONSUMPTION

By Market Sector 1981

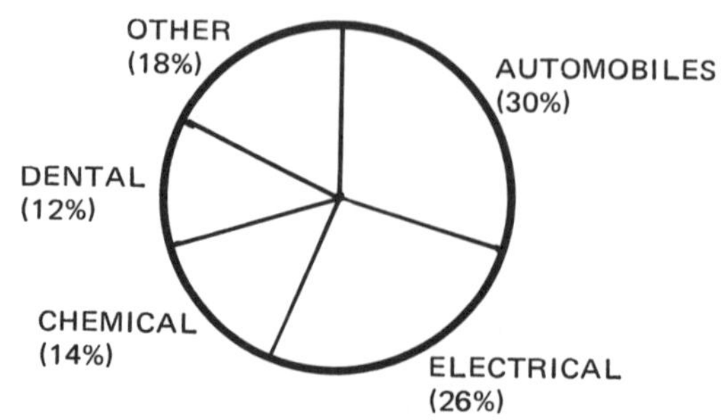


*Source: US Bureau of Mines, 1982*

Mined Production of Platinum Group Metals

| *Countries* | *1981 Output (Thousand Ounces Troy)* | *Percentage of Western World Output* |
|---|---|---|
| USA | 6 | 0.2 |
| Canada | 350 | 9.9 |
| South Africa | 3,100 | 87.8 |
| Others | 74 | 2.1 |
| WESTERN WORLD TOTAL | 3,530 | 100.0 |

(Production in the USSR is currently estimated at 3,250,000 ounces troy a year.)

*Source: USBM, Mineral Commodity Summaries, 1982*

Mined Production of Platinum

| *Countries* | *1981 Output (Thousand Ounces Troy)* | *Percentage of Western World Output* |
|---|---|---|
| South Africa | 2,100 | 84.0 |
| Canada | 150 | 6.0 |
| Others | 250 | 10.0 |
| WESTERN WORLD TOTAL | 2,500 | 100.0 |

(Exports by the USSR are currently estimated at 350,000 ounces troy a year.)

*Source: Shearson American Express, Annual Economic Review of Metal Markets 1981-82.*

The 1981-82 recession has brought production cutbacks for the South African producers of 15-20 per cent, and by Canadian nickel producers of up to 30 per cent. Expansion plans at established mines have been deferred and the exploitation of Zimbabwe's considerable nickel/copper/platinum reserves postponed.

### Technological Developments

The underground platinum mines operate to feasibility limits in terms of depth and efficiency. The USBM is researching methods of exploiting the US low grade Stillwater Complex, but reserves here remain uneconomic.

## TRANSPORTATION

The USA is 85 per cent dependent on PGM concentrate and metal imports whilst the EEC, and Japan are almost wholly dependent on imports. Transportation is therefore a vital link. The mechanics cause few problems since volumes are small and the high unit value ensures that a suitable level of care is taken, but the politics of exporting/importing these metals is more delicate.

## APPLICATIONS

### Consumption Trends

Platinum group metals are consumed in the proportion 47 per cent platinum, 45 per cent palladium, three per cent rhodium, five per cent ruthenium, iridium and osmium. The USA and Japan each consume around one third of annual PGM output of 6.7-6.9 million ounces a year. The USBM expects demand for platinum in the USA to grow by 1.5 per cent a year in the next decade, although industrial consumption of PGMs fell ten per cent in 1981.

### The Principal Markets

A pie chart showing the end-use markets in the USA is given opposite. Western Europe uses a smaller percentage in the automobile industry and Japan uses around 50 per cent of its annual platinum consumption of 1.1 million ounces in jewellery. The major growth area for platinum, palladium, and rhodium is as catalysts in vehicle exhaust systems to nullify potentially harmful combustion products. Catalytic convertors are now required in all new cars sold in the USA, Japan and Australia, but the combination of the effect of the recession on car sales and the technical advances which reduce the weight of PGM needed per car has temporarily halted the healthy growth seen in this sector. Traditional catalytic convertors contained 0.05 oz PGM in the ratio 70Pt/30Pd, but newer three-way covertors contain 0.05 oz Pt/0.02 oz Pd/0.005 oz Rh, or 0.06 oz Pt/0.002 oz Rh.

All the PGMs are used in a wide range of electrical and electronic devices as contacts, thermocouples, and other components. Platinum use is showing a steady expansion in electronic applications, and ruthenium use is expected to grow considerably due to new applications for high quality resistors. Palladium has suffered due to a switch from electromechanical switch gears, with precious metal contacts, in tele-communications to electronic switching devices. Platinum demand for jewellery remains the preserve of the Japanese despite a 1981 advertising campaign in the USA and Europe. Platinum and rhodium have important uses in machinery for producing glass and fibreglass. A possible major new sector is for PGM electrodes in fuel cells. Volumes of PGMs consumed in dental and medical applications are tiny but this forms a stable sector.

### Recycling

Recycling is an important element due to the relative indestructability of PGMs in most applications and their high unit cost. Secondary platinum production is estimated at 200,000 ounces in 1981, 8 per cent of western mined supplies, and secondary palladium production estimated at 275,000 ounces, 28 per cent of western mined supplies. Secondary output is likely to become more important as PGM convertor-equipped cars move into the scrapping stage. Approximately one-third of the catalysts' PGM content is lost during the cars' lifetime and recycled metal could amount to only 30-40 per cent of original PGM consumption with refining complicated by the presence of two or three metals used together in the catalyst. The new technology of fuel cells could provide further recycling possibilities. Obviously scrap recovery is very price sensitive and it is estimated that, at higher price levels, recovery from the chemical industry could jump by 25-30 per cent.

### Substitution

In several chemical and electrical applications the PGMs can be virtually interchanged. Alternatives in electrical and electronic uses are gold, silver and, for resistors and thermocouples, tungsten. Gold is a dental alloy substitute whilst the rare-earths, nickel and vanadium, are possible substitutes for catalytic uses in oil cracking and nitration reactions. Improved engines, fuels and electric cars or reduced environmental awareness could reduce the use of PGM emission control catalysts. Silver and gold are possible jewellery substitutes for platinum.

## INTERNATIONAL TRADE AND WORLD PRICES

### Supply Arrangements

Platinum group metals are marketed via producer/consumer contacts and via the free market. The producer element is governed by the refineries which sell an estimated 90 per cent of PGM supply under one or two year contracts to particular consumers such as the major automobile makers. The two major South African producers effectively fix their producer price for all contracted sales and historically have followed a conservative policy, although, in January 1983, Rustenburg decided to sell a proportion of its material on a free market-based price. Canadian producers and US refiners tend to reflect these prices.

The free market has two effective mechanisms; an informal merchant market for all six metals and terminal markets for platinum and palladium. Trading details are summarised below.

- —platinum: 50 oz, 99.8 per cent minimum purity metal as plates or ingots. Futures contracts tradeable on New York Mercantile Exchange (NYMEX)
- —palladium: 100 oz, 99.8 per cent minimum purity metal as sponge or ingots. Futures contracts tradeable on New York Mercantile Exchange
- —rhodium: 50 oz, 99.9 per cent purity as sponge or powder. Fairly active merchant market
- —iridium: 100 oz, 99 per cent purity as powder or ingots. Very little trading interest
- —osmium: 99.5 per cent purity. Least important of all the PGMs
- —ruthenium: 99.9 per cent powder. Very thin market.

The free market is used by Communist bloc countries and other marginal sources of supply including consumers selling surplus holdings, investment stocks and minor producers. The free market platinum price is obviously volatile and often moves in tandem with the political forces which affect gold. Increasingly, platinum and palladium are being considered as investment vehicles as well as industrial metals.

### Companies

The two major South African producers are Impala Platinum and Rustenburg, other producers include the USSR state owned operations, Inco of Canada, Western Platinum of South Africa, and Cia Mineros Colombianos, Colombia. The main refineries are Johnson Matthey UK, which also markets the entire Rustenburg output, Engelhard in the USA, Impala in South Africa, Degussa of West Germany and Inco in the UK and Norway.

### Prices

Producer prices have been maintained on a sensible up trend for most of the six metals over the last decade with producers ignoring the vagaries of the free market. Free market platinum and palladium saw marked peaks in 1980, following gold's sudden rise. Gold has been at a discount to platinum for most of the last decade and, recently, speculators have traded off their views on the price ratio. The recession has hit the PGMs harder than gold, with platinum trading at a hefty discount to gold during 1981-82.

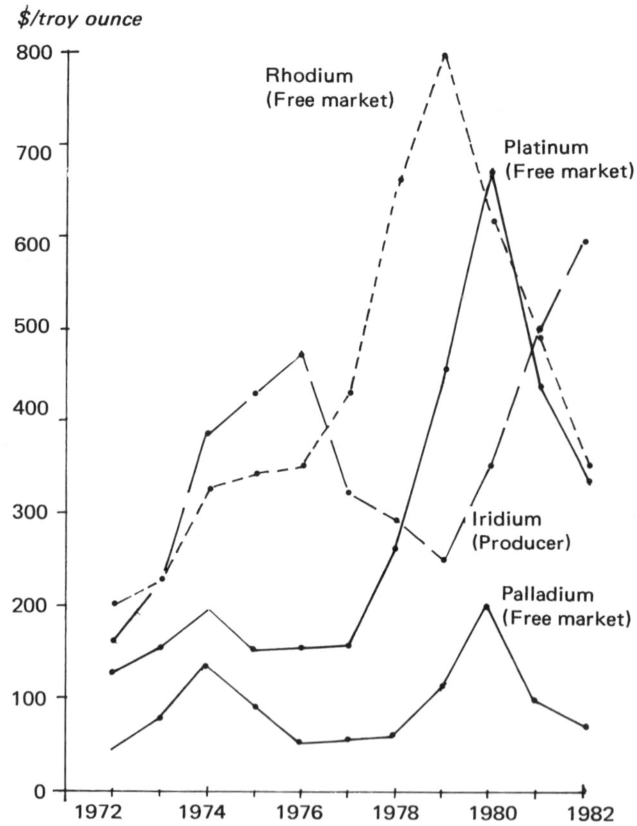


**Platinum Group Metals Prices—Annual Average**

# RARE-EARTH MINERALS (Lanthanides)

| | |
|---|---|
| **FACTOR FOR LIKELIHOOD OF SUPPLY DISRUPTION** | **1.2** |
| **FACTOR FOR COST OF SUCH A DISRUPTION** | **5.5** |
| **TOTAL STRATEGIC FACTOR** | **6.6** |

*The factors for supply disruption and cost of such a disruption are assessed on a scale from 1.0 (lowest) to 10.0 (highest) risk. Multiplied together these factors give the total strategic factor rating. For components of these assessments see table later in this section.*

Rare-earth minerals include a variety of metals which have been discovered relatively recently. Their low strategic factor reflects present use which remains in minor quantities. They are in good supply with no envisaged shortage and the major producers are in the USA and Australia. The rare-earths are mainly used in sophisticated applications such as lasers, catalysts and computer components, and new applications continue to be developed. They often tend to occur in minerals derived as a by-product of titanium and zirconium ores.

### Physical Characteristics

The lanthanides are a group of 15 elements with similar properties which are valued for their interesting responses to magnetic, electric and stress fields giving them special magnetic, piezoelectric or dielectric properties.

## OCCURRENCE AND PRODUCTION

They are generally found as the oxides in a complex ore such as monazite or bastnaesite. Monazite is a phosphate containing 70 per cent combined rare earth oxides (REO) plus 1.5 per cent yttrium oxide. Bastnaesite is a fluoro-carbonate material containing about 75 per cent REO and only 0.05 per cent yttrium oxide. Yttrium is strictly not in the lanthanide group but has closely allied properties and can be treated in the same context.

### World Reserves

These are estimated at over seven million tonnes rare-earth oxide content, sufficient for 440 years of consumption at current levels. Reserves are concentrated in North and South America, Australia and in Asia. The USSR and its satellites possess six per cent of known reserves. No accurate assessment of Chinese reserves is, however, available.

### Production Characteristics

| | |
|---|---|
| *Mining Methods:* | Monazite: depending on method for co-product, usually beach deposits or open cast mining.<br>Bastnaesite: by open cast mining |
| *Processing:* | Rare-earth minerals separated out by gravity and electro-magnetic methods. Either reduced chemically en masse to mischmetal (an alloy with the same proportional mixture of metal as in the original ore), or sophisticated techniques used to extract the different metals individually |
| *Energy Requirements:* | Relatively low and irrelevant when extracted as a by-product |
| *By-Products:* | Monazite: titanium and zirconium ores or tin, occasionally gold, semi-precious stones.<br>Bastnaesite: occasionally strontium and barium |
| *Environmental Factors:* | Minimal, but certain deposits not developed due to ecological considerations |
| *Development Lead Time:* | Three to six years. |

### The Major Producers

| *Countries* | *1981 Mined Output (tonnes)* | *Percentage of Western World Output* |
|---|---|---|
| USA (bastnaesite only) | 15,880 | 57.9 |
| Australia | 7,530 | 27.4 |
| Brazil | 1,090 | 4.0 |
| India | 2,450 | 8.9 |
| Malaysia | 225 | 0.8 |
| Others | 275 | 1.0 |
| WESTERN WORLD TOTAL | 27,450 | 100.0 |

(Communist bloc production is currently estimated at 5,450 tonnes and production in China 3,175 tonnes a year.)

*Source: USBM Mineral Commodities Summaries, 1982.*

## COMPONENTS OF RISK ASSESSMENT

| *Factor* | *Rating* | *Comment* |
|---|---|---|
| **Production Risks** | | |
| Existing capacity | 2 | Adequate |
| Labour disputes | 1 | Unlikely to affect rare-earth mineral output, particularly since monazite is a by-product |
| Violent conflict | 1 | Not a major risk in the USA and Australia, the two main producers |
| Range of primary supply sources | 4 | The USA and Australia produce over 85 per cent of western output. Various minerals contain different amounts of rare-earth compounds and processing capacity to produce individual metals is limited so supply problems are possible when a large demand for one of the least abundant metals is developed |
| Time lags for new supplies | 6 | Up to six years to develop beach sands deposits and viable concentrations of bastnaesite deposits are difficult to find |
| **Transportation Risks** | | |
| Primary | 1 | No problems |
| Secondary | 1 | No problems |
| **Application/Use Risks** | | |
| Total economic impact | 1 | Total world market of 33,000 tonnes per year REO contained |
| Effect on key industries | 8 | High bias towards sophisticated ceramic and glass applications, the improvements of certain steels and use in solid state electronics, magnets and other developing sectors |
| Availability of substitutes | 8 | Alternatives are available in the magnet, ceramics and metallurgical sectors, but tend to be less effective |
| Longer term substitutability | 6 | New uses being developed for rare-earth oxides and a drive to replace them would require ingenuity by materials scientists |
| **Trade Risks** | | |
| Collusive price agreements | 1 | Unlikely due to by-product status, giving a sweetener to the main revenues. Prices are maintained on an upward trend but increases have been kept sensible to encourage usage |
| Embargoes | 1 | Unlikely, except as part of high technology ban to certain governments |

***These components are assessed on a scale from 1 (lowest) to 10 (highest) risk. Grouped together, they form the basis for the assessments of the factors for likelihood of supply disruption, and for cost of such disruption which appear in the first table in this section.***

## WORLD RARE-EARTH METALS RESERVES

Total over 7,730,000 Tons Rare-Earth Oxide Content, not all reserves delineated

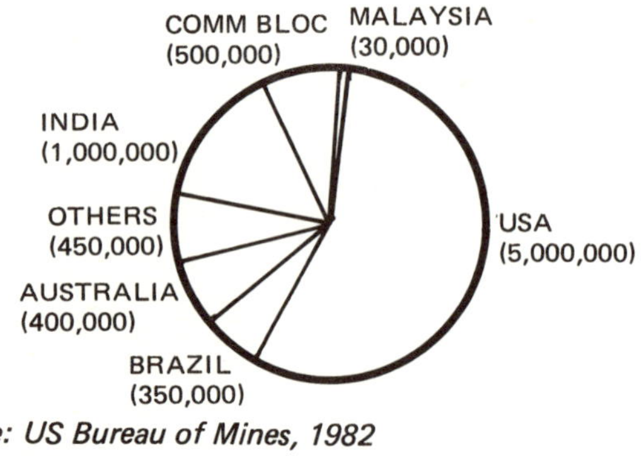


*Source: US Bureau of Mines, 1982*

## US RARE-EARTH METALS CONSUMPTION

By Market Sector 1981
Other Sector includes: Electrical, Nuclear, Magnetic and Research Areas

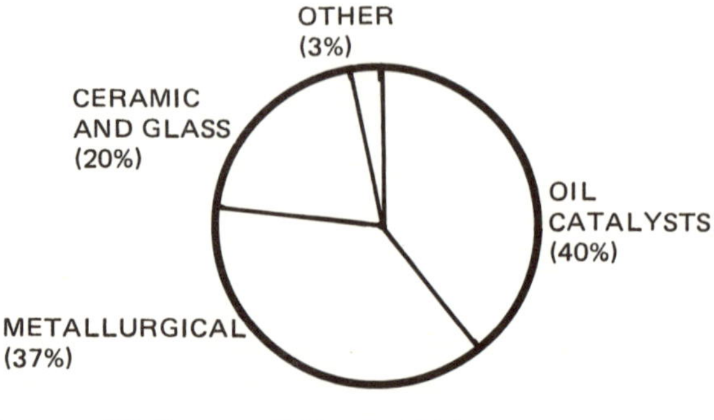


*Source: US Bureau of Mines, 1982*

Output is concentrated in the USA, Australia, China and India. Bastnaesite and monazite are both mined in the USA, but monazite is the main source in Australia (as a zirconium and titanium ore by-product) and in Malaysia and China (as a tin by-product). The rare-earth elements are in abundant supply but various metals contain different proportions of the elements and, when processed, large surpluses may develop in those elements for which there is little demand. Production of the separate metals and of mischmetal is concentrated in the USA, France and Japan, near to the main consuming centres.

### Technological Developments

Complex ion-exchange and solvent extraction techniques for the separation of the various metals continue to be improved. USBM research continues aimed at increasing REO recovery from bastnaesite ore.

## TRANSPORTATION

Few problems encountered in transporting these metals, either as individual metal or as by-product content of ilmenite, rutile or zircon ores.

## APPLICATIONS

### Consumption Trends

The USBM expects demand for rare-earths to increase by six per cent a year during the 1980s with US demand in 1982 at 18,000 tonnes REO. The demand pattern is expected to continue shifting towards individual compounds, metals and special mixtures formulated to meet set requirements. Improved knowledge of rare-earth properties may lead to new industrial applications.

### The Principal Markets

A pie chart showing the end-use markets in the USA is given opposite. Western Europe follows much the same pattern whilst Japan uses only 14 per cent in metallurgical applications and more in ceramics and glass.

Mischmetal and rare-earth silicides improve the physical properties of steels and allow advanced forming techniques to be used. High strength low alloy steels for pipelines and automobiles contain 3 lb rare-earth compounds per ton of steel. The ceramics and glass sector employ lanthanum oxide in optical glasses and ceramic capacitors and cerium oxide as a glass decolonising agent. Miscellaneous uses include cerium and samarium in high-performance permanent magnets, europium oxide as the red phosphor for colour television tubes, and gadolinium or samarium oxides as neutron absorbers. Yttrium and gadolinium compounds are used in solid state lasers. These applications are clearly limited in volume, but important in many rapidly developing areas. The main growth sectors are expected to be phosphors for fluorescent lamps and computer VDU screens plus rare-earth alloy magnets for tachometers and line printers.

### Recycling

Small quantities of samarium recovered from samarium-cobalt permanent magnets and small amounts of europium from scrapped colour televisions.

### Substitution

Substitutes are generally available but usually much less effective.

## INTERNATIONAL TRADE AND WORLD PRICES

### Supply Arrangements

Producers tend to sell directly to consumers with a very limited free market except for occasional periods of shortage when there is active trading in mischmetal. Australia, Brazil, India and Malaysia are the main exporters, with the USA, Europe and Japan as net importers. China is also an erratic exporter of rare-earth minerals. The specifications generally traded are mischmetal, with a defined cerium content, and monazite and bastnaesite concentrates, with a price quoted in $/lb REO contained or per unit gross weight. Other high purity mixed or individual compounds are traded in small volumes only and end-users in the sophisticated applications tend to develop their own precise specifications.

### Companies

The main mining producers are Molycorp of the USA; Allied Eneabba Pty Ltd, of Australia; Indian Rare-Earths Ltd; Nuclebras Monazite and Associados of Brazil; and Westralian Sands of Austria. The main rare-earth oxide and yttrium compound producers are Molycorp, Rhone-Poulenc SA of France, WR Grace & Co of the USA, and Mitsui Mining & Smelting Co of Japan.

### Prices

Producer prices for monazite and bastnaesite have been kept on a steadily increasing trend over the last 20 years with a fair degree of control possible because of the small number of producers.

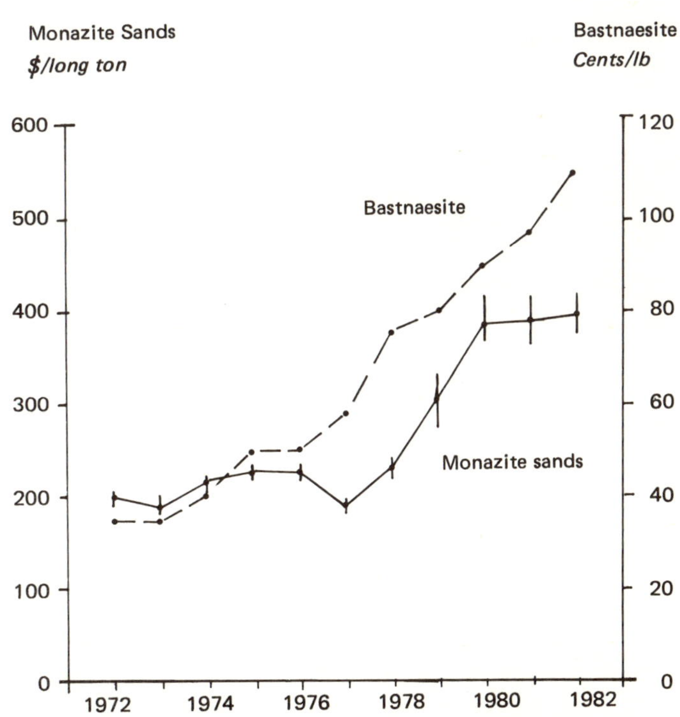


**Rare-Earth Oxide Ores Prices—Annual Averages and High/Lows for Monazite Sands, Annual Averages for Bastnaesite**

# RHENIUM (Re)

| | |
|---|---|
| FACTOR FOR LIKELIHOOD OF SUPPLY DISRUPTION | 3.0 |
| FACTOR FOR COST OF SUCH A DISRUPTION | 3.3 |
| TOTAL STRATEGIC FACTOR | 9.9 |

*The factors for supply disruption and cost of such a disruption are assessed on a scale from 1.0 (lowest) to 10.0 (highest) risk. Multiplied together these factors give the total strategic factor rating. For components of these assessments see table later in this section.*

Rhenium is a little-known minor metal with a moderately low strategic factor. Over 90 per cent of consumption is as a catalyst with platinum in the production of low lead, high octane petrol. Rhenium can only be obtained as a by-product of molybdenum but reserves are large and concentrated along the western backbone of the Americas. Substitutes are available for all applications and demand is expected to remain static.

### Physical Characteristics

High specific density (21); elevated melting point (3,180$^{o}$C); good high temperature strength and high modulus of elasticity; excellent corrosion resistance.

## OCCURRENCE AND PRODUCTION

Rhenium is never found as a major mineral concentration. The main commercial mineral is rhenium sulphide which occurs almost solely with molybdenite in copper and molybdenum porphyries in concentrations of approximately 150 ppm and 100 ppm respectively.

### World Reserves

These are estimated at 7 million lbs of which 74 per cent is split equally between Chile and the USA. Known reserves are sufficient for 400 years of consumption at current levels. The Communist bloc controls just over seven per cent of reserves.

### Production Characteristics

*Mining Methods:* Open cast mining of molybdenum and copper porphyry ore bodies

*Processing:* Rhenium oxides recovered during roasting of molybdenite concentrates. Chemical extraction gives perrhenic acid which is reduced by hydrogen to high purity rhenium metal powder. This is sintered and formed to the required shape.

*Energy Requirements:* Largely irrelevant due to by-product status

*By-Products:* Molybdenum, copper, often also silver and gold

*Environmental Factors:* None unique to rhenium

*Development Lead Time:* Up to 10 years for open cast mine development

### The Major Producers

| *Countries* | *1981 Output (Thousand Pounds)* | *Percentage of Western World Output* |
|---|---|---|
| USA | 8.0 | 31.4 |
| Canada | 4.0 | 15.7 |
| Chile | 8.0 | 31.4 |
| West Germany* | 4.5 | 17.6 |
| Peru | 0.4 | 1.6 |
| Other | 0.6 | 2.3 |
| WESTERN WORLD TOTAL | 25.5 | 100.0 |

(Production in the USSR is currently estimated at 10,000 lbs a year.)

*Treats imported concentrates

*Source: USBM 1982, except for US figure.*

Chile, the USA, Canada, USSR, Peru and Zaire, are all known for their molybdenum and/or copper production. West Germany has no indigenous reserves but recovers rhenium from molybdenite concentrates imported from Chile, Canada and Peru. Chile effectively dominates the market with around 40-45 per cent of annual mined output and is moving towards the integration of its molybdenum industry which will involve building more of its own rhenium extraction facilities.

### Technological Developments

Only 50-60 per cent of the contained rhenium can be recovered from molybdenum porphyry ores and 25-30 per cent from copper ores. Efforts are directed towards increasing these recovery rates by using more sophisticated scrubber systems to remove rhenium oxides from the roaster flue gases. The USSR claimed to have reached 42 per cent recovery from copper ores in 1980 due to improved techniques.

## COMPONENTS OF RISK ASSESSMENT

| *Factor* | *Rating* | *Comment* |
|---|---|---|
| **Production Risks** | | |
| Existing capacity | 1 | Currently in surplus, coinciding with pronounced surpluses for both copper and molybdenum, although this is not always the case |
| Labour disputes | 4 | The copper industries of the USA, Canada and South America are noted for their labour problems |
| Violent conflict | 3 | Chile is relatively unstable and a sizeable disruption there would have major repercussions |
| Range of primary supply sources | 2 | Supply dominated by Chile, Canada, USA and USSR but reserves and resources are huge |
| Time lags for new supplies | 7 | Up to ten year lead time for new open-cast copper and/or molybdenum mines. However rhenium is extracted in the early roasting ore stage so the time lag from mining to its extraction is minimised |
| **Transportation Risks** | | |
| Primary | 1 | Unlikely to affect rhenium due to minor by-product status |
| Secondary | 1 | Transport of finished products containing rhenium suffers little risk |
| **Application/Use Risks** | | |
| Total economic impact | 1 | Minor market of only 35,000-36,000 lbs a year and all applications use relatively little rhenium per unit |
| Effect on key industries | 6 | The effect would be limited to the petroleum industry although demand from this sector has fallen in the recession |
| Availability of substitutes | 1 | Substitutes available in almost all applications so consumption level is very price sensitive, particularly for catalysts |
| Longer term substitutability | 3 | Available in most applications and iridium is already a viable alternative in bimetallic catalysts and others are being considered |
| **Trade Risks** | | |
| Collusive price agreements | 7 | Given the heavy supply dependance on Chile and the mere handful of metal refiners, this is a fairly high risk. It is a high value commodity which quintupled in price over 1979 and the first few months of 1980 |
| Embargoes | 2 | Relatively unlikely |

***These components are assessed on a scale from 1 (lowest) to 10 (highest) risk. Grouped together, they form the basis for the assessments of the factors for likelihood of supply disruption, and for cost of such disruption which appear in the first table in this section.***

## WORLD RHENIUM RESERVES

Total = 7,000,000 lbs Rhenium Content

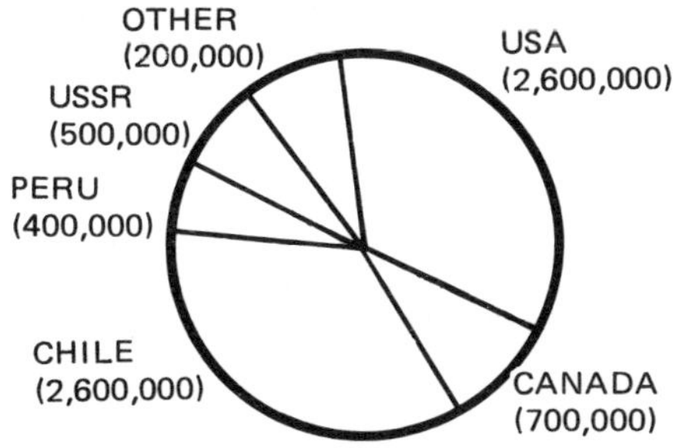


*Source: US Bureau of Mines, 1982*

## US RHENIUM CONSUMPTION

By Market Sector 1981

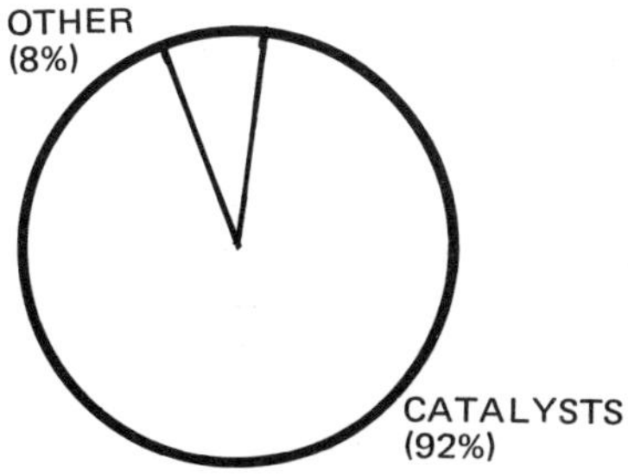


Other sector includes X-ray tubes, thermocouples and electrical contacts

*Source: US Bureau of Mines, 1982*

gases. The USSR claimed to have reached 42 per cent recovery from copper ores in 1980 due to improved techniques.

## TRANSPORTATION

Not a major problem area since largely transported contained in molybdenite concentrates. Chile and West Germany provide the bulk of the rhenium imported by the USA but travel disruptions on these long shipping routes have not been frequent.

## APPLICATIONS

### Consumption Trends

In the OECD countries demand has slackened over the past four to five years with US consumption in 1982 expected to reach 8,000 lbs. Demand is expected to stagnate during the 1980s.

### The Principal Markets

The main applications exploit rhenium's catalytic properties plus its high temperature strength, corrosion resistance and electrical conductivity. 92 per cent of the rhenium used in the USA is for bimetallic catalysts in low-lead or lead-free petrol production. This sector has been hard hit by the recession due to lower petrol consumption and by a stronger anti-environmentalist lobby. The remaining eight per cent is used in a wide variety of applications, generally alloyed with tungsten and/or molybdenum including thermocouples, heating elements, X-ray tubes, flash bulbs, vacuum tubes, electrical contacts and metallic coatings. Western Europe and Japan use a smaller percentage as a catalyst and more as alloys for heating elements, thermocouples, etc.

### Recycling

Recycling is relatively unimportant since only small quantities of molybdenum-rhenium and tungsten-rhenium scrap are recovered and reused.

### Substitution

Iridium is a commercial substitute for rhenium in bimetallic catalysts and other metals being evaluated are gallium, germanium, indium, rhodium, selenium, tin, tungsten and vanadium. Cobalt and tungsten can be used for coatings on X-ray targets, rhodium for thermocouples, tungsten and platinum/ruthenium for coatings on electrical contacts and tungsten and tantalum for electron emitters. The plethora of alternatives for rhenium suggests that the level of demand in its applications is very price sensitive.

## INTERNATIONAL TRADE AND WORLD PRICES

### Supply Arrangements

The free market has traditionally been small and little known although in 1979-80 a brief spurt of activity was seen. Rhenium is sold mainly on a producer/consumer contract basis in the form of 99.99 per cent pure metal powder or as perrhenic acid or perrhenate salts. Prices are quoted in \$/lb or \$/kilo. Chilean output dominates the market. The USA is known to import supplies but the exact amount is not disclosed by the USBM for security reasons. As a by-product, prices tend to be determined on a relatively ad hoc basis.

### Companies

Codelco, Chile's state-owned mining group, operates the largest copper-molybdenum mines and controls around a third or more of annual world rhenium output. The main rhenium metal refiners include Hermann C Starck of West Gemany; Kennecott, M & R Refractory Metals Inc and S W Shattuck Chemical Co of the USA; Utah International Inc in Canada; Molibdenos Y Metales SA Chile; and the USSR state-owned operations. Production is limited to a mere 10-12 companies.

### Prices

Prices have swung wildly over the last decade, registering peaks over \$1,500/lb in 1980. This was due to a rapid increase in demand for platinum-rhenium petroleum reforming catalysts. This sector was hit hard by the 1981-82 recession and this was reflected in a price fall.

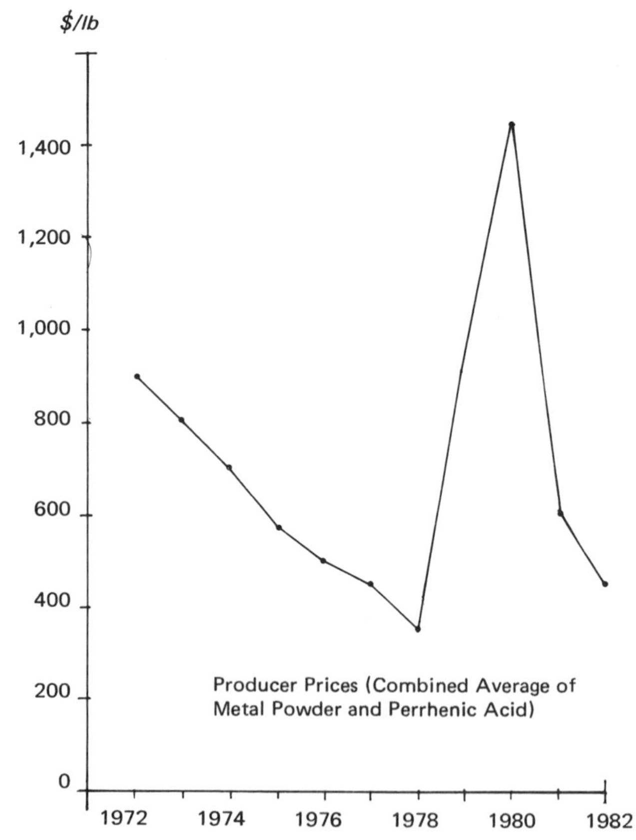


**Rhenium Prices—Annual Averages**

# SELENIUM (Se)

| | |
|---|---|
| FACTOR FOR LIKELIHOOD OF SUPPLY DISRUPTION | 1.2 |
| FACTOR FOR COST OF SUCH A DISRUPTION | 3.8 |
| TOTAL STRATEGIC FACTOR | 4.6 |

*The factors for supply disruption and cost of such a disruption are assessed on a scale from 1.0 (lowest) to 10.0 (highest) risk. Multiplied together these factors give the total strategic factor rating. For components of these assessments see table later in this section.*

Selenium has a relatively low strategic factor since it is essential in very few applications including photocopiers, although it is useful in the glass, chemical and electronic industries. It is a minor by-product of copper but occurs in almost all copper ore bodies (although in low concentrations) and there would only be a danger of selenium supply interruptions if a very large part of copper capacity were shut down.

**Physical Characteristics**

Strictly a semi-metal with fairly low melting point (217°C); behaves as an electrical semi-conductor; controversy continues over the toxic versus the beneficial effects of selenium and the US Environmental Protection Agency has set the maximum drinking water level at 10 micrograms selenium/litre although small amounts are used as a feed additive for farm animals.

## OCCURRENCE AND PRODUCTION

The main selenium minerals are clausthalite PbSe, berzelianite $Cu_2Se$ and silver selenide $Ag_2Se$ but they only occur as a by-product of base metal sulphides. The principal commercial source is slimes from electrolytic copper refineries.

**World Reserves**

These are estimated at 264 million lbs selenium, sufficient for 135 years of consumption at current rates. Western world reserves are concentrated in the USA, Canada, and Chile whilst Communist bloc reserves are known to be large, although not adequately delineated. Coal deposits form an, as yet, untapped resource and they contain an average of 1.5 ppm selenium.

**Production Characteristics**

| | |
|---|---|
| *Mining Methods:* | As for copper or other parent metal |
| *Processing:* | Recovered during the processing of tankhouse sludges for gold and silver. The slimes are treated with sulphuric acid to precipitate out the selenium. Selenium present in concentrations of around 0.1 ppm in certain copper deposits |
| *Energy Requirements:* | Largely irrelevant due to minor by-product status |
| *By-Products:* | Copper, lead, gold, silver, tellurium |
| *Environmental Factors:* | None unique to selenium extraction |
| *Development Lead Time:* | Depends on copper mine plus six months or so lag for sufficient anode slimes to accumulate to allow by-product extraction. |

**The Major Producers**

| *Countries* | *1981 (Thousand Pounds)* | *Percentage of Western World Output* |
|---|---|---|
| USA | 600 | 19.0 |
| Belgium | 130 | 4.1 |
| Canada | 800 | 25.4 |
| Japan | 1,000 | 31.7 |
| Mexico | 180 | 5.7 |
| Sweden | 150 | 4.8 |
| Yugoslavia | 120 | 3.8 |
| Others | 170 | 5.5 |
| WESTERN WORLD TOTAL | 3,150 | 100.0 |

(Production in the USSR is currently estimated at 231,000 lbs a year.)
*Source: USBM 1982, Roskill's Data Book (1982) for USSR output.*

Smelter production is dominated by Canada, the USSR, Japan, the USA and Sweden. Japan extracts selenium from imported copper concentrates, mainly from the USA. Belgium and Yugoslavia are the only European selenium smelters of any consequence, Belgium using imported Central African concentrates. Obviously the major producers tend to be in the copper industry and strikes in

## COMPONENTS OF RISK ASSESSMENT

| *Factor* | *Rating* | *Comment* |
|---|---|---|
| **Production Risks** | | |
| Existing capacity | 1 | Pronounced over-capacity. Producer stocks at record highs |
| Labour disputes | 3 | The North American copper industry is particularly prone to long strikes |
| Violent conflict | 1 | Unlikely to affect the three major producers |
| Range of primary supply sources | 1 | 76 per cent of Western refined output derives from Japan, Canada and the USA, with the remainder from South America or from Central African concentrates |
| Time lags for new supplies | 6 | Unless, as seen in 1982, producer stocks are high, the upper limit of production is controlled by refined copper output |
| **Transportation Risks** | | |
| Primary | 1 | Unlikely to affect free trade in selenium since it is present in such tiny proportions in copper concentrates |
| Secondary | 1 | Unlikely to affect selenium supplies since it forms such a small part of the finished products in which it is used |
| **Application/Use Risks** | | |
| Total economic impact | 2 | Total market volume only 3.15 million lbs a year in the Western World |
| Effect on key industries | 7 | Used in sophisticated applications including electronics, pigments and chemicals |
| Availability of substitutes | 3 | Available in most applications but not in photocopiers where only a change in information technology will affect selenium consumption |
| Longer term substitutability | 4 | Most applications are relatively price insensitive since selenium provides good performance characteristics in its applications |
| **Trade Risks** | | |
| Collusive price agreements | 1 | Unlikely due to relatively active free market and the sizeable number of refined selenium producers scattered over Europe, Japan, the USA, South America and the Communist bloc |
| Embargoes | 1 | Largely irrelevant |

***These components are assessed on a scale from 1 (lowest) to 10 (highest) risk. Grouped together, they form the basis for the assessments of the factors for likelihood of supply disruption, and for cost of such disruption which appear in the first table in this section.***

### WORLD SELENIUM RESERVES

Total = 264 Million lbs

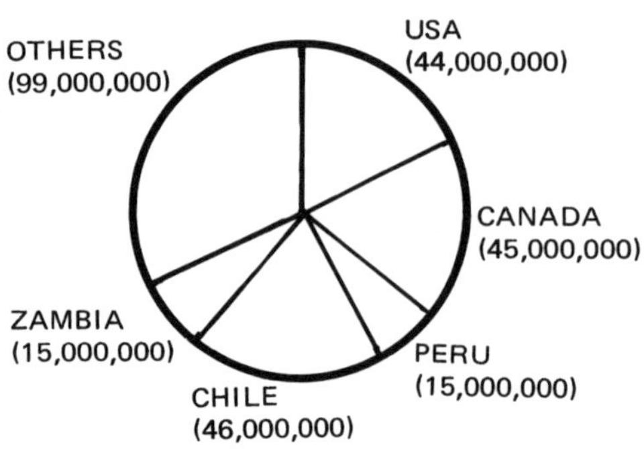


*Source: US Bureau of Mines, 1982*

### US SELENIUM CONSUMPTION

By Market Sector 1981

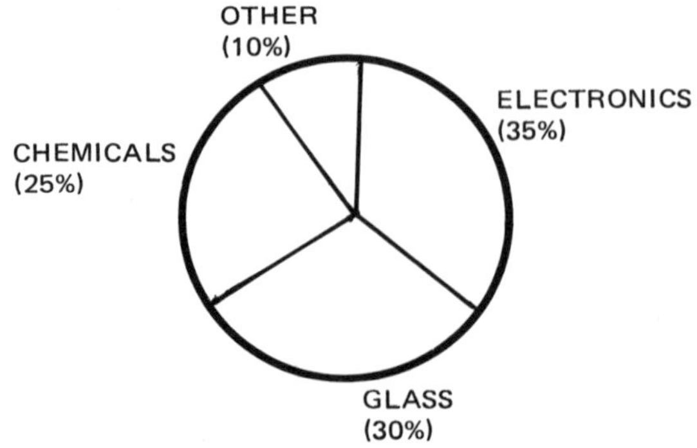


*Source: US Bureau of Mines, 1982*

the copper industry, such as the one in the USA in 1980, play havoc with output. Japan has been steadily increasing its capacity over the last five years and there have been plans for a new 260,000 lb refinery in the UK.

### Technological Developments

Relatively stable extraction technology but purification techniques particularly for electronic applications, continue to improve.

## TRANSPORTATION

Not a major problem since it is present in such tiny quantities in copper concentrates. However US import reliance was 49 per cent in 1981, with 40 per cent of that deriving from Canada. Major trade routes include copper concentrate being shipped from the US to Japan and selenium being exported from Japan, Belgium and Sweden to the rest of Europe and the USA.

## APPLICATIONS

### Consumption Trends

Selenium has a broadly based and sophisticated market where the cost of selenium is minute in comparison with the overall system. Nonetheless, several of these sectors have performed badly and selenium consumption has fallen with the experience in the USA typical. US consumption has fallen from 1,167,000 lbs in 1981 to around 900,000 lbs in 1982. From a 1978 base, demand is expected to increase at an annual rate of 2.8 per cent until 1990. Japan consumes less than half the US figure.

### The Principal Markets

Demand for selenium is based on the need for reliable colouring pigments in glasses and plastics, coupled with the need for semi conductors in a variety of electronic equipment.

A piechart showing the end-use markets in the USA is given opposite. Western Europe follows much the same pattern, whilst in Japan 38-40 per cent of all selenium is used in rectifiers which forms a minor component of the electronics sector elsewhere. Selenium has a unique use in photocopiers, transferring a photographic image by means of a change in static electricity when exposed to light. This area has a major growth potential exceeding that for selenium as a whole, but this depends on the development of information handling technology. The chemicals and pigments sector includes the use of selenides to give highly insoluble, heat-resistant, colourfast pigments in the orange to maroon range, whilst selenium is added to carbon or stainless steels and copper to improve machinability and also to lead battery grids to allow reduction of antimony in maintenance-free batteries.

### Recycling

The production of old selenium is very price sensitive. About 210,000 lbs of selenium amounting to 6-7 per cent of Western world output, was recovered in North America in 1981. This material was almost totally reclaimed from used electronic and photocopier components.

### Substitution

There are no direct substitutes in photocopier technology but a move towards computerised filing systems and microfilm/VDU usage could perhaps undercut this sector. Silicon, germanium and cadmium are suitable alternatives in rectifier and semiconductor applications, cerium oxide in glass manufacturing, tellurium in pigments and rubber compounding and tellurium, bismuth and lead in the production of free-machining metals.

## INTERNATIONAL TRADE AND WORLD PRICES

### Supply Arrangements

Selenium has a sporadically active and volatile free market although the major proportion of material is sold via producer/consumer contracts. The form generally traded is 99.5 per cent purity in five ton lots with a price quoted in $/lb. Metal of greater purity for electronic applications is sold on a producer basis since the merchant trade is unwilling to be involved in this specialised area. The upper limit on output is obviously set by the rate at which copper is mined but there is no effective limit on resources. Producer stocks are currently at record highs and pronounced overcapacity has not yet been affected by the savage cutbacks in North America's copper industry.

### Companies

The major selenium smelters are Noranda's Canadian Copper Refiners, which had a lengthy strike in the summer of 1982; the state owned companies of the USSR and Yugoslavia; Mitsubishi, Nippon Mining and Sumitomo of Japan; Kennecott, Phelps Dodge and Asarco of the USA; plus Boliden of Sweden. Selenium is smelted and produced in its refined state in over 12 countries worldwide.

### Prices

Selenium has a volatile free market price history which peaked in 1974-75 and again in 1977. Prices fell in 1982 to a ten-year low as a copper glut coincided with reduced demand in the vital photocopier and glass production sectors. The weakness of producer control over prices was manifested in the speed of the price fall.

**Selenium Prices—Annual Averages**

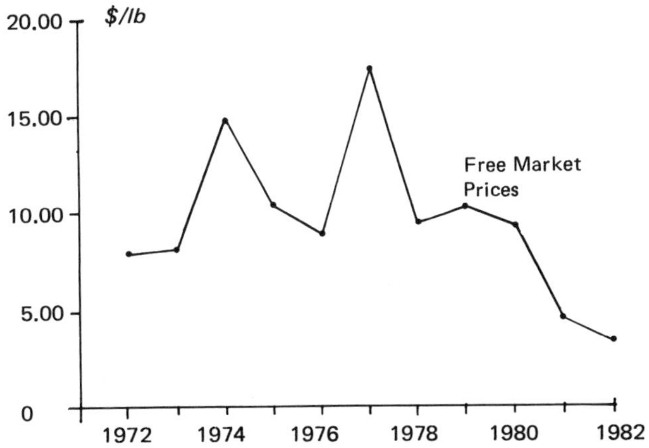

# SILICON (Si)

| | |
|---|---|
| **FACTOR FOR LIKELIHOOD OF SUPPLY DISRUPTION** | **1.0** |
| **FACTOR FOR COST OF SUCH A DISRUPTION** | **4.3** |
| **TOTAL STRATEGIC FACTOR** | **4.3** |

*The factors for supply disruption and cost of such a disruption are assessed on a scale from 1.0 (lowest) to 10.0 (highest) risk. Multiplied together these factors give the total strategic factor rating. For components of these assessments see the table later in this section.*

Silicon is a semi-metal with a very low strategic factor since it is second only to oxygen as the most abundant element in the earth's crust. Over 90 per cent of all silicon is used as the ferroalloy to enhance the properties of cast irons and steels. Only a tiny percentage is used in the high-profile electronics industry and it is only in this sector that political implications are relevant. For the future, silicon's low cost, huge and widespread resources and inertness make it a prime target for research aimed at increasing usage by developing new applications and using it as a substitute for more expensive materials.

**Physical Characteristics**

Low density semi-metal with elevated melting point (1,410°C) and a great affinity for oxygen, giving a wide variety of compounds known as sand, flint and quartz. High purity silicon has useful electrical properties. Bulk silicon is not toxic, but the powdered form causes silicosis, a progressive lung disease similar to that suffered by coal miners.

## OCCURRENCE AND PRODUCTION

The main resource material is silica ($SiO_2$) which occurs in various forms including: quartz, quartzite, sandstone, and the semi-precious gem stones of rose quartz and amethyst.

**World Reserves**

Silica forms 60 per cent of the earth's crust and no meaningful estimate of these effectively limitless and widespread resources has been made. There is a large choice of deposits available for exploitation.

**Production Characteristics**

| | |
|---|---|
| *Mining Methods:* | Open-pit or quarrying techniques |
| *Processing:* | Ore is beneficiated then smelted in submerged arc electric furnaces to silicon or to ferro-silicon if iron is added. Semi-conductor grade silicon produced by careful refining of normal grade material |
| *Energy Requirements:* | Mainly at the smelting stage which uses 6,500-7,500 kwh/ton of ferro-silicon |
| *By-Products:* | None |
| *Environmental Factors:* | Containment of air pollution from the electric furnaces is a problem. Pollution control equipment in the USA now accounts for about 20 per cent of capital costs and 10 per cent of operating costs of new plants |
| *Development Lead Time:* | Relatively short for new quarries but up to six years for new smelters. |

**The Major Producers**

| *Countries* | *1981 Output (Thousand Tons)* | *Percentage of Western World Output* |
|---|---|---|
| USA | 470 | 20.9 |
| Canada | 120 | 5.3 |
| France | 160 | 7.1 |
| Italy | 70 | 3.1 |
| Japan | 250 | 11.1 |
| Norway | 330 | 14.7 |
| South Africa | 170 | 7.6 |
| Spain | 110 | 4.9 |
| Others | 570 | 25.3 |
| WESTERN WORLD TOTAL | 2,250 | 100.0 |

(Production in the USSR is currently estimated at 500,000 tons and Communist bloc production at 750,000 tons a year.)

*Source: US Bureau of Mines, 1982*

Industrially viable reserves of silica are widespread and the final decision is made on the basis of such factors as accessibility to low cost power sources, transportation costs and ready availability to the main markets. The important bulk ferro-alloy producers possess ample electric power or fossil fuels, raw materials and a close

## COMPONENTS OF RISK ASSESSMENT

| *Factor* | *Rating* | *Comment* |
|---|---|---|
| **Production Risks** | | |
| Existing capacity | 1 | In current over-capacity due to the slump in the steel industry |
| Labour disputes | 1 | No historical record of major problems in the mining or conversion to ferro-alloy sectors. Problems tend to occur further downstream in steel making |
| Violent conflict | 1 | Unlikely to affect this element |
| Range of primary supply sources | 1 | Wide range of producers, including the USA, EEC, Japan and South Africa. South Africa dominates supplies of the higher energy grades |
| Time lags for new supplies | 1 | Short for new quarries, up to six years for new smelter |
| **Transportation Risks** | | |
| Primary | 1 | Mines located at distances less than 300 miles from smelter due to cost of transport |
| Secondary | 1 | Few problems associated with transporting silicon metal or its ferro-alloys |
| **Application/Use Risks** | | |
| Total economic impact | 10 | High volume product with three million tonnes a year silicon content mined plus huge volume of sand and gravel (mainly silica) also used |
| Effect on key industries | 2 | Over 90 per cent of silicon employed as an additive to cast irons, steels and aluminium alloys with fairly low profile end uses |
| Availability of substitutes | 6 | Substitutes available for silicon's deoxidising applications but none which include its alloying properties. Germanium a useful substitute in electronic industry |
| Longer term substitutability | 6 | Silicon is cheap and readily available and research is progressing to find new uses for it in ceramics and other solid state devices rather than to replace it |
| **Trade Risks** | | |
| Collusive price agreements | 1 | Cut-throat competition between producers is only alleviated by US and EEC protectionism in the steel industry |
| Embargoes | 1 | Only possible in the high-technology electronics industry where exports of silicon chips or devices to Communist countries is strictly monitored |

*These components are assessed on a scale from 1 (lowest) to 10 (highest) risk. Grouped together, they form the basis for the assessments of the factors for likelihood of supply disruption, and for cost of such disruption which appear in the first table in this section.*

## US SILICON CONSUMPTION

By Market Sector 1981

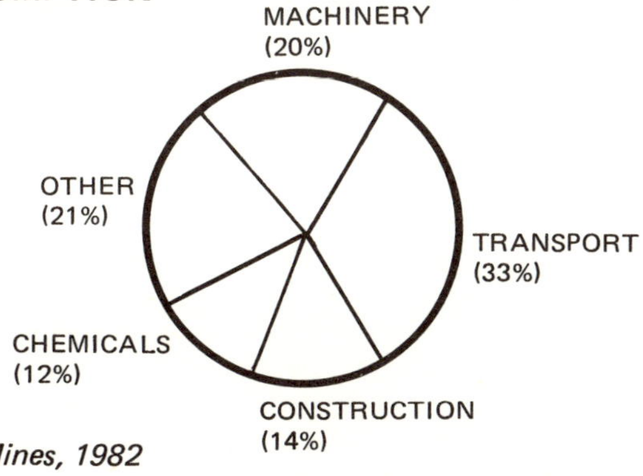


*Source: US Bureau of Mines, 1982*

proximity to steel and other user industries. Norway, Canada and the USA are notable western producers. Price competition from South Africa has already undercut the EEC and US producers whilst in Asia the whole structure of the market is changing. Japan's ferro-silicon production is on the downturn and the Philippines and Taiwan have stepped into the breach aided by their lax pollution regulations and low labour costs. Trimetallic master alloys, such as ferro-silico-chromium and ferro-silico-manganese, are becoming more widely used in the steel industry and obviously their production depends on the cost and reliability of the chromium or manganese supplies. This favours the South African or Communist bloc producer to an even greater extent. Amongst western producers, both Norway and South Africa are heavily biased towards production of the higher energy content, and thus higher silicon content, ferro-alloys. The production of electronic grade silicon is a highly specialised sector and is integrated with the production of solid state devices using silicon.

### Technological Developments

The production of silicon and its ferro-alloys is highly energy intensive and efforts have been directed towards reducing the energy required. The computerised control of process variables and the re-use of heat contained in waste gases have improved the situation. The techniques of producing pure electronic grade silicon crystals undergo constant refinement.

## TRANSPORTATION

This forms an important cost factor in the production of this high volume, low unit cost material. In general, silica mines and quarries are located within 300 miles of a smelter. Major silicon and ferro-alloy exporters include South Africa, Canada and Norway with transport by ship. Problems tend to be a consequence of dock and other labour disputes.

## APPLICATIONS

### Consumption Trends

Demand for silicon metals and alloys is closely related to activity in the iron and steel and aluminium industries and the USBM expects demand to grow by three per cent a year on average until 1990. Consumption stagnated in the late 1970s and fell by one to two per cent during the 1981-82 recession.

### The Principal Markets

A pie chart showing the end-use markets of silicon in the USA is given on page 129. Over 90 per cent is used in the metallurgical industry, with the bulk of this as the ferro-alloy which acts as a deoxidant and also as an alloying agent to produce specialised irons and steels. Ferro-silicon is used as a reducing agent in the manufacture of magnesium or nickel. The largest use of silicon metal is as an addition to aluminium castings to improve their wear resistance. Silicon is used to make silanes which find a wide variety of uses as silicone resins, lubricants and water repellant compounds. Silica is used as reinforcing filler in the manufacture of silicone rubber whilst high purity silicon dioxide has a range of uses in the chemical and pharmaceutical industry. Standard grade silicon carbide is an important abrasive but it is the high purity silicon carbide and nitride ceramics which have a great potential as the engineering material of the future for precision components in engines and turbines. The electronics industry uses only one per cent of silicon, as single crystals which form the basis for semi-conductor devices and most integrated circuits. Here the cost of the silicon is virtually irrelevant when considering the cost of producing an integrated circuit of guaranteed reliability. The sector could see a major boost from the advent of commercially reliable photovoltaic power systems.

### Recycling

Only small quantities of silicon are recycled in any form due to the dissipative nature of its metallurgical, chemical and electronic uses plus its low unit cost.

### Substitution

Aluminium, magnesium or zirconium can replace ferro-silicon as a steel deoxidant, but are more expensive and do not double up as useful alloy additions. Germanium is the main substitute for silicon in semi-conductor and other solid state devices and here the main criterion is that of performance.

## INTERNATIONAL TRADE AND WORLD PRICES

### Supply Arrangements

Silicon and its ferro-alloys are mainly sold directly from producer to consumer with very little leeway for a free market. Price competition between producers remains cut-throat with a recent increase in protectionism. There have been many closures of smaller, older and less efficient units of production, but the USA and EEC have tried to protect their own producers by a variety of plans including the Davignon scheme and recommendations from the US International Trade Commission. The conflict extends to a variety of steel related elements including chromium, manganese and vanadium. Although the trade war continues, the Third World countries plus South Africa, which have low energy and labour costs and an indigenous iron industry, look set to succeed. Silicon can be traded either as the element of 98-98.5 per cent purity, ferro-silicon alloys containing 45 per cent, 75 per cent or 90 per cent silicon with a specified maximum impurity level, or as silico-manganese ferro-alloys containing 65 per cent manganese and 14-25 per cent silicon. Very high purity electronic grade silicon is rarely traded since it is generally made and used in situ due to stringent handling requirements.

## Companies

The main ferro-silicon producers are Union Carbide Corp of the USA, Ohio Ferro-Alloys Corp of the USA, Foote Mineral Co of the USA, Cie. Universelle d'Acetylene et d'Electrometallurgie of France and A/S Elkem Spigerverket, Norway. Leading silicon metal producers include Ohio Ferro-Alloys, Alabama Metallurgical Corp of the USA, Samancor of South Africa, Union Carbide Canada Ltd, and Elkem Spigerverket.

## Prices

Silicon and its ferro-alloy prices have historically moved in approximate parallel, since the end-uses overlap and both are high-energy products. Silicon saw a peak in 1974 coinciding with the oil prices, but prices weakened in 1977-78 and again in 1981-82 as the recession struck.

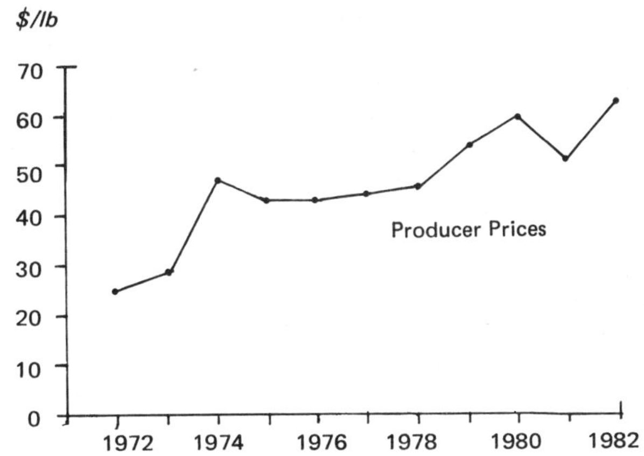


**Silicon Metal Prices—Annual Averages**

# SILVER (Ag)

| | |
|---|---|
| **FACTOR FOR LIKELIHOOD OF SUPPLY DISRUPTION** | **3.5** |
| **FACTOR FOR COST OF SUCH A DISRUPTION** | **3.3** |
| **TOTAL STRATEGIC FACTOR** | **11.6** |

*The factors for supply disruption and cost of such a disruption are assessed on a scale from 1.0 (lowest) to 10.0 (highest) risk. Multiplied together these factors give the total strategic factor rating. For components of these assessments see the table later in this section.*

Silver is a precious metal with a moderately low strategic factor. It is prized as an investment medium and is also used in the photographic and electronics sectors. Silver production is spread over 50 countries and a supply shortage is unlikely given the huge above-ground stocks which regularly help plug the deficit between annual mined output and consumption. Production is concentrated in relatively stable areas and India has the largest available stocks, exceeding three years' world mined output.

### Physical Characteristics

Lustrous, soft, malleable metal with extremely high thermal and electrical conductivities and melting point of 962$^{o}$C; silver's halide salts are light sensitive, the basis for almost all film emulsions.

## OCCURRENCE AND PRODUCTION

Sometimes occurs in the free state but the most widespread sources are the complex sulphides including argentite ($Ag_2S$), cerargyrite (AgCl) and hessite ($Ag_2Te$).

### World Reserves

These are estimated at 8.4 billion ounces silver content, sufficient for 25 years of mining at current levels. Most silver reserves are in lead, zinc, and copper ores and are widely distributed geographically. Communist countries possess just under a quarter of viable reserves.

### Production Characteristics

| | |
|---|---|
| *Mining Methods:* | Only 20 per cent of silver derives from mines operated solely for their silver content, usually underground mines. For the remaining 80 per cent, methods depend on the parent metal ore body |
| *Processing:* | Silver removed from lead/zinc concentrates during smelting process. Silver from copper ores recovered as part of anode slime during electrolysis. The silver recovered is electrolytically refined to 99.9 per cent purity |
| *Energy Requirements:* | Largely irrelevant until final electrolytic refining process due to by-product status |
| *Co-Products:* | Lead, zinc, copper, gold, selenium, tellurium, cadmium, thallium |
| *Environmental Factors:* | The treatment of waste generated in the processing of silver from base metals has required controls to protect local water supplies |
| *Development Lead Time:* | Depending on parent metal. Five to eight years for primary underground silver mines. |

### The Major Producers

The USA, Canada, Peru and Mexico dominate Western world output and account for just under half world production with the Communist bloc, mainly Russia, providing 23 per cent. Production in Mexico and the USA is over 50 per cent derived from primary operations whilst elsewhere by-product silver is dominant. The high price of silver during 1979-80 encouraged the exploration and development of new resources worldwide with continued developments in Mexico and marginal increments in the USA, Peru, Australia and Southern Africa. Predictions of silver output depend heavily on the health of the base metal mining sector since decisions to change the output of lead/zinc or copper concentrates are not made on the basis of the silver price or demand. 1981-82 saw cutbacks in US silver production in particular, firstly as copper mines producing by-product silver closed and then later as higher grade underground silver mines also began closing. Refined silver is mainly produced within the country of origin as a consequence of extracting and purifying the parent lead/zinc or copper.

### Technological Developments

The USBM is investigating new flotation techniques for higher recovery from ore, heap-leach cyanidation and associated methods for the treatment of anode slimes.

## COMPONENTS OF RISK ASSESSMENT

| *Factor* | *Rating* | *Comment* |
|---|---|---|
| **Production Risks** | | |
| Existing capacity | 1 | Currently in over-capacity with cutbacks enforced by primary and by-product silver producers in North America |
| Labour disputes | 4 | Some history of shortages caused by strikes and labour problems, particularly during North America copper strikes |
| Violent conflict | 6 | Over 30 per cent output centred in South America, whilst silver's market profile is notably susceptible to political news |
| Range of primary supply sources | 1 | Despite 50 per cent of current mined output in Mexico, Peru, the USA and Canada, there are huge above ground refined stocks, available at the right price |
| Time lags for new supplies | 3 | Six to ten years for by-product silver but up to four years for small primary silver mines |
| **Transportation Risks** | | |
| Primary | 1 | Few problems in transporting silver since it is a low volume commodity |
| Secondary | 1 | The high unit cost of refined silver promotes smuggling and theft |
| **Application/Use Risks** | | |
| Total economic impact | 2 | Relatively small with world mined output at 340-360 million ounces per year |
| Effect on key industries | 4 | Heavy bias towards the luxury industries of photography, jewellery, sterling ware and commemorative coins |
| Availability of substitutes | 5 | Substitutes are available but silver retains its luxury uses due to its appearance and a long tradition and keeps its industrial markets on a cost and performance basis |
| Longer term substitutability | 5 | Complete replacement of silver could be achieved but would need extensive public education |
| **Trade Risks** | | |
| Collusive price agreements | 8 | Speculator interest in silver dates back to Biblical times, but the most recent example was the 1979-80 Bunker Hunt market manipulation |
| Embargoes | 1 | Unlikely, although Peru attempted a limited sales stoppage in 1982 in protest at prevailing low prices |

*These components are assessed on a scale from 1 (lowest) to 10 (highest) risk. Grouped together, they form the basis for the assessments of the factors for likelihood of supply disruption, and for cost of such disruption which appear in the first table in this section.*

## WORLD SILVER RESERVES

Total = 8,435 Million Troy Ounces Silver Content

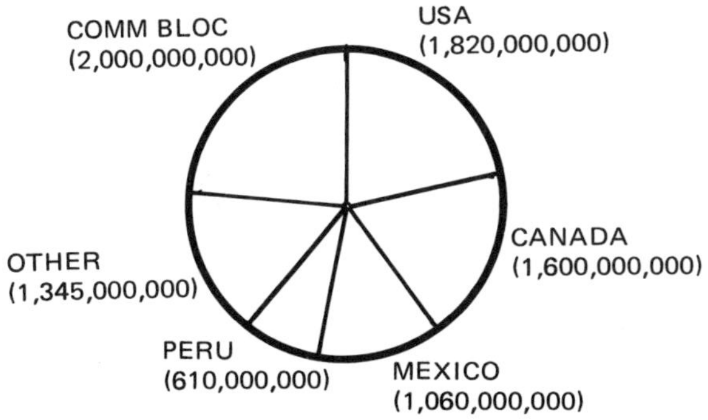


*Source: US Bureau of Mines, 1982*

## US SILVER CONSUMPTION

By Market Sector 1981

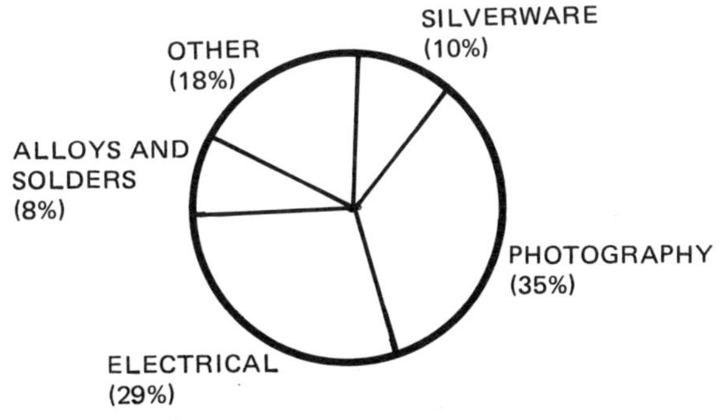


*Source: US Bureau of Mines, 1982*

**The Major Producers**

New Production

| *Countries* | *1981 (million troy oz)* | *Percentage of Western World Output* |
|---|---|---|
| USA | 36.0 | 13.6 |
| Canada | 38.7 | 14.7 |
| Mexico | 45.2 | 17.1 |
| Peru | 40.1 | 15.2 |
| Australia | 27.0 | 10.2 |
| Other | 77.0 | 29.2 |
| WESTERN WORLD TOTAL | 264.0 | 100.0 |

(Communist bloc production is currently estimated at 84.0 million troy ounces a year.)

*Source: USBM, Mineral Commodity Summaries 1982.*

Secondary Production

| | *1981 (million troy oz)* |
|---|---|
| From US Government | 2.1 |
| Stocks of Foreign Governments | 2.0 |
| Demonetised coin | 12.0 |
| Indian stocks | 33.5 |
| Old Scrap | 105.0 |
| Dishoarding (additions to private stocks) | (49.3) |
| WESTERN WORLD TOTAL | 105.3 |

*Source: Handy & Harman, 1982.*

## TRANSPORTATION

The USA was 50 per cent dependent on silver imports in 1981, with the main trade routes from Mexico and Canada into the USA consuming centres. Japan and Europe are also dependent on imports mainly from Australia, South America and material smuggled out of China and India. As a prized high unit cost metal, disruption of deliveries has mainly been due to theft and government controls rather than to dock strikes etc.

## APPLICATIONS

### Consumption Trends

Over 50 per cent of all silver is used in the luxury sectors of amateur photography, sterling ware, coins and commemorative articles, which are particularly susceptible to a change in the level of freely disposable income. Between 1977 and 1979 Western world annual consumption stabilised at 445-475 million ounces, but then fell to 360-370 million ounces in 1980 and 1981. The USBM estimate demand for silver to increase by an average 2.9 per cent a year until 1990.

### The Principal Markets

A pie chart showing the end-use markets in the USA, which consumes 30 per cent of the total in Europe and around 50 per cent in Japan is given on page 133. The health of the photographic industry is closely linked to living standards in the industrialised countries and is controlled by a small number of large corporations. The sector is split 40:40:20 between the amateur photography, X-ray photography and graphic arts sectors as far as silver usage goes. The silver intensity of photographic products has been reduced to almost half the level of a decade ago, but the increasing penetration of colour photography and the shift in preference towards faster films is reducing the potential for further cuts in silver content. Nonetheless, the Kodak disc film camera, which uses five per cent less silver per frame and Sony's Mavica video tape camera, a totally silver-free system, could pose a future threat. Electrical and electronic products absorb around 25-30 per cent of all silver, 15 per cent of which is used in batteries and the remainder as contacts and conductors, where it is often alloyed to reduce cost and increase strength. The jewellery sector fluctuates in popularity from year to year and, in a recession, may benefit from a down-market move away from gold. Sterling and plated ware has had two lean years since purchases are a sensitive function of silver prices. Brazing alloys for aircraft and steam turbine assemblies and dental and medical applications also use small amounts of silver.

### Recycling

Silver recovery in the USA totalled 37 per cent of consumption in 1981, of which a large proportion is derived from scrap film. One troy ounce of silver produces about 470 rolls 110-12 colour film, but despite this apparently low usage intensity there are viable chemical techniques for recovering silver from all types of scrap film and paper. The dishoarding of silver sterling and plated ware is a price sensitive phenomenon but the widespread availability of private silver stocks is a major factor against the successful manipulation of the market.

### Substitution

Silver is rarely used in coinage today, having been replaced on grounds of cost by nickel and other alloys. Aluminium and rhodium substitute for silver in mirrors and other reflecting surfaces. Stainless steel is the main alternative to silver cutlery and flatware. Almost all silver's uses are very cost sensitive and the metal has to work hard to retain its industrial markets.

## INTERNATIONAL TRADE AND WORLD PRICES

### Supply Arrangements

Most newly mined silver is sold on long term contracts between major producers and consumers whilst the remainder is still channelled through banks and bullion brokers, a throwback to the time before silver severed its physical links with the monetary system. The free market pricing mechanism predominates with the New York Commodities Exchange leading in terms of futures contracts traded, the London Metal Exchange and the London Silver Bullion Market. Prices are quoted in $/c or in £/p for a troy ounce of 0.999 fineness material. There are also silver futures markets in Canada, Australia and Hong Kong, but the USA markets attract such a high volume of business that normal trading activity elsewhere is swamped.

### Companies

The main silver mining organisations are the state-owned organisations of the USSR and Poland; Industrial Minerá Mexico SA of Mexico; Mount Isa Mines Ltd of Australia; Centromin-Peru of Peru; and Asarco Inc of the USA. The main silver refiners in the Western world are Asarco Inc; Industrias Penoles SA de CV of Mexico, Centromin-Peru, and Noranda of Canada. London's three major silver bullion brokers are: Mocatta & Goldsmid, Samuel Montagu and Sharps Pixley.

### Prices

The silver market has always attracted speculators, partly because it tends to follow gold as a political indicator whilst its relatively low unit cost has meant that it is available to a wider investment public. The Bunker Hunt attempt to corner the world silver market in 1979-80 was a potentially dangerous reminder of this, when silver prices brushed the $50/oz level.

**Silver Prices—Annual Averages, Highs and Lows**

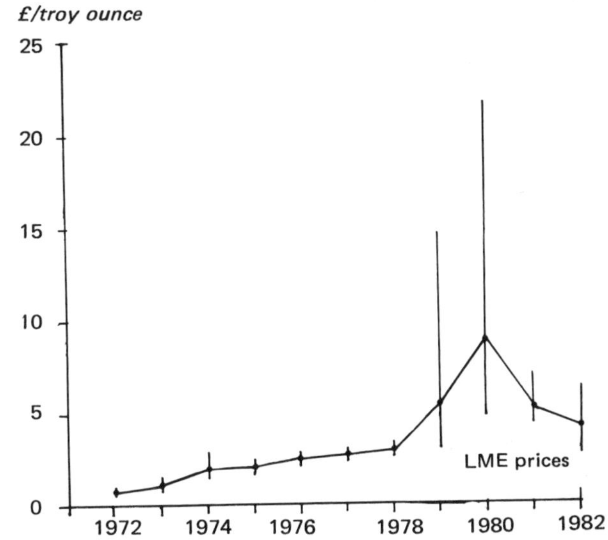

# TANTALUM (Ta)

| | |
|---|---|
| **FACTOR FOR LIKELIHOOD OF SUPPLY DISRUPTION** | **2.2** |
| **FACTOR FOR COST OF SUCH A DISRUPTION** | **6.2** |
| **TOTAL STRATEGIC FACTOR** | **13.6** |

*The factors for supply disruption and cost of such a disruption are assessed on a scale from 1.0 (lowest) to 10.0 (highest) risk. Multiplied together these factors give the total strategic factor rating. For components of these assessments see the table later in this section.*

Tantalum has a moderately high strategic factor due to its importance in the electronics and electrical industries. Usage is concentrated in the high-growth sectors of these industries and, although substitutes are available, they tend to be less efficient and are only viable on a cost comparison basis. Reserves are concentrated in South East Asia and Australia, the areas which jointly produce 60 per cent mined output, and there is a growing trend to processing concentrates and slags within the country of origin rather than exporting material for treatment in the USA and EEC.

### Physical Characteristics

Extremely high melting point (2,996°C), high specific density (16.6), good conductor of heat and electricity. Tantalum pentoxide ($Ta_2O_5$) is prized for its high dielectric constant and chemical inertness which guarantee its use as a capacitor.

## OCCURRENCE AND PRODUCTION

The main source is a series of minerals containing varying proportions of tantalum and columbium as the pentoxides. Tantalum oxide is also present in the pyrochlore mineral series and in tin-bearing ores.

### World Reserves

These are estimated at 48 million lb tantalum content (22,000 tonnes), sufficient for over 50 years of exploitation at current levels. Reserves are concentrated in South East Asia, Australia and parts of Africa. Estimates of Communist bloc reserves are not reliable.

### Production Characteristics

| | |
|---|---|
| *Mining Methods:* | Tantalite-columbite minerals by open-pit or underground methods. Alluvial tin deposits containing tantalum by dredging or open-pit methods |
| *Processing:* | Tantalum mineral concentrates and tin slags chemically processed to the pure oxide or complex fluoride (K-salt). These salts can be reduced to tantalum metal powder which can be consolidated by sintering and melting |
| *Energy Requirements:* | Moderately high |
| *By-Products:* | Columbium, tin, beryllium, zirconium, rare-earths, tungsten |
| *Environmental Factors:* | No major health hazards since fumes, gases and dust generated by extraction plants can be easily controlled |
| *Development Lead Time:* | Six to eight years for new tantalite/columbite mine, up to six years for tin mines although immediately accessible stock of tin slags do exist. |

### The Major Producers

| *Countries* | *1981 Output (Thousand Pounds)* | *Percentage of Western World Output* |
|---|---|---|
| Australia | 160 | 16.3 |
| Brazil | 250 | 25.5 |
| Canada | 250 | 25.5 |
| Malaysia | 10 | 1.0 |
| Mozambique | 60 | 6.1 |
| Nigeria | 60 | 6.1 |
| Thailand | 130 | 13.3 |
| Zaire | 25 | 2.6 |
| Other | 35 | 3.6 |
| WESTERN WORLD TOTAL | 980 | 100.0 |

Includes production from concentrate only. Production of tantalum in tin slags estimated at 1.23 million pounds in 1981 by the Tantalum Producers International Study Centre. The tin slags have a tantalum content of 1-30 per cent.

*Source: US Bureau of Mines. Mineral Commodity Summaries, 1982.*

## COMPONENTS OF RISK ASSESSMENT

| *Factor* | *Rating* | *Comment* |
|---|---|---|
| **Production Risks** | | |
| Existing capacity | 1 | Major over-capacity problems in 1982 with a processors' inventory-to-shipments ratio of 2.4 to 1 (normal stocks are 1.2-1.4 years of demand) |
| Labour disputes | 1 | Few problems with labour disputes due to large number of small individual operations |
| Violent conflict | 2 | South East Asia, Canada and Australia are relatively safe sources of supply |
| Range of primary supply sources | 7 | Australia and South East Asia possess 58 per cent of reserves and produce almost 60 per cent of mined output although this is decreasing as stocks of old tin slags become depleted |
| Time lags for new supplies | 8 | Six to eight years for new tantalite mines, up to six years for new tin mines although old stocks of tin slags (not all of which are viable) are available |
| **Transportation Risks** | | |
| Primary | 5 | Transported mainly as the ore/tin slags. Long shipping lines from Australia and South East Asia to main refining centres in USA, Japan and EEC |
| Secondary | 1 | Few problems in transporting tantalum as contained in finished electrical or electronic goods |
| **Application/Use Risks** | | |
| Total economic impact | 1 | Small market, with demand at only 2.5 million lb (1,135 tonnes) $Ta_2O_5$ each year |
| Effect on key industries | 8 | 50 per cent tantalum used as capacitors in electronic applications, the remainder as the carbide for cutting tool tips and in super-alloys for heavy duty aerospace engines and gas turbines |
| Availability of substitutes | 6 | Tantalum's uses are very price sensitive. Trend towards using cheaper alumina or ceramic capacitors even though performance slightly inferior. Similar trends for carbide tool tips |
| Longer term substitutability | 7 | The excessive price increases in the 1975-80 period caused consumers to use substitutes and to reduce tantalum requirements per unit application and the process is continuing |
| **Trade Risks** | | |
| Collusive price agreements | 5 | Possible, considering small number of major producers although the presence of an active free market in ore, reduces the likelihood of successful price fixing |
| Embargoes | 1 | Unlikely since most end-products, particularly capacitors, mainly used in consumer goods rather than in specialised military applications so an embargo would have little point |

*These components are assessed on a scale from 1 (lowest) to 10 (highest) risk. Grouped together, they form the basis for the assessments of the factors for likelihood of supply disruption, and for cost of such disruption which appear in the first table in this section.*

## WESTERN WORLD TANTALUM RESERVES

Total = 48 Million lbs Tantalum Content

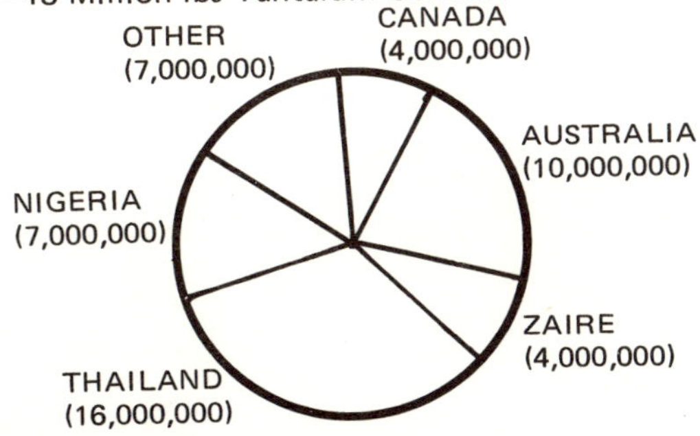


*Source: US Bureau of Mines, 1982*

## US TANTALUM CONSUMPTION

By Market Sector 1981

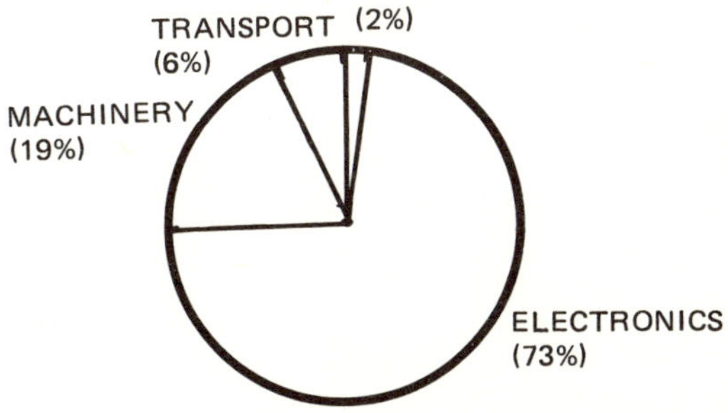


*Source: US Bureau of Mines, 1982*

Tantalum mineral concentrates are produced in Canada, Brazil, Australia, Nigeria, Mozambique and the USSR, whilst tantalum-bearing tin slags originate in Thailand, Malaysia, Spain and Nigeria. There are few tantalite mines with an output exceeding 75,000 lb $Ta_2O_5$ (tantalum pentoxide) and the mines are usually small one-man operations. Asia and Australia currently account for almost 60 per cent of world output, but this is expected to decline somewhat over the next five years, due to the depletion of old stocks of tantalum-bearing slags and the new production coming onstream. China's tantalum production level is unknown but exports are estimated at 8,000-10,000 lbs a year $Ta_2O_5$, and recent US-China defence talks have identified tantalum as one of the three metals which the Chinese will supply to the US aerospace industry. The Soviet bloc is expected to remain an importer of 50,000-100,000 lbs a year. The processing of tantalite-bearing raw materials is concentrated in Germany, Belgium, the USA and Japan, although there is now a trend towards upgrading of the raw materials in producer countries, particularly in Thailand and Australia. Worldwide plant production capacity is around 3.5 million lb $Ta_2O_5$, although the 1981-82 recession caused Canadian, Australian and the South East Asian producers to cut back.

### Technological Developments

Research continues to improve recovery from low-grade (0.5 per cent) tantalum-bearing tin slag.

## TRANSPORTATION

Major exporters include Australia, Brazil, Canada, Thailand and Zaire, with the EEC, Japan and the USA as the main importers. Transport is mainly in the form of concentrate or tin slags and, although supply lines are long, few problems have been recorded.

## APPLICATIONS

### Consumption Trends

Use of tantalum saw rapid growth rates during the last decade, plateauing at 3.35 million lb $Ta_2O_5$ equivalent in 1980. This fell in 1981 whilst 1982 demand fell below 2.3-2.5 million lb $Ta_2O_5$ due to a slump in the capacitor industry. The USBM predicts demand in the USA to increase at an annual rate of about four per cent through 1990. The Tantalum Producers International Study Centre predicts that demand will rise to 3.5 million lb $Ta_2O_5$ equivalent by 1985 with some substitution causing a reduction in demand after 1986, assuming a sustained price of $60-80/lb contained oxide.

### The Principal Markets

A piechart showing the end-use markets in the USA is given on page 137. The capacitor industry, which accounts for 50 per cent of consumption worldwide and for about 65 per cent in the USA, is vital to the health of this market. The future looks relatively bright because the world electronics market is expected to grow from $176 billion in 1981 to $818 billion in 1990, with a compound growth rate of 18 per cent a year. Unit sales of tantalum capacitors bear a linear relationship to sales of integrated circuits, in a 1:8 ratio in the USA, suggesting that tantalum capacitor demand will see the same startling rise over the period. Tantalum carbide (TaC) is used to provide tool cutting edges in the farm machinery, steel and automotive sectors and this sector registered a 33 per cent fall in demand during 1981. Future demand for TaC depends on the demand for steel cutting grades of cemented carbide cutting tools and on the effort put into reducing the tantalum content in those tools. A growth rate of seven to eight per cent is expected in this sector over the 1982-86 period. Tantalum is an invaluable additive to superalloys for aerospace structures, jet engines and gas turbine parts due to its high melting point and corrosion resistance. Tantalum now makes up four per cent (6.0-6.5 lb Ta) of the weight of the turbine blades of a typical commercial aircraft engine and has not yet reached its full potential in the military aerospace sector. Chemical plant and other machinery applications include tantalum in such uses as heat exchangers, heating elements, condensers, coolers, valves, etc, and accounts for 15 per cent of the total. This sector is expected to show only modest growth but has the advantage of relative price insensitivity.

### Recycling

The recovery of scrap provided only five per cent of US tantalum demand in 1981. Most end-uses are dissipative: cemented tool tips wear away and capacitors form too small a part of an overall device to be viably recycled. The tantalum in alloys could be reclaimed but, as yet, little progress has been made with recycling super-alloys which are a complex mixture of up to ten elements.

### Substitution

Applications in the capacitor and tool tip sectors are particularly price sensitive. There is a constant drive towards higher charge powders for capacitors, thus using less $Ta_2O_5$ per unit, whilst alumina and ceramics provide cheaper, if less effective, capacitors for consumer goods and other such areas. Glass, titanium, zirconium, columbium and platinum provide good, but often more costly, alternatives in corrosion resistant equipment. Columbium is an all purpose substitute in the tool tip and super-alloys areas. The tool tip sector is suffering, however, from use of coated carbide tools, the increase of mixed carbides, and the redesign of tools to reduce the tantalum content. Tungsten, rhenium, molybdenum, hafnium, and columbium are alternatives in high temperature applications.

## INTERNATIONAL TRADE AND WORLD PRICES

### Supply Arrangements

Tantalum is generally traded in the form of tantalite ore containing 30 or 60 per cent tantalum pentoxide with a price fixed on the basis of the oxide content. A high percentage of all ore is sold on the basis of the Metal Bulletin free market price quotation, but there is no publicly quoted price for tantalite-bearing tin slags. The Metal Bulletin price attempts to reflect the bulk of ore transactions between producers, consumers and merchants in all countries. Tantalum powder and compounds are rarely traded and only then between producer and consumer. Over the past five years, speculator interest in this market has increased leading to price volatility.

### Companies

Major tantalite ore producers include Tantalum Mining Corp of Canada Ltd, Greenbushes Tin of Australia, Bisichi-Jantar Co of Nigeria, Societe des Mines du Rwanda SA of Rwanda and Bhuket Union Thai Minerals of Thailand. Major producers of tantalite-bearing tin slags are Kamativi Tin of Zimbabwe, Zairetain of Zaire and Cia de Estanho Minas Brazil of Brazil.

### Prices

Between 1975 and 1980 the free market tantalite ore price increased by a factor of ten, to top the $120/lb oxide contained level. This was due to rapid growth in demand resulting from the steady constant-dollar ore prices realised over the past 20 years. By mid-1981, prices fell back down through $100 and then down below $40 in the spring of 1982. The heritage of skyrocketing ore prices came home to roost giving an example of the classic cycle where drastic price rises spurred marginal capacity into production and also forced consumers to look for alternatives and for ways of reducing the tantalum content of their products.

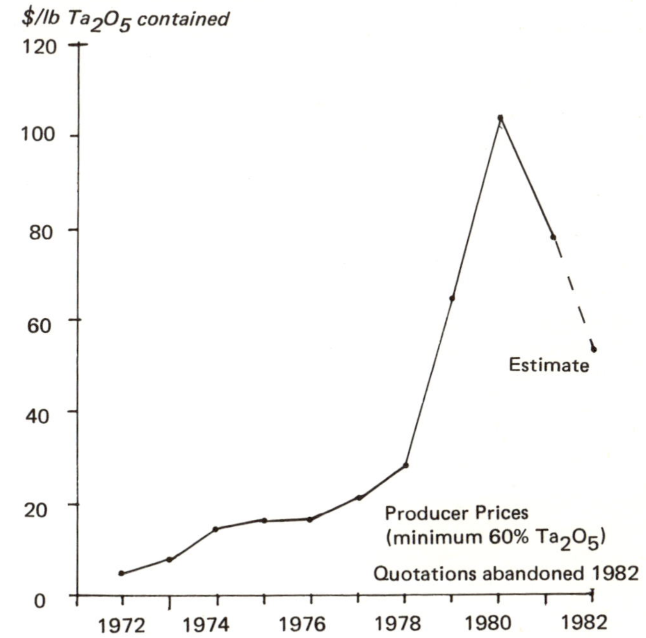


**Tantalite Ore Prices—Annual Averages**

# TELLURIUM (Te)

| | |
|---|---|
| **FACTOR FOR LIKELIHOOD OF SUPPLY DISRUPTION** | **1.2** |
| **FACTOR FOR COST OF SUCH A DISRUPTION** | **3.2** |
| **TOTAL STRATEGIC FACTOR** | **3.8** |

*The factors for supply disruption and cost of such a disruption are assessed on a scale from 1.0 (lowest) to 10.0 (highest) risk. Multiplied together these factors give the total strategic factor rating. For components of these assessments see the table later in this section.*

Tellurium has a low strategic factor since it is a minor metal with only limited importance in the electronic, rubber and metallurgical industries. It is a copper by-product but reserves are large and output is concentrated in the relatively trouble free areas of North America and Japan. Adequate substitutes are available in most applications.

### Physical Characteristics

A semi-metal with a melting point of 450°C and a specific density of 6.24; electrically it is a semi-conductor.

## OCCURRENCE AND PRODUCTION

Tellurium is found associated with gold, silver, copper, lead, mercury and bismuth ores and the most common mineral is sylvanite; a gold/silver tellurium complex. The principal commercial source is slimes from electrolytic copper refineries.

### World Reserves

These are estimated at 71 million lbs tellurium content, sufficient for over 235 years of consumption at current levels. Reserves are concentrated in the USA, Canada and Peru although the Communist bloc is estimated to hold over half the world's reserves. Coal deposits contain around 0.015 ppm tellurium but this is, as yet, an untapped resource.

### Production Characteristics

| | |
|---|---|
| *Mining Methods:* | As for parent metal—usually copper |
| *Processing:* | Recovered during the processing of tankhouse sludges for gold and silver. Chemically refined. Most economic copper ores contain around 0.004 ppm tellurium |
| *Energy Requirements:* | Largely irrelevant due to minor by-product status |
| *By-Products:* | Copper or lead, gold, silver, usually selenuim |
| *Environmental Factors:* | None unique to tellurium extraction |
| *Development Lead Time:* | Depends on parent metal. |

### The Major Producers

| *Countries* | *1981 Plant Output (Tonnes)* | *Percentage of Western World Output* |
|---|---|---|
| USA | 80.0 | 27.1 |
| Canada | 45.0 | 15.3 |
| Fiji | 23.0 | 7.8 |
| Hong Kong | 45.0 | 15.3 |
| India | 0.5 | 0.2 |
| Japan | 80.0 | 27.1 |
| Peru | 21.0 | 7.1 |
| Other | 0.5 | 0.2 |
| WESTERN WORLD TOTAL | 295.0 | 100.0 |

(Production in the USSR is currently estimated at 40 tonnes a year.) Independently rounded.

*Source: USBM and Roskill's Metal Databook, 1982.*

Mined output is dominated by the major copper producers of Canada, USA, Peru, Japan and the USSR. Refined output is dominated by Hong Kong, where imported tankhouse slimes are used, the USSR, Japan, Canada and the USA. The USSR is a major exporter to the West of around 30 tonnes or more tellurium each year. In the long term more stringent environmental regulations coupled with a change in the geographical pattern of copper refining could alter the relative importance of the current producers.

### Technological Developments

Relatively stable production technology.

## TRANSPORTATION

Tellurium is present in such tiny quantities in copper that transport risks are minimal until the final pure material is reached. It is a low volume, largely unexciting metal and transport stoppages are unlikely. A relatively

## COMPONENTS OF RISK ASSESSMENT

| *Factor* | *Rating* | *Comment* |
|---|---|---|
| **Production Risks** | | |
| Existing capacity | 1 | Current over-capacity in the copper industry has meant production cutbacks but this has coincided with a decline in demand for the tellurium by-product |
| Labour disputes | 2 | Poor labour record in North America and Peru |
| Violent conflict | 1 | Unlikely to affect tellurium |
| Range of primary supply sources | 3 | Relatively limited |
| Time lags for new supplies | 4 | Only present as 0.004 ppm in some copper ores and there is a time lag between copper mining and the accumulation of tank house sludges |
| **Transportation Risks** | | |
| Primary | 1 | Largely irrelevant as a very minor constituent of copper concentrates |
| Secondary | 1 | Tellurium does not have a high risk trade profile |
| **Application/Use Risks** | | |
| Total economic impact | 1 | Small volume market of only 330-340 tonnes a year |
| Effect on key industries | 5 | Important in rubber and car component (as an additive to steels and cast irons) industries. Could see major growth in electronic industries |
| Availability of substitutes | 3 | Available in almost all applications and consumption levels are very price sensitive |
| Longer term substitutability | 3 | Largely possible |
| **Trade Risks** | | |
| Collusive price agreements | 1 | The market is dominated by several large producers but prices are forced to remain at a sensible level due to fierce competition by substitutes |
| Embargoes | 1 | The only major international body is the Selenium/Tellurium Development Institute but an embargo is an unlikely and self defeating possibility |

*These components are assessed on a scale from 1 (lowest) to 10 (highest) risk. Grouped together, they form the basis for the assessments of the factors for likelihood of supply disruption, and for cost of such disruption which appear in the first table in this section.*

### WORLD TELLURIUM RESERVES

Total over 71 Million lbs Tellurium Content, not all reserves delineated

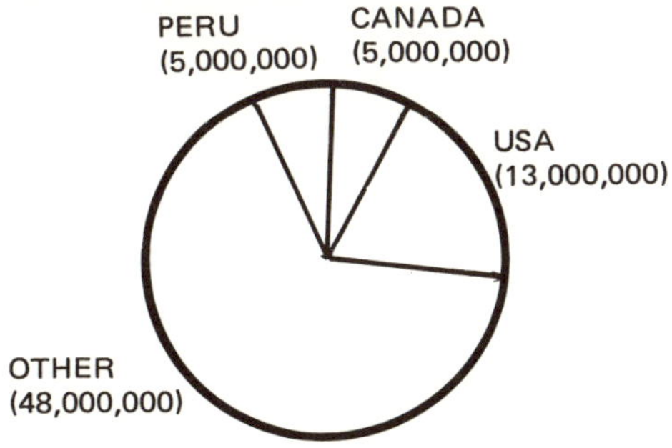


*Source: US Bureau of Mines, 1982*

### US TELLURIUM CONSUMPTION

By Market Sector 1981

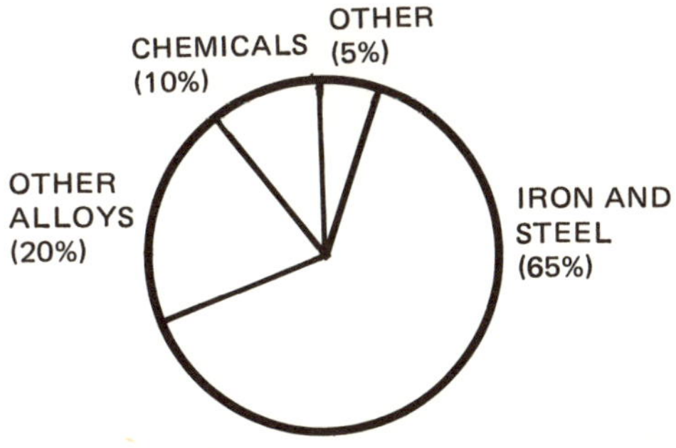


*Source: US Bureau of Mines, 1982*

new trend is the export of copper concentrates, bearing tellurium, from the USA to Japan for treatment.

## APPLICATIONS

### Consumption Trends

Tellurium has a small market with US consumption fluctuating in the 400-500,000 lb a year range. Demand growth is expected to average only one per cent a year over the next decade.

### The Principal Markets

85 per cent of tellurium is used in the metallic state as an alloy additive and a pie chart showing the end-use markets in the USA is shown on page 141. Tellurium is used in its elemental state as a free machining agent in steel and copper alloys and as a carbide stabiliser to provide stronger and tougher cast irons. Tellurium chemicals are used as curing agents and accelerators in rubber compounding principally for the automobile industry. Tellurium dioxide is used for tellurite glass, an infra-red filter, and some catalysts contain tellurium. These uses are, at best, stable. Bismuth and lead tellurides are used in electronic applications such as infra-red detectors, photo-electric cells and thermo-electric junctions. Research is being directed at the possible use of cadmium telluride in photovoltaic solar cells but many other alternatives are also being considered. The solid state uses of tellurides may prove to be the fastest growth area for this minor metal.

### Recycling

Virtually insignificant due to dissipative uses as a minor additive to many alloy systems.

### Substitution

Selenium, bismuth and lead can be substituted in metallurgical applications: selenium and sulphur in rubber compounding applications and selenium and germanium in electronic applications. It is not essential in any of its uses.

## INTERNATIONAL TRADE AND WORLD PRICES

### Supply Arrangements

Almost all tellurium is sold on a producer basis with virtually no free market interest. It is traded as metal of 99.0-99.5 per cent purity in the form of lump, powder or stick with a price quoted in \$/lb. All its applications are very price sensitive and consumption levels depend on the price differential between tellurium and its substitutes, which forces producers into a relatively competitive situation.

### Companies

Metal output is dominated by Metals Refiners (Asia) of Hong Kong; the state-owned USSR organisations; Mitsubishi Metal Corp of Japan; Noranda's Canadian Copper Refiners Ltd (Canada); and Asarco in the USA. Emperor Gold Mining Co of Fiji and Centromin-Peru follow shortly behind.

### Prices

Producer prices moved steadily upwards in the 1973-78 period and then stabilised. A decrease in demand resulted in producers abandoning their list price in order to compete for orders on an unrestricted free market price basis. A low level of consumer demand has meant that prices are now quoted on a nominal basis only.

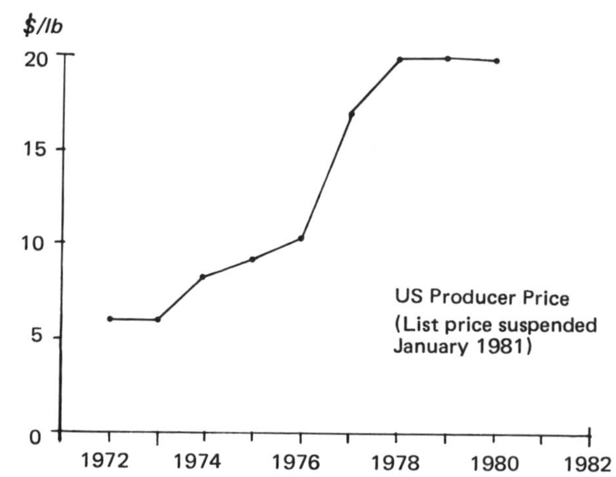


**Tellurium Prices—Annual Averages**

# TIN (Sn)

| | |
|---|---|
| FACTOR FOR LIKELIHOOD OF SUPPLY DISRUPTION | 4.2 |
| FACTOR FOR COST OF SUCH A DISRUPTION | 5.2 |
| TOTAL STRATEGIC FACTOR | 21.8 |

*The factors for supply disruption and cost of such a disruption are assessed on a scale from 1.0 (lowest) to 10.0 (highest) risk. Multiplied together these factors give the total strategic factor rating. For components of these assessments see the table later in this section.*

Tin has a moderately high strategic factor mainly because of the many applications to which it is put in improving the characteristics of other materials. 35-40 per cent of tin is used in the packaging industry with the remainder as solders, bronzes, etc. Tin is unusual in that it is one of the few commodities successfully controlled by an international trading agreement. Output is dominated by three South East Asian countries plus Bolivia, whereas consumption is based in the USA, Western Europe and Japan in an unusually clear division of interests. Tin is a fairly low volume metal in the mature phase of its commodity life cycle, but has an important combination of properties which guarantee its market.

**Physical Characteristics**

Soft, weak, white metal with a slightly bluish tinge; relatively low melting point (232°C) and moderate specific density (7.0); highly corrosion resistant.

## OCCURRENCE AND PRODUCTION

The only economic ore is cassiterite containing tin dioxide ($SnO_2$), but small quantities are recovered from the complex sulphides.

**World Reserves**

These are estimated at ten million tonnes tin, sufficient for 40 years of consumption at current levels. Reserves are scattered irregularly along a belt surrounding the Pacific Ocean with the larger, richer deposits on the Asiatic portion of the ring. The Communist bloc controls a quarter of world reserves.

**Production Characteristics**

*Mining Methods:* Deposits found as primary lodes—mined by underground methods, or as secondary alluviai/eluvial deposits—mined by bucket line dredging and gravel pump methods

*Processing:* Placer deposits upgraded by sorting processes to 65-70 per cent tin, smelted and carbon reduced to tin metal. Lode deposits require more beneficiation but only give 40-60 per cent tin concentrates which need roasting and acid leaching prior to smelting

*Energy Requirements:* Moderate at 200-210 million Btu's/tonne

*By-Products:* Tungsten, antimony, lead and silver, plus heavy minerals such as ilmenite, monazite and zircon

*Environmental Factors:* No major problems specific to this mineral

*Development Lead Time:* Up to eight years for new underground or open cast mines; up to three years for new mechanical dredges at placer deposits and almost instantaneous for one-man gravel-pump operations.

**The Major Producers**

| *Countries* | *1981 Output (Tonnes)* | *Percentage of Western World Output* |
|---|---|---|
| Malaysia | 59,938 | 29.7 |
| Indonesia | 35,268 | 17.5 |
| Thailand | 31,474 | 15.6 |
| Bolivia | 29,801 | 14.8 |
| Brazil | 7,315 | 3.6 |
| UK | 3,869 | 1.9 |
| Australia | 12,083 | 6.0 |
| Others | 22,252 | 10.9 |
| WESTERN WORLD TOTAL | 202,000 | 100.0 |

(Production in the USSR estimated at 16,000 tonnes. *Source: World Bureau Metal Statistics February 1983.)*

*Source: International Tin Council publications.*

Tin is mined almost exclusively in the developing countries most of which have economies which are highly dependent on export earnings from tin. Growth in the world tin industry has been relatively low in comparison with the main tonnage metals and world production has only risen

## COMPONENTS OF RISK ASSESSMENT

| *Factor* | *Rating* | *Comment* |
|---|---|---|
| **Production Risks** | | |
| Existing capacity | 1 | Currently in over-capacity with mid-1982 producer stocks at 70,000 tonnes, 35 per cent of world annual tin metal output |
| Labour disputes | 6 | Troubled labour record in Bolivia particularly, and also in Australia. The small one-man operations in South East Asia are run by their owners, and do not face such problems |
| Violent conflict | 3 | Problems in Bolivia and Zaire are possible. The South East Asian producers are fairly stable |
| Range of primary supply sources | 7 | 63 per cent of output concentrated in Malaysia, Thailand and Indonesia |
| Time lags for new supplies | 7 | Up to eight years for new open-cast or underground mines, but small dredging operations are very flexible. |
| **Transportation Risks** | | |
| Primary | 6 | Tin concentrates or metal shipped from South East Asia, Bolivia, and Australia to the main consuming centres of Europe, Japan and the USA often liable to delays since transport lines are long |
| Secondary | 2 | Few problems encountered in transport of tin or its products |
| **Application/Use Risks** | | |
| Total economic impact | 6 | Moving from being a 'tonnage' to a 'kilo' commodity with world consumption of all metal at 200,000 tonnes per year |
| Effect on key industries | 5 | Heavy bias towards the growth packaging, electrical and construction industries |
| Availability of substitutes | 3 | As a high unit cost metal, its applications are very price sensitive and substitutes are available, particularly aluminium and rival non-metals in packaging |
| Longer term substitutability | 3 | Tin's share of the beverage can market is falling, alloys and solders are being developed with less tin and the metal can only retain its markets at a competitive price |
| **Trade Risks** | | |
| Collusive price agreements | 10 | Prices largely controlled by a successful International Trading Agreement and is susceptible to a price support exercise as happened in 1981-82. The major producers began negotiations in 1982 for a Tin Producers Association |
| Embargoes | 1 | Unlikely |

*These components are assessed on a scale from 1 (lowest) to 10 (highest) risk. Grouped together, they form the basis for the assessments of the factors for likelihood of supply disruption, and for cost of such disruption which appear in the first table in this section.*

## WORLD TIN RESERVES

Total = 10 Million Tonnes Tin Content

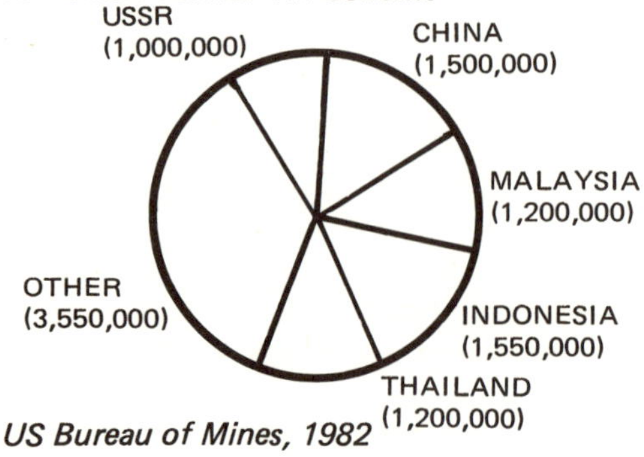


*Source: US Bureau of Mines, 1982*

## US TIN CONSUMPTION

By Market Sector 1981

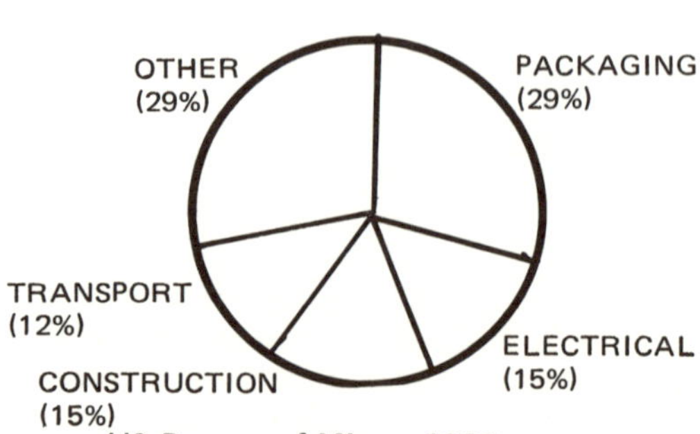


*Source: US Bureau of Mines, 1982*

40 per cent in the last 40 years. There is also a global trend towards the exploitation of progressively leaner ores, such as alluvial deposits which may contain as little as 0.01 per cent tin. Malaysia and Thailand derive most of their tin from alluvial deposits by dredging and gravel pump mining with a heavy reliance on the flexible, and often illegal, one-man operations. Bolivia has one of the highest tin mining cost structures since it has complex vein deposits situated underground in remote rocky areas. The state mining concerns of China and Russia are also major producers and have a long history of distorting Western markets with their variable trade patterns.

The world tin industry has historically been relatively unstructured with small privately owned, labour intensive mines providing at least half of each of the four main countries' output. This is now changing with the formation of larger groups of companies and an increasing state interest, particularly in Malaysia. The degree of vertical integration is increasing although tin smelters have made little effort to move into areas of fabrication, such as tinplate production. Tin producers have continued their policy of installing new smelting capacity in an attempt to produce a higher value product, in direct competition to the older custom smelters in the industrialised countries where most of the tin is used.

### Techological Developments

Research is directed towards improving tin recovery, currently averaging between 50 and 65 per cent, from complex lode areas. There has been much interest in improved smelting processes to recover the 10-25 per cent tin content of old slags and produce cleaner slags.

## TRANSPORTATION

An important link in the chain lies between Penang, the main physical tin market for South East Asia, and the consuming centres of the USA, Europe and Japan. Concentrates tend to be transported by ship and the distances involved are quite large. Other main routes are those from Bolivia and Brazil to the USA and Europe plus that from Australia to Japan. A large proportion is exported as concentrate, with the EEC refining almost three times as much tin as it mines because of traditional colonial ties with South East Asia. Malaysia, however, refines 10,000 tonnes more tin a year than it mines, mainly from other nearby producers.

## APPLICATIONS

### Consumption Trends

Tin's useful combination of properties makes it suitable for a wide variety of applications but its high unit cost makes it a prime target for attempts at substitution. Total Western world primary consumption levelled at 185,000 tonnes from 1977-79, but fell sharply to 163,000 tonnes in 1981 as a result of the recession. The USBM envisages tin consumption increasing by less than one per cent a year when averaged over the next decade.

### The Principal Markets

About 40 per cent of all tin is used to coat steel sheet to produce cheap, light corrosion resistant tinplate for use mainly in packaging. Tinplate production has been steadily increasing for many years, but present-day cans use only about 20 per cent of the tin and 65 per cent of the steel employed for cans in 1945. A tonne of finished tinplate in the USA contains only 4.0-4.2 kg tin. Tinplate output is shifting from traditional USA and West European locations to the developing countries of Brazil, Mexico and Malaysia. About 70 per cent of tinplate cans go to the food and beverages industries in the ratio 2:1, whilst the remainder is used for chemicals, paints, etc.

Solders use another 25-27 per cent of tin output, mainly in the electrical industry and recently for electronics where automated mass soldering devices enable the basic soldering operations to be carried out sequentially by a single machine. Solders, bronze and brass alloys are also used in the construction and transportation industries. Organic tin compounds are a growing end-use for application as pesticides and plastics stabilisers. Inorganic tin chemicals are used in the production of polyurethane and soap stabilisers amongst other minor uses. Development interest has recently been focused on tin powder metallurgy and more sophisticated tin alloys. The pie chart opposite shows the end-use markets in the USA with Western Europe and Japan following much the same pattern. The newly developing countries including Brazil, Thailand, Mexico and Malaysia use over half their tin in tinplate.

### Recycling

In 1981 world consumption of recycled tin metal totalled 9,100 tonnes, five per cent of all total consumption. Recycled scrap in the USA accounted for 13 per cent of consumption. Old scrap derives mainly from the chemical detinning of tinplate scrap and also as the alloyed form, where reclamation does not involve separating the alloy into its component metals. The potential for tin reclamation is very large given a moderate 35 per cent tinplate recovery rate would provide another 25,000 tonnes a year of tin to the market.

### Substitution

Tin's end-uses are particularly price sensitive and substitutes are competitive in almost every application. Tin consumption in the can-making sector is under a two-pronged attack from:

- competition from other types of packaging material including glass, plastic, metal/plastic laminates and paper
- the reduction in tin use in can production due to thinner tin coatings, tin-free steels and replacement by aluminium cans.

Aluminium now controls three quarters of the major US beverage can market and, elsewhere, is increasing its market share. Recently solders have been developed containing proportionately less tin whilst modern PVC water pipes, instead of copper, and aluminium car radiators do not use solders, thus reducing the size of the sector twofold.

## INTERNATIONAL TRADE AND WORLD PRICES

### Supply Arrangements

A series of International Tin Agreements have been negotiated between producing and consuming countries since 1956. These have been relatively successful in stabilising supplies and preventing wide price fluctuations using a buffer stock and export controls to maintain the price between agreed ranges. The majority of tin is produced in South East Asia and prices fixed daily for physical tin at the Penang market are used as a world producer price. The Penang price is quoted in Malaysian dollars per kilo. There are also two LME tin contracts for five tonnes, high grade (purity exceeding 99.85 per cent) or standard grade tin (minimum purity 99.75 per cent), which provide a useful free market for both hedgers and speculators. An unusual feature of the world market is the existence of a large US stockpile of which 35,000 tons is considered to be surplus to requirements, to be liquidated at a maximum rate of 10,000 tons a year. The USSR is an erratic importer of around 15,000 tonnes a year, from the LME, and China exports 2,000-3,000 tonnes a year of tin.

### Companies

The Western world's largest tin mining group is Malaysia Mining Corporation, 72 per cent state owned, which has additional interests in Thailand and Nigeria. Other major tin mining companies are the state-owned operations of the USSR and China; the state-owned PT Tambang Timah of Indonesia; the state-owned Comibol Group of Bolivia; and Renison, a Consolidated Gold Fields subsidiary, of Australia. Major tin smelters include Thailand Smelting and Refining Co; Datuk Keramat Smelting of Malaysia; and Malaysia Mining Corporation, via its holding in the Straits Trading Company operating in Malaysia. PT Tambang Timah, Enaf of Bolivia, and the Communist bloc state owned operations are also major smelters.

### Prices

1981-82 saw an audacious market support exercise which raised prices to historic highs and expertly manipulated supplies in the face of poor physical demand. The market believed the Malaysian government to be responsible, but with no concrete proof, as a consequence of the producers failing to win an International Tin Agreement price increase in June 1981.

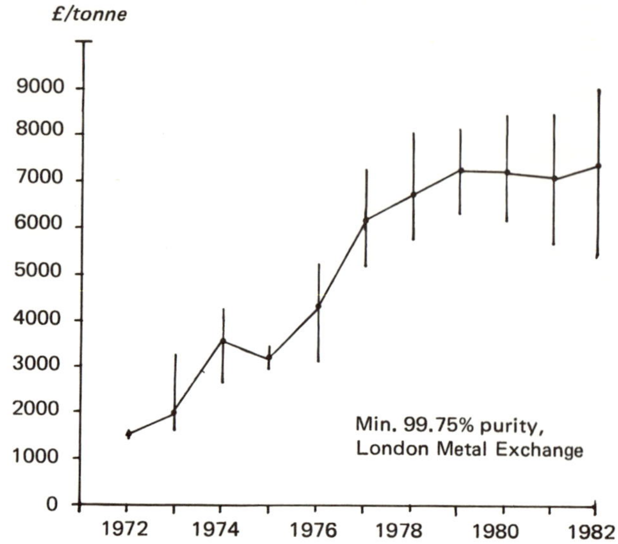


**Tin Prices—Annual Averages, Highs and Lows**

# TITANIUM (Ti)

| | |
|---|---|
| **FACTOR FOR LIKELIHOOD OF SUPPLY DISRUPTION** | **2.3** |
| **FACTOR FOR COST OF SUCH A DISRUPTION** | **7.7** |
| **TOTAL STRATEGIC FACTOR** | **17.7** |

*The factors for supply disruption and cost of such a disruption are assessed on a scale from 1.0 (lowest) to 10.0 (highest) risk. Multiplied together these factors give the total strategic factor rating. For components of these assessments see the table later in this section.*

Titanium has a relatively high strategic factor due to its importance in the aerospace, military and chemical industries whilst mined output in the USA, the EEC and Japan is virtually nil. Australia dominates the titanium ore market and the importance of other Southern Hemisphere producers underlines the somewhat precarious nature of supplies to the USA and Europe. Titanium is virtually irreplacable in many of its applications and supplies can be affected by political manipulations since it is refined and used in countries other than those in which ore deposits are found.

### Physical Characteristics

Useful combination of strength; high melting point (1,668°C); low specific density (4.51) and excellent corrosion resistance.

## OCCURRENCE AND PRODUCTION

Present in many ores of which the most commonly exploited are ilmenite, containing 50-60 per cent titanium dioxide ($TiO_2$) and rutile, 95 per cent titanium dioxide.

### World Reserves

These are estimated at 130 million tonnes rutile and 650 million tonnes ilmenite, sufficient for around 200 years of exploitation at current levels. Australia dominates world reserves of both ores. Communist countries possess two per cent of both ilmenite and rutile reserves.

### Production Characteristics

| | |
|---|---|
| *Mining Methods:* | Rutile and ilmenite mainly mined from beach sand deposits with only a few per cent mineral content using bucket ladder or suction dredges |
| *Processing:* | Rutile is fully oxidised, converted to the chloride and reduced with magnesium to titanium sponge. Ilmenite mainly used for titanium dioxide pigment |
| *Energy Requirements:* | High, 400-420 million Btu's per tonne titanium sponge, most of which is consumed during reduction |
| *By-Products:* | Zirconium, hafnium, rare-earths |
| *Environmental Factors:* | Ecological land-use conflicts and problems of waste water disposal cause most concern |
| *Development Lead Time:* | Up to five years for beach sand deposits. |

### The Major Producers

Australia produces 64 per cent of world rutile and 26 per cent of world ilmenite output, with Norway, South Africa and Canada as other important producers. Ilmenite supplies about 85 per cent of titanium raw materials despite its lower titanium dioxide content. The main barrier to the more extensive use of rutile is the relatively limited supply plus extensive environmental problems in Australia. Processes have been developed to use ilmenite directly to give pigment or to blend it with rutile to give titanium sponge. Ilmenite can also be upgraded to synthetic rutile by increasing the titanium dioxide content. Sponge metal capacity has grown rapidly in the last five years and is concentrated in the USA, Japan and the USSR. There is an increasing degree of integration between sponge production, conversion to ingot, and fabrication into products. Titanium dioxide pigment is produced by the sulphate process, using ilmenite or slag as feed material, or by the chloride process which uses high-grade ilmenite, synthetic or natural rutile and produces less waste material.

### Technological Developments

The emphasis is on reducing the energy consumed in producing sponge and the Showa Denko/Ishizuka Laboratory joint venture in Japan has developed a new process using 30 per cent less energy than that required by conventional methods.

## COMPONENTS OF RISK ASSESSMENT

| *Factor* | *Rating* | *Comment* |
|---|---|---|
| **Production Risks** | | |
| Existing capacity | 1 | Titanium ore and sponge producers currently operating at well below rated capacity |
| Labour disputes | 2 | Unlikely to affect titanium output at any stage since there is no centralised union as in the North American copper or nickel sectors |
| Violent conflict | 3 | A possibility in Sierra Leone, South Africa and the Communist bloc, and would disrupt production |
| Range of primary supply sources | 7 | Australia provides 64 per cent of world rutile and 26 per cent of world ilmenite output with only the USSR, South Africa, and Sierra Leone as other major producers |
| Time lags for new supplies | 7 | Up to five years for exploitation of beach sand deposits but it is difficult to find the right deposits |
| **Transportation Risks** | | |
| Primary | 5 | Long supply lines for ore transport from Southern Hemisphere producers to consumers in the USA, EEC and Japan |
| Secondary | 1 | Few problems transporting titanium sponge |
| **Application/Use Risks** | | |
| Total economic impact | 6 | Total market size of around 90,000 tonnes a year sponge and 2.5-2.7 million tonnes a year pigment |
| Effect on key industries | 10 | 60 per cent of titanium metal vital for aircraft construction and engine components. Remainder of sponge used in marine applications and chemical industry |
| Availability of substitutes | 8 | Virtually no substitute available for dioxide pigment or for sponge in aircraft and space use due to its high strength/density ratio |
| Longer term substitutability | 8 | Alternative alloys could be developed or fibre reinforced materials but at the cost of much engineering research |
| **Trade Risks** | | |
| Collusive price agreements | 3 | Only a handful of producers but a small free market and the unpredictability of Communist bloc trade tends to limit the possibility of artificially fixed prices |
| Embargoes | 2 | Unlikely, except as part of high technology engine or other end-use |

*These components are assessed on a scale from 1 (lowest) to 10 (highest) risk. Grouped together, they form the basis for the assessments of the factors for likelihood of supply disruption, and for cost of such disruption which appear in the first table in this section.*

## WORLD ILMENITE RESERVES

Total = 727 Million Tons Concentrate

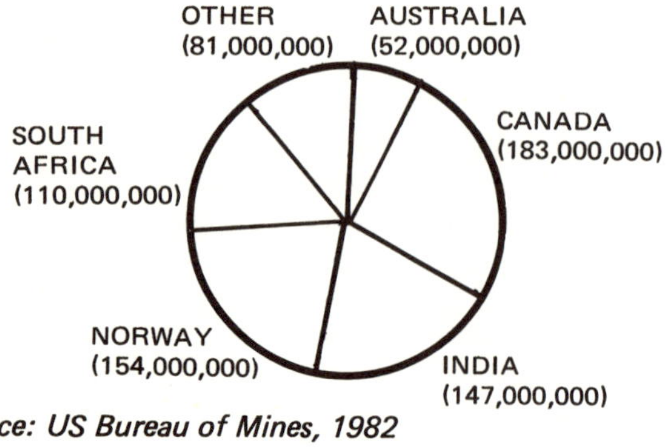


*Source: US Bureau of Mines, 1982*

**The Major Producers**

Rutile

| *Countries* | *1981 Output (Thousand Tonnes)* | *Percentage of Western World Output* |
|---|---|---|
| USA* | 30 | 6.7 |
| Australia | 275 | 61.0 |
| India | 8 | 1.8 |
| Sierra Leone | 68 | 15.1 |
| South Africa | 53 | 11.8 |
| Sri Lanka | 16 | 3.5 |
| Other | 1 | 0.2 |
| WESTERN WORLD TOTAL | 451 | 100.0 |

(Communist bloc production is currently estimated at 10,000 tonnes a year.) Independently rounded.

**Figure withheld by USBM, estimated from imports and consumption data.*

Ilmenite

| *Countries* | *1981 Output (Thousand Tonnes)* | *Percentage of Western World Output* |
|---|---|---|
| USA* | 550 | 11.4 |
| Australia | 1,350 | 28.1 |
| India | 200 | 4.2 |
| South Africa | 420 | 8.8 |
| Norway | 910 | 19.0 |
| Canada | 960 | 20.0 |
| Other | 410 | 8.5 |
| WESTERN WORLD TOTAL | 4,800 | 100.0 |

(Communist bloc production is currently estimated at 460,000 tonnes a year.)

**Figure withheld by USBM, estimated from imports and consumption data.*

Titanium Sponge

| *Countries* | *1981 Output (Tonnes)* | *Percentage of Western World Output* |
|---|---|---|
| USA(e) | 27,000 | 49.3 |
| UK | 2,360 | 4.3 |
| Japan | 25,400 | 46.4 |
| WESTERN WORLD TOTAL | 54,760 | 100.0 |

(Communist bloc production estimated at 42,650 tonnes.)

*Source: US Bureau of Mines 1982, except for USA.*

## TRANSPORTATION

An important stage in the chain from ore to fabricated products or titanium dioxide. The USA has an almost total import reliance for titanium source materials, mainly from Australia and South Africa, with long supply lines in all cases which rely on the free passage of shipping through various global choke points.

## APPLICATIONS

### Consumption Trends

Approximately half all titanium source materials are converted to the dioxide for use as pigments and paper fillers due to its opacity and brightness and this sector is expected to grow by two per cent a year during the 1980s. Titanium sponge consumption saw rapid growth in the 1970s but also a 20 per cent slump in the 1982 recession, despite a USBM prediction of a six per cent annual growth rate from 1978 to 1990.

### The Principal Markets

The demand growth for titanium and its alloys is based on their high strength density ratio making them particularly useful in weight-sensitive applications.

A pie chart showing the end-use markets in the USA is given on page 150. About 60 per cent of all titanium metal is used for load bearing components in military and civilian aircraft construction, strictly on a performance/cost basis. In 1981 aircrafts contained an average of 13.5 tonnes titanium against only 5.2 tonnes in 1968. The necessity of replacing old, noisy and fuel-inefficient aircraft worldwide over the next few years should ensure that this sector continues to show healthy growth. Titanium alloys are useful for chemical processing equipment and in marine applications, such as Russia's well publicised non-magnetic titanium nuclear submarines, due to its corrosion resistance. Titanium is also an additive in steel and other alloy systems to improve hardness and strength.

The military aerospace industry has been given a boost by recent shifts of emphasis within NATO and the manufacture of 100 B-1 bomber aircraft during the next five years could use between 10-20 per cent of current US titanium supply. Superalloys for jet engines continue to be developed with up to 10 per cent titanium content and it is clear that the health of the titanium market and the aircraft industry will continue to be closely linked in the foreseeable future.

### Recycling

All uses of titanium dioxide are totally dissipative. Around two per cent of titanium metal consumption is provided by scrap, mainly recycled aerospace construction alloys such as the 90 per cent titanium/six per cent columbium/ four per cent vanadium alloy. Superalloys or specialised steel alloys containing titanium are rarely recycled back into their component metals.

### Substitution

There is no adequate substitute for titanium in aircraft and space use despite its functional cost disadvantage of 20 per cent even at 1982 depressed prices. In industrial uses, high-nickel steel is a possible alternative. There is no cost effective substitute for titanium dioxide pigment.

## INTERNATIONAL TRADE AND WORLD PRICES

### Supply Arrangements

Within Europe, Japan and the USA, sponge and pigment producers tend to set their own prices in each country after taking account of Australian or South African ore prices. Erratic exports from the USSR and China, however, tend to keep prices at a general consensus. There is only a small free market in titanium sponge and, during boom times, in aircraft alloy scrap. Titanium is traded in various forms including 99.3 per cent purity sponge, rutile or ilmenite of specified $TiO_2$ content, and certain widely used alloys. Increased Japanese shipments of sponge to Europe and the USA is a relatively new phenomenon.

### Companies

Major titanium bearing ore producers include Allied Eneabba, Associated Minerals Consolidated, Consolidated Rutile, Westralian Sands, and Union Corporation of Australia; Sierra Rutile of Sierra Leone; and Qit-Fer et Titane of Canada and South Africa. Major titanium sponge producers include Oregon Metallurgical, Timet and International Titanium of the USA; Osaka Titanium, Toho Titanium and Showa Denko of Japan; plus the integrated state-owned operations of the USSR. The 1981-82 recession caused many titanium sponge producers to cut back sharply with the USA and Japanese companies operating at under 70 per cent capacity. Despite these harsh measures, inventories of sponge and fabricated products continued to rise and several major consumers and aircraft manufacturers, appeared on the free market to liquidate excess inventories.

### Prices

The titanium market has moved both up and down since 1978. During 1979 and part of 1980 a rapid increase in civil and military aircraft orders and an expansion in industrial demand provided the base for firm demand. The USSR meanwhile halted sponge exports, probably due to construction of its nuclear submarines. Prices soared as a consequence of panic buying in the West. During 1981 sponge prices weakened considerably and then slumped dramatically by 60 per cent in 1982. The prospects of a price renewal depend more than anything else on the state of the aircraft industry.

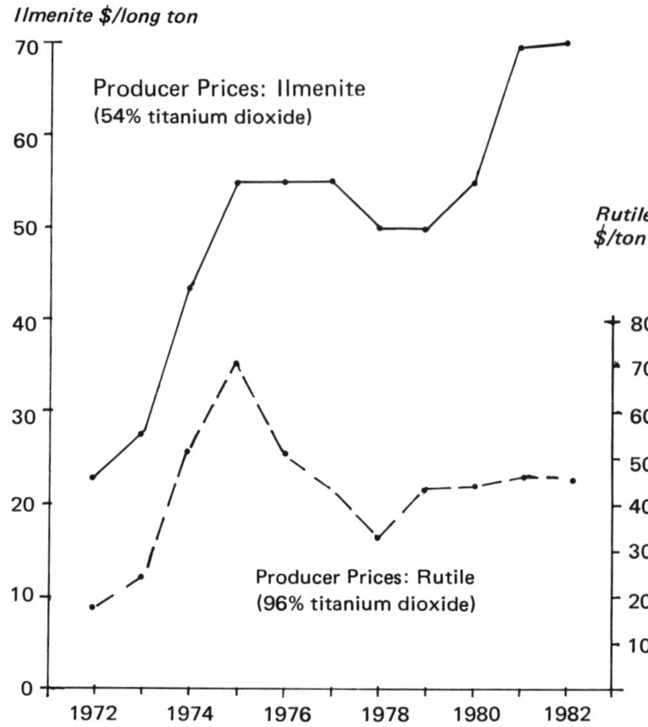


**Titanium Ore Prices—Annual Averages**

## US TITANIUM DIOXIDE CONSUMPTION

By Market Sector 1981

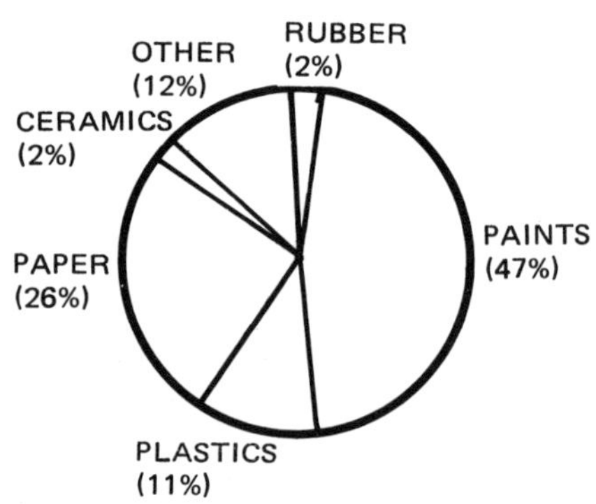


*Source: US Bureau of Mines, 1982*

## US TITANIUM SPONGE CONSUMPTION

By Market Sector 1981

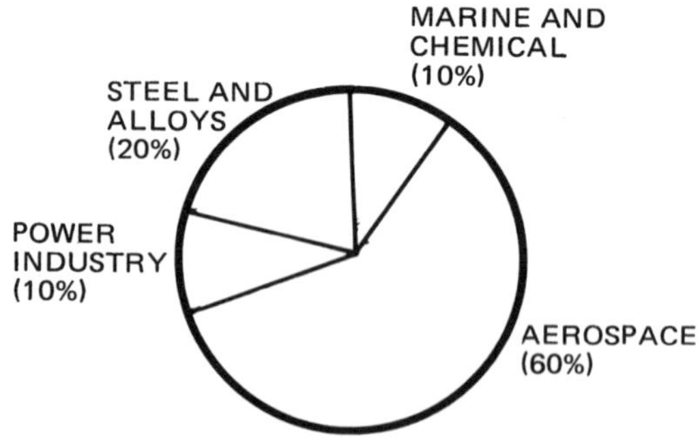


*Source: US Bureau of Mines, 1982*

# TUNGSTEN (W)

| | |
|---|---|
| **FACTOR FOR LIKELIHOOD OF SUPPLY DISRUPTION** | **2.4** |
| **FACTOR FOR COST OF SUCH A DISRUPTION** | **5.8** |
| **TOTAL STRATEGIC FACTOR** | **13.9** |

*The factors for supply disruption and cost of such a disruption are assessed on a scale from 1.0 (lowest) to 10.0 (highest) risk. Multiplied together these factors give the total strategic factor rating. For components of these assessments see the table later in this section.*

Tungsten has only a moderate strategic factor despite the heavy dependence on China and North Korea for mined supplies. It is a fairly low tonnage metal and substitutes are available in most applications. It is valued for its useful combination of strength and wear resistance, particularly at high temperatures, and is mainly used in metal working and construction machinery. New projects are coming onstream to reduce the West's dependence on erratic Chinese exports and major supply disruptions are unlikely.

### Physical Characteristics

Highest melting point of all commercially important metals (3,410°C); high specific density (19.4) and hardness: good conductor of electricity.

## OCCURRENCE AND PRODUCTION

The main ores are the wolframites, containing $WO_3$ (tungsten trioxide) with iron and manganese oxides and the scheelites containing $WO_3$ with calcium oxides.

### World Reserves

These are estimated at 2.9 million tonnes tungsten content, sufficient for over 50 years of consumption at current levels. The majority of reserves are concentrated in Communist countries and China possesses 47 per cent of world reserves. Wolfram ores account for around 70 per cent of the total, but their occurrence in vein deposits is extremely erratic and reserve estimates are subject to more than the usual error.

### Production Characteristics

*Mining Methods:* Mainly small underground mines with ores grading one per cent $WO_3$

*Processing:* Ore beneficiated and converted chemically to tungsten powder via ammonium paratungstate (APT) as intermediate. Powder converted to the carbide, the ferro-alloy or to pure metal products

*Energy Requirements:* Moderate for one per cent $WO_3$ ores but planned exploitation of 0.1-0.2 per cent $WO_3$ ores will require far more energy at the beneficiation stage

*By-Products:* (i) Molybdenum: USA, USSR, Korea. (ii) Bismuth: Korea. (iii) Copper: Peru, Korea. (iv) Tin and bismuth: Australia, Bolivia, Japan, Thailand.

*Environmental Factors:* Pulmonary health hazards a concern in occupational exposure

*Development Lead Time:* Up to eight years for underground mines.

### The Major Producers

| *Countries* | *1981 Output (Tonnes)* | *Percentage of Western World Output* |
|---|---|---|
| USA | 3,175 | 11.5 |
| Canada | 2,725 | 9.9 |
| Mexico | 225 | 0.7 |
| Bolivia | 3,175 | 11.5 |
| Brazil | 1,180 | 4.3 |
| Australia | 3,315 | 12.0 |
| Portugal | 2,500 | 9.0 |
| Austria | 1,360 | 4.9 |
| Burma | 680 | 2.4 |
| Korea | 2,725 | 9.9 |
| Thailand | 1,590 | 5.8 |
| Other | 4,990 | 18.1 |
| WESTERN WORLD TOTAL | 27,640 | 100.0 |

(Production in China is currently estimated at 13,600 tonnes, production in the USSR 8,800 tonnes and total Communist bloc production 24,725 tonnes a year.)

*Source: USBM, Mineral Commodity Summaries, 1982.*

About half world production comes from the USSR, China and North Korea with the balance shared by many countries. The USA, Bolivia, South Korea, and Australia account for approximately one third. There are very few

## COMPONENTS OF RISK ASSESSMENT

| *Factor* | *Rating* | *Comment* |
|---|---|---|
| **Production Risks** | | |
| Existing capacity | 1 | Currently in over-capacity with major cutbacks by producers in Canada, the USA and Australia. There is also a heavy overhang of material in both Europe and the Far East |
| Labour disputes | 1 | Unlikely to affect the bulk of output, which derives from the Communist countries |
| Violent conflict | 5 | Output concentrated in below-average stability countries whilst demand/prices would increase during a war |
| Range of primary supply sources | 6 | China provides 26 per cent of world mined output with another 18 per cent from the USSR and North Korea |
| Time lags for new supplies | 5 | Up to eight years to develop new underground mines but wolfram ore bodies occur erratically and may be difficult to assess |
| **Transportation Risks** | | |
| Primary | 6 | Transport routes are long and exports from China are erratic in timing and volume. Main importers (EEC, Japan and the USA) are long distances from major producers |
| Secondary | 1 | Few problems in transporting tungsten products or ammonium para-tungstate |
| **Application/Use Risks** | | |
| Total economic impact | 5 | Market volume totals around 51,000-53,000 tonnes a year |
| Effect on key industries | 7 | Mainly used as carbide for tool tips and to improve the properties of tool steels, superalloys and special armour plating and heavy duty steels |
| Availability of substitutes | 6 | Substitutes are available and often cheaper. Molybdenum is an alternative alloy constituent and titanium or tantalum carbides used for tool tips |
| Longer term substitutability | 6 | Extensive engineering changes required to effectively replace tungsten, particularly unlikely in the electrical sector |
| **Trade Risks** | | |
| Collusive price agreements | 3 | UNCTAD talks on the stabilisation of tungsten prices failed in 1981 and the producer countries' working committee has had limited success in balancing supply and demand and stabilising prices |
| Embargoes | 1 | Unlikely, China needs foreign exchange generated by exports and limits set elsewhere on tungsten trade are not expected |

***These components are assessed on a scale from 1 (lowest) to 10 (highest) risk. Grouped together, they form the basis for the assessments of the factors for likelihood of supply disruption, and for cost of such disruption which appear in the first table in this section.***

## WORLD TUNGSTEN RESERVES

Total = 6,400 Million lbs Tungsten Content

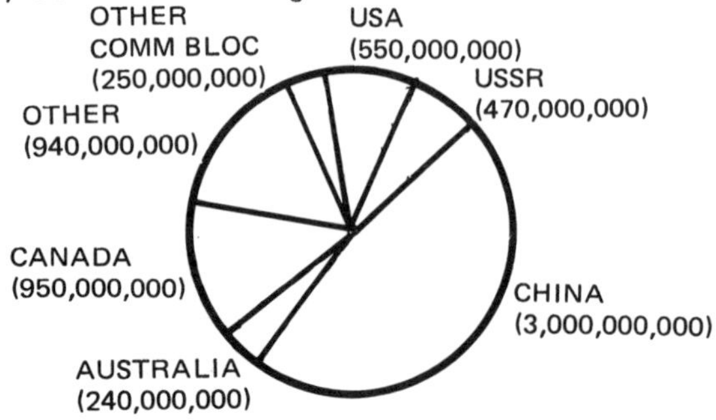


*Source: US Bureau of Mines, 1982*

## US TUNGSTEN CONSUMPTION

By Market Sector 1981

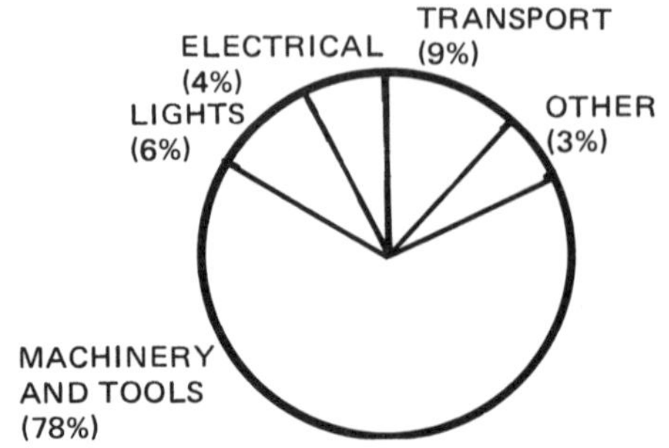


*Source: US Bureau of Mines, 1982*

large tungsten bearing mineral operations with the average mine producing under 1,000 tonnes per day of ore. The four major Western mines are all scheelite with wolfram mining operations concentrated in the Communist bloc, South East Asia and Bolivia. Expansion programmes in the US, Australian and Canadian mining industries would appear to rule out any shortage problems for the foreseeable future. The USA has over a quarter of world APT production capacity and planned increases will prevent a bottle-neck in the supply line. No new major projects in the Communist world are known about. The USA is barred from fully developing its existing resources due to its low grade ores and environmental factors but, in South Africa a new deposit was found in 1981 which could make it self-sufficient within the decade.

### Technological Developments

Effort is concentrated on the efficient exploitation of low grade ores and the USBM is researching methods to recover tungsten from low grade brines and hot springs.

## TRANSPORTATION

The major exporters are China, the USSR, Australia, Korea and Canada, mainly in the form of tungsten-bearing concentrates. Few major disruptions have been recorded. The USA is 55 per cent import dependent and Western Europe and Japan are even more heavily reliant on imports.

## APPLICATIONS

### Consumption Trends

Tungsten consumption rose at an annual rate of five per cent in the 1970s and the USBM predicts that US demand will continue to grow at this rate over the next decade. 1982 however, saw a 20-30 per cent reduction in consumption in most countries, due mainly to the poor state of the world steel industry.

### The Principal Markets

Tungsten is valued for its strength and abrasion resistance, especially at high temperatures and it is used in four main forms: the carbide, as an alloy constituent, as the pure metal and as particular chemicals. A pie chart showing the end-use markets in the USA is given opposite. Tungsten consumption in cutting and wear-resistant applications, dominated by the use of carbide for tool tips, uses around three quarters of world supplies. This sector is, however, sensitive to the level of oil and gas drilling activity, and the health of the machine tool industries, neither of which have been performing well recently. Tungsten is an important constituent (15-20 per cent) of tool steels and a vital part of superalloys and heavy duty steels for armour plating and mining equipment. The main growth areas for superalloys are gas turbines and jet engines. The electrical industry employs tungsten wire as the filament in light bulbs, heating elements in furnaces and as components in cathode ray and X-ray tubes. Non-metallurgical applications include chemicals for dyes, luminescent compounds in X-ray and video screens. Other tungsten compounds have useful catalystic properties. The health of the tungsten carbide sector is the major issue for this metal and it performs well in non-recessionary times. Demand forecasts could, however, be lowered by the increased use of coated carbide inserts rather than solid carbide tips.

### Recycling

Use of tungsten in the electrical, lighting and chemical sectors is dissipative, but scrap, mainly recovered in the form of alloy steels, provided 15 per cent of US consumption in 1981. The steels are rarely separated into their component metals and the tungsten is recycled as an integral part of the alloy.

### Substitution

Titanium or tantalum carbides can be substituted for tungsten in some wear-resisting applications. Molybdenum in tool steels and superalloys provides an attractive alternative and in many cutting tool applications, bulk ceramics or carbide/nitride coatings provide economical substitutes. Alternatives are available in all, except the most arduous applications, and are often preferred on the basis of lower density and cost.

## INTERNATIONAL TRADE AND WORLD PRICES

### Supply Arrangements

The tungsten market is complicated by its reliance on erratic sales from China, particularly centred around the time of the Cantung Trade Fairs and the uncertainty of import volumes required by East European countries. Tungsten is unusual in that a high proportion is sold from producer to consumer via an active and relatively volatile free market. Forms traded include tungsten ore, containing a minimum of 65 per cent $WO_3$ with prices quoted in \$/metric tonne unit (22 lb) of contained $WO_3$, (\$/short ton unit in the USA) or as tungsten powder of 99.5 per cent purity.

Two price indicators have been developed: the International Tungsten Indicator and the Metal Bulletin wolframite price, to provide a more realistic measure of tungsten prices on which to base supply contracts. Continuing sales of surplus tungsten ore and concentrate by the GSA form another important supply source and are the cause of many complaints by producers.

The recession has caused many western producers to cut back on operations and nine producing countries have formed an international working committee to study measures to stabilise tungsten prices. China has been repeatedly blamed for undermining the market by increasing exports of low-priced concentrate and the Working Committee has met with little success so far.

**Companies**

The major mining companies are the state-owned organisations of China, North Korea, and the USSR; Canada Tungsten Mining Corporation; Korea Tungsten Mining Co of South Korea; and Peko Wallsend of Australia. Major APT producers include Union Carbide, General Electric and Amax, all of the USA.

**Prices**

Prices fell to their lowest level in real terms since 1973 during 1982. At $100/m.t.u., for contained $WO_3$, prices fell close to average production costs, a large fall from a 1977 peak of $178/m.t.u.

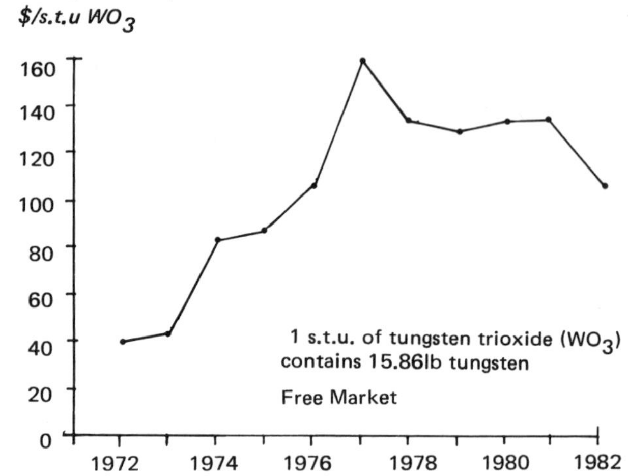


**Tungsten Ore Prices—Annual Averages**

# URANIUM (U)

| | |
|---|---|
| **FACTOR FOR LIKELIHOOD OF SUPPLY DISRUPTION** | **3.3** |
| **FACTOR FOR COST OF SUCH A DISRUPTION** | **4.3** |
| **TOTAL STRATEGIC FACTOR** | **14.2** |

*The factors for supply disruption and cost of such a disruption are assessed on a scale from 1.0 (lowest) to 10.0 (highest) risk. Multiplied together these factors give the total strategic factor rating. For components of these assessments see the table later in this section.*

Uranium has a moderately high strategic factor due to its necessity in the production of nuclear energy by fission methods. Its military applications use only a small proportion of supplies. Output is currently concentrated in the USA, Canada, Australia and South Africa, although China is known to be increasing production. Over the 1980-82 period, a marked slowdown in reactor construction due both to the recession and to environmental pressures, created a growing level of world uranium stocks. This suggests that a supply shortage for the EEC, Japan and USA, is unlikely to occur within the next decade.

**Physical Characteristics**

Natural uranium consists mainly of the U-238 isotope with only one per cent of the easily fissionable U-235 isotope. Thermal reactors use fuel enriched uranium in which the amount of U-235 has been artificially raised to two to three per cent. Fast reactors use plutonium, produced initially as a by-product from thermal reactors. Plutonium results from U-238 which has absorbed neutrons, but without splitting. Uranium has a high specific density (19) and a moderately high melting point (1,133$^{o}$C).

## OCCURRENCE AND PRODUCTION

Uranium is present in several ores as the oxide $U_3O_8$.

**World Reserves**

These are estimated at about 2.6 million tonnes uranium in the form of 'reasonably assured resources' while a further 2.5 million tonnes are considered more speculative according to the Uranium Institute, London. This is sufficient for over 125 years of consumption at current levels. Three factors militate against accurate assessments of reserves; commercial secrecy, lack of adequate exploration programmes and the difficulty of detecting uranium when it lies more than a few tens of metres deep. Uranium is also present in seawater in tiny quantities.

**Production Characteristics**

| | |
|---|---|
| *Mining Methods:* | Underground and open-pit |
| *Processing:* | Crude 'yellow-cake' ($U_3O_8$) concentrates refined into uranium hexafluoride, then enriched to three per cent U-235 |
| *Energy Requirements:* | Moderate with ore grading levels of 1,000 ppm or more |
| *By-Products:* | Phosphate, copper, molybdenum |
| *Environmental Factors:* | Problems of disposing of tailings, particularly if industry continues to process increasingly lean ores |
| *Development Lead Time:* | Up to six years for green field projects. |

**The Major Producers**

| *Countries* | *1980 Output (Tons)* | *Percentage of Western World Output* |
|---|---|---|
| Australia | 1,561 | 3.5 |
| Canada | 7,050 | 16.0 |
| France | 2,634 | 6.0 |
| Nigeria | 4,129 | 9.4 |
| South Africa and Namibia | 10,225 | 23.2 |
| USA | 16,810 | 38.0 |
| Other | 1,691 | 3.9 |
| WESTERN WORLD TOTAL | 44,100 | 100.0 |

*Source: Roskill's Metals Databook, 3rd Edition, 1982.*

Output is dominated in the western world by the USA, South Africa, Australia and Canada. Within Europe, France is the largest producer and is also the most committed to nuclear power. A number of large discoveries were made over a decade ago, during which the market has gone from boom to bust. Australia has great potential with two large projects, Jabiluka and Olympic Downs, likely to come onstream within the next two decades. Production in the USA is concentrated in Utah, but exploratory operations have been carried out all over the country. Few details are available from the Communist bloc, but China is now believed to process 1,000 tonnes a year. Depleted

## COMPONENTS OF RISK ASSESSMENT

| *Factor* | *Rating* | *Comment* |
|---|---|---|
| **Production Risks** | | |
| Existing capacity | 1 | Heavy over-capacity worldwide and stocks continue to grow |
| Labour disputes | 2 | Unlikely to affect yellow-cake production or uranium enrichment |
| Violent conflict | 1 | Production of $U_3O_8$ concentrates unlikely to be affected particularly by violent conflict since mainly produced in Australia and North America |
| Range of primary supply sources | 2 | Supply highly inelastic in the short term. All the main producing countries are part of the Western industrialised world |
| Time lags for new supplies | 4 | Up to eight years for green field projects |
| **Transportation Risks** | | |
| Primary | 2 | Few problems transporting yellow-cake |
| Secondary | 10 | Often politically and physically hazardous to transport enriched uranium or spent nuclear fuel. Depleted uranium can be handled with safety |
| **Application/Use Risks** | | |
| Total economic impact | 3 | Fairly small market of 30-40,000 tonnes a year uranium oxide output producing only 5,000 tonnes or less enriched uranium (in the Western world) |
| Effect on key industries | 7 | Nuclear energy is an important energy option for many countries with few fossil fuel resources. Its military applications are obviously vital |
| Availability of substitutes | 7 | No direct substitutes available for nuclear energy production. However, many other forms of energy generation available, often at lower cost |
| Longer term substitutability | 3 | Hydrogen fusion reaction offers possibility of apparently less hazardous nuclear power generation in the future |
| **Trade Risks** | | |
| Collusive price agreements | 1 | The concentration of resource ownership among relatively few companies and the barriers to new entrants both offset by the need to make nuclear energy competitive with oil/coal-fired smelters |
| Embargoes | 9 | The access of developing countries to enriched uranium is considered to be under strict control although well publicised exceptions have occurred. |

*These components are assessed on a scale from 1 (lowest) to 10 (highest) risk. Grouped together, they form the basis for the assessments of the factors for likelihood of supply disruption, and for cost of such disruption which appear in the first table in this section.*

## FREE WORLD URANIUM CONSUMPTION (TONNES URANIUM)

By Market Sector 1981

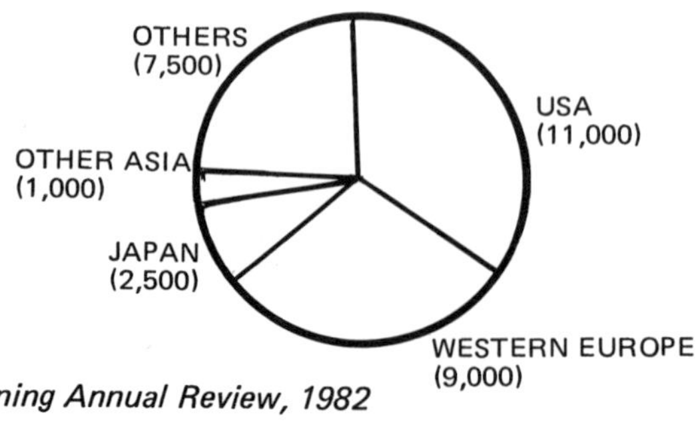


*Source: Mining Annual Review, 1982*

uranium is generated during the process of enriching uranium for nuclear fuel, and about 4.5 tons of depleted product is produced for each ton of enriched material.

### Technological Developments

Chiefly aimed at methods to utilise leaner grades of ore and towards more efficient enrichment procedures at the hexafluoride stage.

## TRANSPORTATION

The principal routes are from Australia to Japan and the UK; from Canada to the USA and Europe; and from South Africa to Europe. The chief problems are those of security since the actual volume of material to be transported is relatively small.

## APPLICATIONS

### Consumption Trends

For all practical purposes, uranium's only use is as nuclear fuel and its market depends on the growth potential for nuclear energy. Only 28,000 tonnes uranium were required for use in 1981 although 41,000 tonnes were contracted by the uranium enrichment factories. The 13,000 tonnes remainder was surplus to requirements and stocks will continue to grow over the 1981-1990 period. This is despite the forecast that nuclear energy will increase in importance and provide 17 per cent of electricity production worldwide in 1985, as opposed to eight per cent in 1981. The USBM predicts that demand for depleted uranium will grow by 8.6 per cent a year through 1990.

### The Principal Markets

Uranium's only major market is that of the energy industry. The 1981-82 recession has forced a radical change into the approach of energy analysts who have had to revise their energy growth rate forecasts markedly downwards. Non-Communist world energy demand rose by 3.5-5.5 per cent a year in the 1970s even though energy use per unit of Gross Domestic Product fell by almost 13 per cent between 1973 and 1980 in International Energy Agency countries. In 1982, an oil glut developed whilst a mixture of price and environmental pressures slowed the growth of the nuclear energy sector.

The International Energy Agency forecasts that nuclear power output will grow by 170 per cent in the coming decade and by a further 65 per cent in the 1990s. The rate of nuclear reactor construction has slowed worldwide due to the effects of recession and high capital costs for construction. Use in the Communist bloc in this sector is considered to top 30,000 tonnes a year.

Over 90 per cent of depleted uranium is used in ordnance applications and for radiation shielding applications due to its high density. The remaining 10 per cent is used in armour-piercing shells and as counterweights and ballast for aircraft. Depleted uranium can be handled and fabricated relatively safely and methods to use it for energy and other non-energy purposes are being investigated.

### Recycling

Fast reactors utilise plutonium which is originally produced as a by-product of thermal reactors. The plutonium core can also be surrounded by an uranium blanket as in a fast breeder reactor, to give more plutonium. Disposal of spent fuel from nuclear power plants is a major problem, both technically and politically.

### Substitution

The development of nuclear reactors using hydrogen fusion reactions as a commercial proposition is still for the future. Currently available alternatives include the traditional methods of energy generation; oil, gas, coal and HEP plus the newer forms including solar, wind, ocean, biomass and geothermal sources. The issues include those of cost, availability and political suitability. However, the uranium price forms only 6-10 per cent of the unit cost of nuclear electricity so demand is relatively insensitive to fuel cost. Lead, tungsten, and other dense metals substitute for depleted uranium in non-nuclear applications.

## INTERNATIONAL TRADE AND WORLD PRICES

### Supply Arrangements

There is no free market in uranium since supply contracts are rigidly monitored by governmental agencies. Prices are quoted in $/lb $U_3O_8$, and sales contracts are arranged on a long term basis. The usually traded form is yellow-cake since enriched uranium is not an easy material to transport. Depleted uranium is safe to handle but supplies greatly exceed demand and trade is minimal.

### Companies

Leading producers are the state-owned organisations of the USSR and China, Rossing Uranium of Namibia, Rio Algom of Canada, Kerr-McGee Corp of the USA, Soc. des Mines de l'Air, Nigeria, Gulf Minerals Canada, and Denison Mines of Canada.

**Prices**

Contract prices for forward delivery, up to ten years in advance, generally give a reasonable indication of market conditions for uranium oxide but contracts are only reported in the USA, and the move towards captive production means that a large proportion of transactions do not have a known market price. Spot transactions account for ten per cent of the total market but most utilities resort to this source of supply only in emergency. US spot prices rose more than fourfold from mid 1974 to mid 1976 but then remained nearly constant for three years before falling sharply in real terms. NUEXCO, an organisation funded solely by subscribers, provides market and price information on uranium.

**Uranium Prices—Annual Averages**

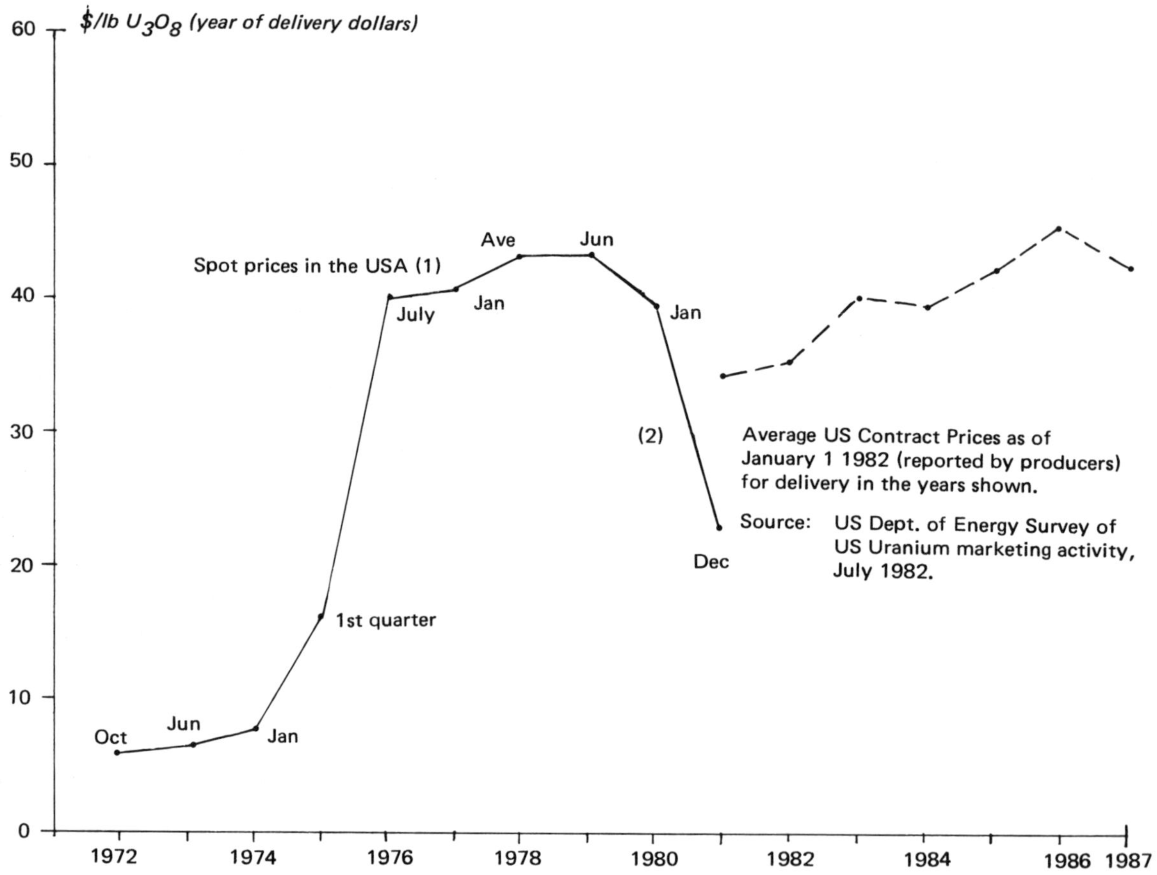


(1) Source: C.S. Buckley, G.S. Mackerron, A.J. Surrey (The International Uranium Market, Energy Policy, 1980 (June)).

(2) Jan 1980, Dec 1981, prices reported by NUEXCO
NUEXCO: Nuclear Exchange Corporation, USA

# VANADIUM (V)

| | |
|---|---|
| FACTOR FOR LIKELIHOOD OF SUPPLY DISRUPTION | 2.7 |
| FACTOR FOR COST OF SUCH A DISRUPTION | 6.5 |
| TOTAL STRATEGIC FACTOR | 17.6 |

*The factors for supply disruption and cost of such a disruption are assessed on a scale from 1.0 (lowest) to 10.0 (highest) risk. Multiplied together these factors give the total strategic factor rating. For components of these assessments see the table later in this section.*

Vanadium has a moderately high strategic factor due to its importance in the production of many steels and sophisticated non-ferrous alloys. South Africa supplies almost half the non-Communist mined output and the USSR is the second largest producer. It is a co-product of iron, uranium and crude oil with an increasing number of primary producers establishing their own conversion facilities at source. This raises a possible future problem of shortage in West European facilities, and elsewhere, to convert concentrates to ferro-vanadium.

**Physical Characteristics**

Moderate specific density (6.11), fairly high melting point (1,900$^{o}$C) malleable metal; reactive metal essential as a trace element for some organisms.

## OCCURRENCE AND PRODUCTION

Vanadium is present as the oxide ($V_2O_5$) as a minor constituent of iron or phosphate or uranium ores and also in crude oils.

**World Reserves**

These are estimated at 18.5 million tonnes contained vanadium, sufficient for over 430 years of consumption at current levels. Vanadium ores are widely distributed throughout the continents with the USSR and South Africa possessing the most extensive resources. In most cases the vanadium constitutes less than two per cent of the total deposit.

**Production Characteristics**

| | |
|---|---|
| *Mining Methods:* | Depends on parent metal or oil |
| *Processing:* | Vanadium pentoxide is recovered from the ores and residues from the extraction of the parent metal and is then converted to ferro-vanadium in an electric furnace or refined to pure oxide. Vanadium also extracted from residue of crude oil and tar sands and from the alumina clay of Arkansas, USA |
| *Energy Requirements:* | 490 million Btu's to produce one ton of ferro-vanadium of which 85-90 per cent is expended on producing the pentoxide from low grade ores |
| *Co-Products:* | Iron, titanium, uranium, oil |
| *Environmental Factors:* | Vanadium is a health hazard at industrial concentrations and other problems include disposal of mine wastes |
| *Development Lead Time:* | Depends on parent product. |

**The Major Producers**

| *Countries* | *1981 Output (tonnes V content)* | *Percentage of Western World Output* |
|---|---|---|
| USA | 4,800 | 24.3 |
| Australia | 590 | 3.0 |
| Finland | 2,450 | 12.4 |
| South Africa | 10,900 | 55.2 |
| Others | 1,000 | 5.1 |
| WESTERN WORLD TOTAL | 19,740 | 100.0 |

(Production in China is currently estimated at 4,990 tonnes and production in the USSR 10,880 tonnes a year.)

*Source: USBM Mineral Commodity Summaries, 1982.*

South Africa produces more than 30 per cent of total world production and over half of non-Communist world output. Other notable producers include the USA, Australia, Finland, and the Communist bloc. Caribbean oils are growing in importance as source materials with Venezuela and Mexico providing the bulk of output in

## COMPONENTS OF RISK ASSESSMENT

| *Factor* | *Rating* | *Comment* |
|---|---|---|
| **Production Risks** | | |
| Existing capacity | 1 | Currently in 24 per cent over-capacity due to recession, but a supply/demand balance likely to be maintained after that |
| Labour disputes | 2 | South Africa and the USSR tend to keep their labour forces under control |
| Violent conflict | 2 | Vanadium production unlikely to suffer primary disruption |
| Range of primary supply sources | 7 | South Africa produces around 50 per cent of non-Communist mined world supply and 50 per cent of pentoxide supplies |
| Time lags for new supplies | 9 | Processing plants find it difficult to adapt to alternative source materials and obtaining other supplies would be uncertain |
| **Transportation Risks** | | |
| Primary | 3 | Transported as slags, pentoxide or ferro-vanadium with proportional costs decreasing in the same order. Transport routes to the main consuming centres are long |
| Secondary | 1 | Relatively few problems in transporting alloy steels containing vanadium |
| **Application/Use Risks** | | |
| Total economic impact | 4 | World annual mined output of 35,000-37,000 tonnes a year |
| Effect on key industries | 6 | Usage of steels containing vanadium (the major use) is concentrated in the growth oil and gas industry and in the chemical and transportation industries |
| Availability of substitutes | 7 | Substitutes include molybdenum, titanium, columbium and others for alloying applications and choices are often made on grounds of cost and reliability of supplies |
| Longer term substitutability | 6 | Would need extensive changes in the metallurgy of established alloys, both ferrous and titanium based, but could be done given time |
| **Trade Risks** | | |
| Collusive price agreements | 6 | South African producers are considered to collaborate with the USSR, to give an effective stranglehold on the market which has, so far, been used to establish sensible prices and promote consumption |
| Embargoes | 3 | A possible political weapon but unlikely as yet |

*These components are assessed on a scale from 1 (lowest) to 10 (highest) risk. Grouped together, they form the basis for the assessments of the factors for likelihood of supply disruption, and for cost of such disruption which appear in the first table in this section.*

## WORLD VANADIUM RESERVES

Total = 40,800 Million lbs Vanadium Content

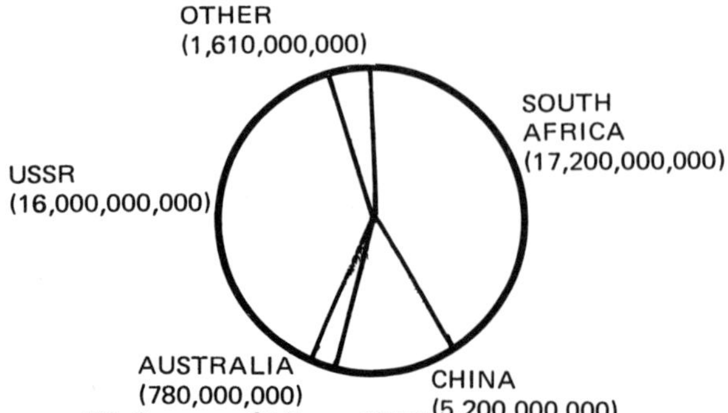


*Source: US Bureau of Mines, 1982*

## US VANADIUM CONSUMPTION

By Market Sector 1981

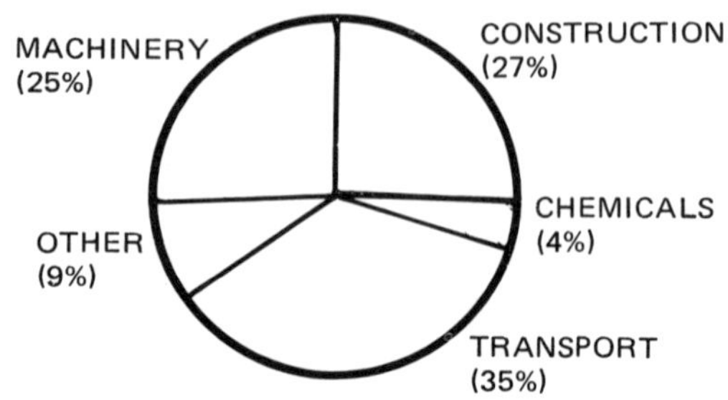


*Source: US Bureau of Mines, 1982*

this sector. South Africa, however, effectively controls the market and the growing integration between mining and ferro-vanadium production puts the country at an advantage given its reserves of cheap labour and coal. Weakening demand in 1981-82 exposed a world over-capacity of 24 per cent in this relatively small market. Operations effectively ceased in Australia and Finland and even in South Africa cutbacks were made with Highveld Steel & Vanadium obliged to reduce production (by an undisclosed amount).

### Technological Developments

The USSR claims to have developed a new process for making ferro-vanadium which uses less energy and costs one third of the conventional process but no details are available. Other developments include Engelhard's new extraction technique from crude oils plus improvements in ore beneficiation techniques.

## TRANSPORTATION

This is a major issue for this mineral both in terms of cost and security. Vanadium forms such a minor constituent in most ores that the high cost of transporting concentrates containing vanadium must be reduced by locating processing plants as near to the mine site as possible. The pentoxide is a widely traded transitional product, but the increase in transport costs plus the development of reduction techniques increasingly allow the circumvention of the pentoxide on the route to higher value-added ferro-alloys. The heavy reliance on South Africa for supplies poses security questions for the long shipping routes to the USA, Europe and Japan. The Communist bloc is a major, if erratic, exporter.

## APPLICATIONS

### Consumption Trends

The non-Communist world market for vanadium grew at an annual rate of 2.5 per cent during the 1965-79 period. The latest USBM forecast, which may well prove optimistic, foresees the demand increasing at 3.6 per cent a year until the end of the century. In 1982, vanadium production capacity totalled 18,000 tonnes a year more than requirements but this is considered to be a short term imbalance.

### The Principal Markets

The chief use for vanadium is as an alloying agent for iron and steel and in the production of iron and steel. A pie chart showing the end-use markets in the USA is given opposite. Vanadium can only be conveniently added to steels in the form of a ferro-alloy due to the high reactivity of the pure element at the elevated temperatures necessary in steel making but, once there, it increases strength, fabricability, and wear resistance. High-speed tool steels for metal working, corrosion-resistant steels for the chemical industry and heavy-duty rail steels all contain vanadium. Its use in high-strength low-alloy steels for oil and gas pipelines forms a major growth area whilst there is a large potential market in vanadium for weight-saving dual-phase steels in automobile manufacture to improve fuel consumption. Vanadium is a possible substitute in alloys containing the more expensive molybdenum and nickel.

Titanium alloys for aerospace applications contain four per cent vanadium to improve properties and this sector accounts for ten per cent of US consumption. Vanadium chemicals are used as catalysts in sulphuric acid manufacture and in gas desulphurisation reactions. Vanadium compounds also provide a wide range of colouring agents.

### Recycling

Small quantities of spent catalysts containing vanadium are reprocessed and some tool steel and other alloys are recycled for their vanadium content.

### Substitution

Various metals are interchangeable with vanadium in alloy steels including columbium, manganese, titanium, tungsten, and particularly molybdenum. It is often a simple question of choosing the cheapest alloying agent with the most reliable supply lines. Platinum is an alternative in the catalyst sector.

## INTERNATIONAL TRADE AND WORLD PRICES

### Supply Arrangements

The EEC, Japan and the USA, are heavily dependent on imported vanadium source materials, usually as the pentoxide or ferro-vanadium. Britain produces ferro-vanadium from slags originating in South Africa and the USSR, but the South Africans adopt a pricing policy favourable to their own converting facilities. Non-South African supplies are restricted by the production economics of other residues and also the time-lag necessary for plants to be converted to cope with alternative feed stocks.

Vanadium is traded as the ferro-alloy (50-60 per cent vanadium), the pentoxide (minimum 98 per cent purity) and as slags of various grades and origins. The pentoxide is the only form in which there is a sizeable free market and the major merchant in this sector is Philipp Brothers. Most vanadium is sold using long-term producer/consumer contracts and the major South African producers have a vested interest in maintaining sensible producer prices. Few disruptions in vanadium supply have been noted, but vanadium appears on the vulnerability lists compiled by West German, UK, French and Japanese authorities and the GSA has a stockpile goal of 15.4 million lbs.

**Companies**

The major mine producers are the state organisations of China and the USSR; Highveld Steel & Vanadium Corporation of South Africa; Rautaruukki Oy of Finland; Agnew Clough of Australia; and Union Carbide Corporation of the USA. South Africa possesses half the West's vanadium pentoxide capacity, two-thirds of which is owned by Highveld Steel & Vanadium. Major ferro-vanadium producers are Highveld, Rautaruukki; Gesellschaft fur Elecktrometallurgie GmbH of Germany; Treibacher of Austria; and Nippon Denko of Japan. As the major consumers are steel producers with few interests in mining or conversion processes there is substantial trade in both primary and intermediate vanadium products.

**Prices**

Vanadium product prices have been maintained on a broad uptrend over the past decade by the effective producer control of the market. Highveld Steel & Vanadium has remained the chief arbiter of prices due to its virtual monopoly of Western world output. Exports from China (about 3,600 tonnes vanadium pentoxide in 1982) have recently been an unpredictable factor in the market place, acting to depress prices in a time of recession. Prices realised for vanadium alloys and chemicals have shown a tendency to move in tandem with vanadium pentoxide prices.

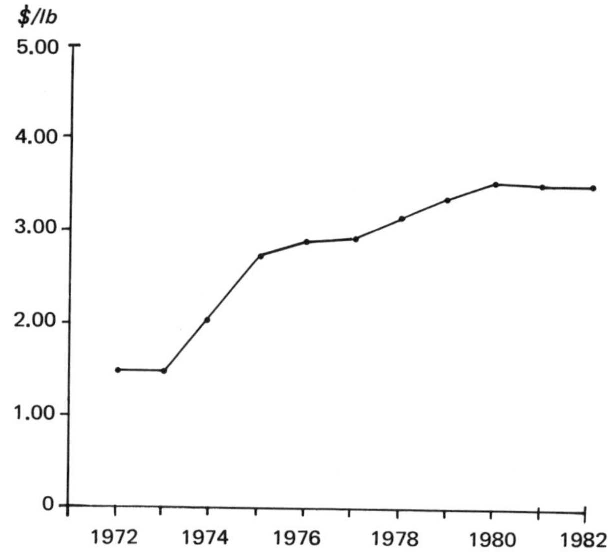


**Vanadium Pentoxide Producer Prices—Annual Averages**

# ZINC (Zn)

| | |
|---|---|
| **FACTOR FOR LIKELIHOOD OF SUPPLY DISRUPTION** | **3.1** |
| **FACTOR FOR COST OF SUCH A DISRUPTION** | **4.8** |
| **TOTAL STRATEGIC FACTOR** | **14.9** |

*The factors for supply disruption and cost of such a disruption are assessed on a scale from 1.0 (lowest) to 10.0 (highest) risk. Multiplied together these factors give the total strategic factor rating. For components of these assessments see the table later in this section.*

Zinc is an industrial metal with a moderate strategic factor. It is versatile and useful in many industries as a component of various alloy systems and as a protective coating to prolong the life of steel structures. It stands fourth amongst the metals in terms of world annual consumption by volume. Reserves are large and the industry is more concerned with developing and maintaining markets than with supply disruptions since 60 per cent of mined output derives from Europe, Japan, Canada and the USA.

**Physical Characteristics**

Bluish-white metal with specific density of 7.13 and low melting point (420°C); chemically active, allowing it to be used to sacrificially protect ferrous alloys against corrosion; as a trace element, it is essential in animal and plant nutrition.

## OCCURRENCE AND PRODUCTION

Zinc is present in many types of ores and the most important mineral is sphalerite (ZnS).

**World Reserves**

These are estimated at 240 million tonnes zinc content, sufficient for over 40 years of exploitation at current rates. Reserves are widespread but concentrated in the USA, Canada and Australia. The Communist bloc possesses around 10 per cent of world reserves.

**Production Characteristics**

*Mining Methods:* A few open pit mines in Australia, Brazil, Canada and Thailand, but mainly by underground methods

*Processing:* Beneficiation process depending on ore type, then roasting followed by electrolytic deposition or distillation methods to reduce to zinc metal. Second electrolytic stage produces pure zinc

*Energy Requirements:* Moderate energy consumption of 70-75 million Btu's per tonne

*By-Products:* Varying amounts of copper, lead, silver plus cadmium, germanium, gallium, indium, thallium, and sometimes gold

*Environmental Factors:* Tailings and mine water do not present massive problems since mine tailings often sold as crushed rock and agricultural limestone

*Development Lead Time:* Up to ten years for greenfield underground mine projects.

**The Major Producers**

Canada dominates world mined output, followed by Australia, Peru and the USA. The Communist bloc produces just over a quarter of the world total. Slab zinc production is dominated by Western Europe, which refines almost twice as much zinc as it mines, Canada, the USA and Australia plus Japan which is now the largest metal producer. Developing countries currently account for only 25-30 per cent of mined output but almost half planned new mine capacity is located in the Third World.

Zinc is unusually dependent on its co- and by-products: only one-third of world zinc production derives from ores that are not associated with lead, whilst two-thirds of world silver and one-tenth of world copper are co-products from the same ore bodies which provide the zinc and lead. The industry is only partially integrated with a large number of toll smelters, particularly in Europe and Japan, which buy concentrates and produce slab zinc.

**Technological Developments**

The USSR has developed a new zinc production method, the Kivcet-CS process which is claimed to combine the functions of sintering, blast furnacing and slag fuming in one smelter unit with the concommitant cost and recovery advantages.

## TRANSPORTATION

Japan imports about two-thirds of its concentrate requirements and Europe, about half. Canada exports approxi-

## COMPONENTS OF RISK ASSESSMENT

| *Factor* | *Rating* | *Comment* |
|---|---|---|
| **Production Risks** | | |
| Existing capacity | 1 | Currently in over-capacity with harsh mining and smelting cutbacks in Europe and North America |
| Labour disputes | 6 | Poor history of labour relations in the zinc mining and smelting industry of the USA, Eire and Canada |
| Violent conflict | 1 | Unlikely to affect zinc output |
| Range of primary supply sources | 3 | Deposits identified in over 50 countries but over two-thirds of mined output from North America, Europe and Australia |
| Time lags for new supplies | 4 | Up to ten years for greenfield underground mines. Some open cast mines in Canada but deposits often located in areas where mine construction is hampered by weather etc |
| **Transportation Risks** | | |
| Primary | 6 | The USA, Japan and Europe depend on imports from Peru, Canada and Australia—long shipping routes |
| Secondary | 3 | Transport of zinc metal tends to be less of a problem |
| **Application/Use Risks** | | |
| Total economic impact | 7 | Total market volume of 6.0-6.1 million tonnes a year slab zinc. Ranks fourth of all metals in tonnage consumed a year |
| Effect on key industries | 5 | Heavily dependent on construction and transportation industries for use in galvanised steel and die castings but both use sectors declining in the industrialised world |
| Availability of substitutes | 5 | Available but usually at higher cost |
| Longer term substitutability | 5 | Zinc is losing the battle in the automobile component sector but is fighting hard against aluminium/zinc alloy galvanising substitutes |
| **Trade Risks** | | |
| Collusive price agreements | 5 | Has already occurred and could do so again given judicious manipulation of the LME price but such action contravenes US and EEC regulations |
| Embargoes | 1 | Unlikely |

*These components are assessed on a scale from 1 (lowest) to 10 (highest) risk. Grouped together, they form the basis for the assessments of the factors for likelihood of supply disruption, and for cost of such disruption which appear in the first table in this section.*

## WORLD ZINC RESERVES

Total = 240 Million Tonnes Zinc Content

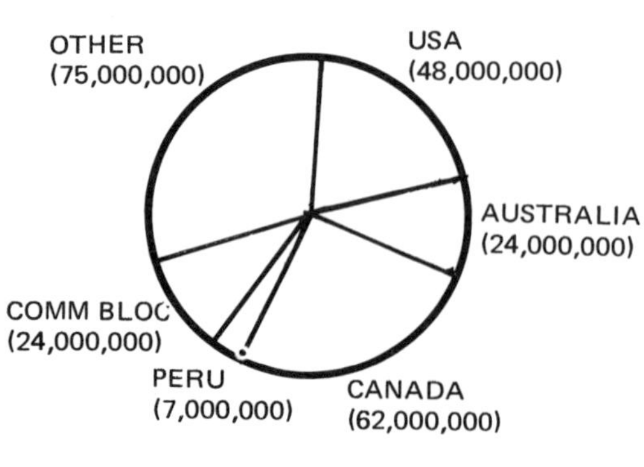


*Source: US Bureau of Mines, 1982*

## US ZINC CONSUMPTION

By Market Sector 1981

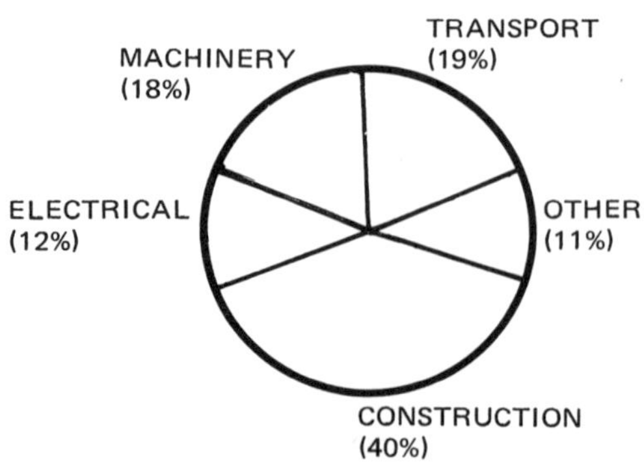


*Source: US Bureau of Mines, 1982*

### The Major Producers

World Mine Production

| Countries | 1981 Output (Thousand Tonnes) | | Percentage of Western World Output | |
|---|---|---|---|---|
| EUROPE | | 989 | | 22.3 |
| AFRICA | | 254 | | 5.7 |
| ASIA | | 382 | | 8.6 |
| Japan | 242 | | 5.4 | |
| AMERICA | | 2,307 | | 52.0 |
| Canada | 1,096 | | 24.7 | |
| Mexico | 204 | | 4.6 | |
| Peru | 497 | | 11.2 | |
| USA | 335 | | 7.6 | |
| AUSTRALIA | | 504 | | 11.4 |
| WESTERN WORLD TOTAL | | 4,436 | | 100.0 |

(Communist bloc production is currently estimated at 1,644,000 tonnes a year.)

World Slab Production

| Countries | 1981 Output (Thousand Tonnes) | | Percentage of Western World Output | |
|---|---|---|---|---|
| EUROPE | | 1,843 | | 40.5 |
| Belgium | 250 | | 5.5 | |
| France | 257 | | 5.6 | |
| West Germany | 367 | | 8.1 | |
| AFRICA | | 208 | | 4.6 |
| ASIA | | 827 | | 18.2 |
| Japan | 670 | | 14.7 | |
| AMERICA | | 1,369 | | 30.1 |
| Canada | 619 | | 13.6 | |
| USA | 373 | | 8.2 | |
| AUSTRALIA | | 301 | | 6.6 |
| WESTERN WORLD TOTAL | | 4,548 | | 100.0 |

(Communist bloc production is currently estimated at 1,698,000 tonnes a year.)

*Source: World Bureau of Metal Statistics, June 1982.*

mately 40 per cent of its output as concentrate, thus delineating the main shipping routes. The main exporters of slab zinc are Canada, Belgium, France and Germany and the USA remained 67 per cent import dependent in 1981. The major shipping route is from Australia to Japan for zinc concentrates. Disruptions tend to be a consequence of dock union strikes rather than deliberate attempts to hinder zinc deliveries.

## APPLICATIONS

### Consumption Trends

Zinc consumption has stagnated since 1973-74 and fallen incrementally in the Western world since 1979. These overall figures mask the ten per cent a year growth in consumption in the less developed countries particularly in the newly industrialised Asian countries. Western world consumption in the first half of 1982 was six per cent down from 1981's annualised total of 4.24 million tonnes. The USBM predicts that US demand will grow by 1.1 per cent a year over the next decade.

### The Principal Markets

Almost half of the zinc used in the USA is for galvanising steel, providing it with a corrosion resistant coating, which is then used in the construction and transportation industries. Automobile makers also require zinc-based alloys as cast components but this application has declined from 60 lb of zinc diecastings on average per US car in the early 1970s to 25 lb in 1981 as the trend towards lighter cars has gained momentum. Zinc is used in electrical equipment and in other machinery mainly as the versatile range of brasses (copper/zinc alloys). Other zinc alloys provide corrosion inhibitive properties and are applied as a plating to different base alloys.

A pie chart showing the end-use markets in the USA is given opposite. Japan uses a greater proportion of zinc (56 per cent), in galvanising, whereas the UK uses only 30 per cent and West Germany 33 per cent for this process, with proportionately more consumed in brasses and other alloys. The use of zinc as a sacrificial anode to protect underwater steel structures and as metallised corrosion resistant paints are two of the few growth areas for zinc in Western markets.

### Recycling

Production of secondary slab zinc represented just five per cent of US consumption in 1981. This does not include the zinc content of remelted brass, which undergoes no further treatment for re-use. Worldwide production of secondary zinc, including brasses, totalled almost 20 per cent of all slab zinc output. The use of zinc in galvanising and other corrosion resistant applications is, by definition, dissipative.

### Substitution

Aluminium and magnesium alloys are now often preferred alternatives in diecasting. Plastic coatings, paints, cadmium plating can often replace zinc in some anti-corrosion applications whilst several new aluminium-zinc alloys have been developed as an alternative to the traditional galvanising process and these obviously require less zinc. Aluminium, magnesium, titanium and zirconium are significant competitors in the chemicals and pigments sector.

## INTERNATIONAL TRADE AND WORLD PRICES

### Supply Arrangements

The bulk of the non-Communist world's supply of zinc concentrate and metal is sold under direct supply contracts arranged on an annual basis between producers or smelters and consumers. The vertically integrated producers of North America use a price quoted in c/lb whilst the European producer price, in $/tonne, is used throughout the rest of the world. Europe has a large number of non-integrated toll smelters. These tend to get squeezed between the cost of concentrates and the price achieved when these are weak. There is an active LME contract for 98 per cent purity zinc in 25 tonne lots, which only sees a small amount of Western output but it is the market place for East/West trade. Weekly or monthly average LME prices are often a reference point in producer/consumer concentrate or metal contracts. There is a UK Zinc Development Association plus a loose association of US and European zinc smelters but the latter's aim of supporting prices has been thwarted by the threat of anti-cartel action by the US government and the EEC Commission plus the rivalry between integrated and non-integrated smelters.

**Companies**

A large number of companies are involved in mining zinc ores and the industry is fragmented. The major producers are the state-owned operations of the USSR and Poland; Cominco of Canada; Tara Exploration & Development of Eire; Centromin-Peru; and New Broken-Hill Consolidated of Australia. The leading smelters are the state-owned operation of the USSR, Poland and China; Mitsui Mining & Smelting of Japan; Cominco, Canadian Electrolytic Zinc; Preussag AG Metall of West Germany; EZ Zinc Co of Australasia and Vieille-Montagne SA of Belgium.

**Prices**

Zinc prices peaked in 1973 in the wake of the oil crisis, at least partly due to a price support exercise by European smelters. But prices have performed poorly during 1981-82 with heavy discounting from published prices and causing sharp production cutbacks in Western Europe and North America. In autumn 1982, a selection of European smelters applied to the EEC to be treated as a special case, allowing cooperation in a much-needed rationalisation of over-capacity. The original 200,000 tonnes a year rationalisation plan has foundered due to reluctance on behalf of the French, Italian and Greek governments.

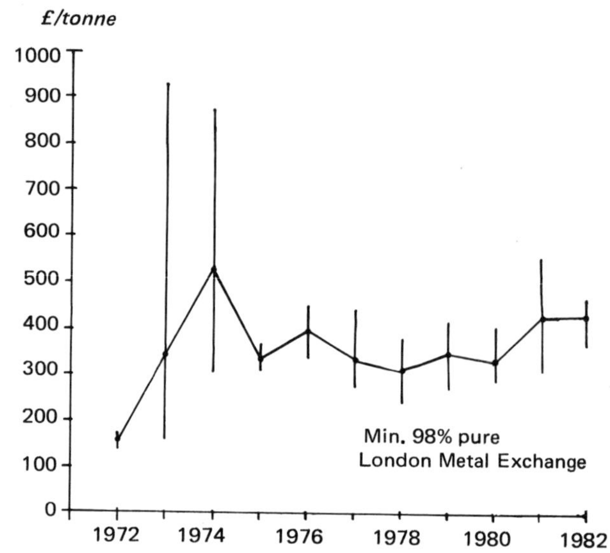


**Zinc Prices—Annual Averages, Highs and Lows**

# ZIRCONIUM (Zr)

| | |
|---|---|
| FACTOR FOR LIKELIHOOD OF SUPPLY DISRUPTION | 1.4 |
| FACTOR FOR COST OF SUCH A DISRUPTION | 4.2 |
| TOTAL STRATEGIC FACTOR | 5.9 |

*The factors for supply disruption and cost of such a disruption are assessed on a scale from 1.0 (lowest) to 10.0 (highest) risk. Multiplied together these factors give the total strategic factor rating. For components of these assessments see the table later in this section.*

Ninety per cent of zirconium is used in foundry sands, refractories and ceramics with the remainder employed in chemical processing equipment and nuclear applications. Reserves and output are both concentrated in Australia and there are virtually no supply problems. Substitutes are available for most applications.

### Physical Characteristics

High melting point (1,852°C); moderate specific density (6.49); strength and corrosion resistance at high temperatures; low neutron absorption cross-section and great resistance to severe bombardment by neutrons and fission fragments.

## OCCURRENCE AND PRODUCTION

Zirconium is generally found as zircon (zirconium silicate) in sand deposits.

### World Reserves

These are estimated at 48 million short tons, sufficient for over 75 years consumption at current levels. Reserves are concentrated in South Africa, Australia, the USA and India. World resources exceed 60 million tons since phosphate sand and gravel deposits may, in the future, yield substantial amounts of zircon.

### Production Characteristics

*Mining Methods:* Dredging and wet-milling from sand deposits involving recovery of only a few per cent minerals

*Processing:* Raw zircon sand chemically treated to give zirconium sponge or furnace treated to give zirconium dioxide

*Energy Requirements:* Moderate

*By-Products:* Titanium ores, hafnium (nuclear grade zirconium is the only commercial source for hafnium)

*Environmental Factors:* Ecological lobby against the large-scale mining of beach sands

*Development Lead Time:* Five to eight year time scale.

### The Major Producers

| *Countries* | *1981 Output (Short Tons)* | *Percentage of Western World Output* |
|---|---|---|
| USA | 98,000 | 14.3 |
| Australia | 450,000 | 66.3 |
| India | 12,000 | 1.8 |
| South Africa | 105,000 | 15.5 |
| Others | 14,000 | 2.1 |
| WESTERN WORLD TOTAL | 679,000 | 100.0 |

*Source: USBM 1982, and USA output estimated from Roskill's Metal Databook, 1982.*

Mined output dominated by Australia, South Africa, the USA and India whilst Communist output is large but not accurately known. The USA imports about half its needs, 90 per cent in the form of zircon from Australia. Europe also depends on these sources. Zirconium metal production plants are concentrated in the USA, Japan and Europe.

### Technological Developments

Mainly stable production technology. Mitsui and Ishizuka Research claim to have developed a new method for separating hafnium from zirconium sand to give pure zirconium sponge at half traditional costs.

## TRANSPORTATION

Mainly by ship from Australia and South Africa. No major problems noted.

## COMPONENTS OF RISK ASSESSMENT

| *Factor* | *Rating* | *Comment* |
|---|---|---|
| **Production Risks** | | |
| Existing capacity | 2 | Current over-capacity during 1982 with most zirconium sponge producers operating at only one-third capacity |
| Labour disputes | 1 | Unlikely to affect beach sand mining |
| Violent conflict | 1 | Unlikely in Australia, which produces 70 per cent of world output |
| Range of primary supply sources | 4 | Major exporters are South Africa, Australia and India |
| Time lags for new supplies | 5 | Only certain deposits available, mainly on the coastlines of South Africa, Australia, India and the USA (Florida) |
| **Transportation Risks** | | |
| Primary | 2 | Long shipping routes but not a likely disruption target |
| Secondary | 1 | Subject to few risks as zircon or zirconia. Possible target as nuclear fuel cans |
| **Application/Use Risks** | | |
| Total economic impact | 4 | Western World Market volume of 650-700,000 short tonnes a year |
| Effect on key industries | 3 | Unexciting but stable market base in foundry sands, refractories etc— ten per cent used as the metal in nuclear applications and to improve the qualities of special steels |
| Availability of substitutes | 5 | Available in non-metallic applications but tend to be inferior in performance. Difficult to use substitutes for zirconium metal due to engineering problems |
| Longer term substitutability | 4 | Reasonably feasible given sufficient time |
| **Trade Risks** | | |
| Collusive price agreements | 2 | Small number of producers but over-supply and the availability of substitutes tempers producers' greed |
| Embargoes | 1 | Unlikely except as part of ban on high technology exports to Communist or non-approved regimes |

*These components are assessed on a scale from 1 (lowest) to 10 (highest) risk. Grouped together, they form the basis for the assessments of the factors for likelihood of supply disruption, and for cost of such disruption which appear in the first table in this section.*

---

## WORLD ZIRCONIUM RESERVES

Total = 48 Million Tons

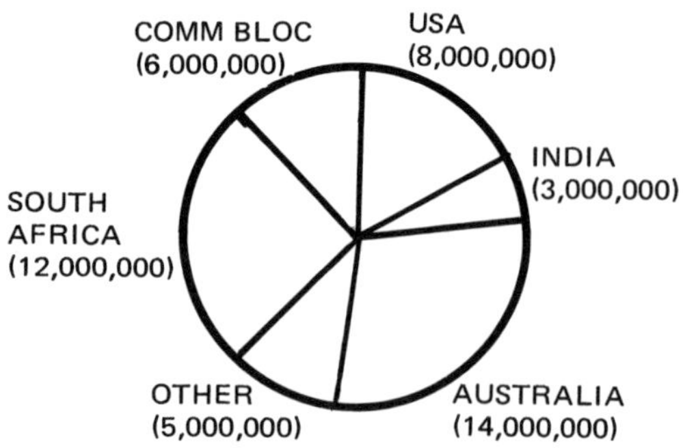


*Source: US Bureau of Mines, 1982*

## US ZIRCONIUM CONSUMPTION

By Market Sector 1981

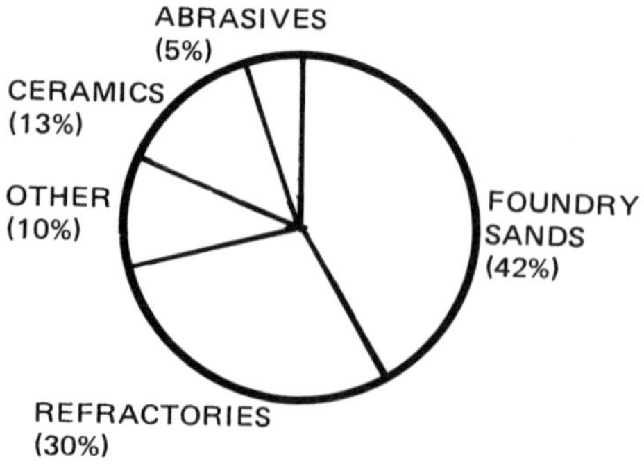


*Source: US Bureau of Mines, 1982*

## APPLICATIONS

### Consumption Trends

The USA consumes 25-30 per cent of all zirconium with an annual demand level of 170,000 tonnes zircon. The USBM expects consumption to increase by four per cent a year during the 1980s. 90 per cent of all zirconium is used as zircon and zirconia (zirconium dioxide).

### The Principal Markets

These are based on the heat and thermal shock resistance of zirconium ceramics. Zircon provides high quality foundry sands and long-life refractory materials for steel production. Zirconia is a useful ceramic for the electronics industry and as an abrasive polishing compound for metal components and lenses. Zirconium metal is an important, but minor, alloying element in steels and magnesium/aluminium alloys. Extra-purity zirconium, with the one per cent co-product hafnium removed, is used as zircalloy for nuclear fuel cans in water and steam cooled reactors.

### Recycling

Insignificant, it is not viable to repurify used refractories or foundry sands.

### Substitution

Chromite and aluminium silicates may be substituted in foundry applications, chromite/magnesite refractories provide slightly inferior substitutes.

## INTERNATIONAL TRADE AND WORLD PRICES

### Supply Arrangements

The free market is minimal as the few manufacturers tend to sell directly to consumers. Price competition is not therefore intense. It is normally a quiet market but it did peak in 1975 coinciding with wide interest in the field of nuclear energy. There is now a marked over-supply. Zirconium is traded either as zircon sand with a 66-76 per cent zirconia content and specified impurity content or as zirconium oxide of ceramic grade. Prices are generally quoted in $/ton. Zirconium metal of high purity is sold on a controlled basis to the nuclear or specialised alloy industries but their specifications are stringent and often deemed proprietary.

### Companies

Major producers include the state-owned companies of the Communist bloc; Associated Minerals, Allied Eneabba and Consolidated Rutile of Australia; Richards Bay Minerals of South Africa; and E I du Pont de Nemours of the USA. The major zirconium sponge producers are Teledyne Wah Chang Albany Corporation, Western Zirconium of the USA; PUK of France; and Zirconium Industries and Nippon Mining Co of Japan.

### Prices

Zirconium prices peaked in the wake of the oil crisis and the upsurge in demand for nuclear power plants using zirconium fuel cans. Prices declined and then stabilised as the move towards nuclear energy slowed. Prices are unlikely to recover substantially within the next five years given the ready availability of supplies.

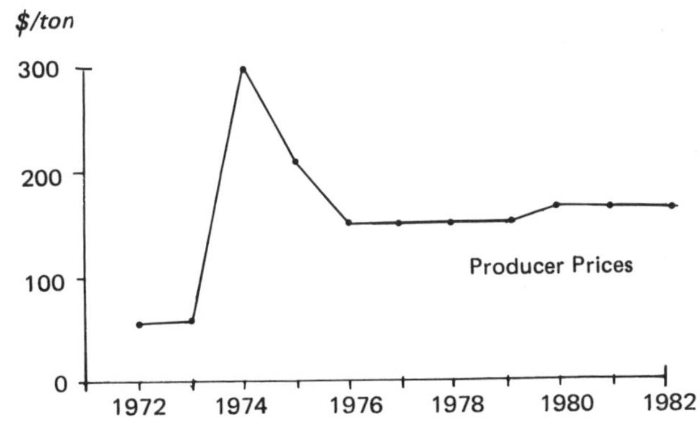


**Zircon (zirconium silicate) Prices—
Annual Averages**

# SECTION THREE:

# THE COUNTRIES

# AUSTRALIA

## COUNTRY RISK FACTOR 4.6

*This factor is assessed from 1.0 (lowest) to 10.0 (highest) risk. For components of this assessment see later table in this section.*

Australia's development as a leading minerals producer and exporter has largely come about since 1960. Production is concentrated on high volume raw materials such as bauxite—of which it produces almost a third of the world supply—coal, manganese, lead, zinc, copper, iron ore, titanium sands and nickel. It is also an important source of gold, tin, tungsten, cadmium and antimony. Large uranium reserves have been proved but their development has been fraught with political and environmental constraints.

### MINERAL PRODUCTION

The major developments since 1960 have been in iron ore, where it now outclasses the USA; bauxite, producing about 30 per cent of non-Communist world output; and nickel where it ranks joint second in production terms with New Caledonia. Unless world energy patterns change again, coal will probably dominate as the chief export by the end of the century. Production of traditional raw materials such as copper, lead and zinc have undergone major expansion alongside the more strategically sensitive tin, tungsten and titanium sands.

Massive new reserves of a wide range of minerals continue to be discovered but their development and financing are becoming increasingly bound up with political considerations. The uranium industry's failure to develop during the energy crises of the 1970s is an extreme but relevant case.

### THE MINING INDUSTRY

Of the dozen or so major companies which provide the framework of Australia's minerals industry, over half are either majority owned by multinationals or have substantial foreign shareholdings. British companies such as RTZ and Consolidated Gold Fields provided the impetus for early development but the build-up of the bauxite and aluminium industry brought in North American and European companies such as Alcoa and Alusuisse. Strengthening trade links have also attracted some Japanese equity. The industry is highly export-oriented and being so firmly attached to the high-volume sector provides the obvious basis for developing the infrastructure of the country. The large land mass and small population density ensure that mining will remain the most important commercial sector domestically whilst the broad base of mineral resources will require the consuming world to continue to encourage its development.

### INVESTMENT AND DEVELOPMENT

Like Canada, Australia has a totally inadequate capital resources base to fund its mineral development programme. Its need for foreign investment creates some conflict within a stated policy of local control of resource ventures. The investment climate has also been chilled by world recession, falling prices and a marked deterioration in labour relations. Political differences at State level have also had an effect, prompting the deferral, or even mid-stream cancellation, of major projects, particularly in the high-risk aluminium sector.

Geologically, the country is a prime investment target. Massive, sub-surface deposits of excellent quality bauxite, coal, copper, uranium and even gold and diamonds continue to be discovered. The large landmass dictates that infrastructure costs are daunting but even these are not the prime consideration. The ability to attract development capital rests upon these factors: political stability, particularly with respect to the rules governing foreign investment; the Japanese connection; competition with other sources in the non-Japanese sector, and, labour relations within Australia.

## COMPONENTS OF RISK ASSESSMENT

| *Factor* | *Rating* | *Comment* |
|---|---|---|
| **Labour** | | |
| Strike incidence | 6 | Volatile in ports and iron ore mining |
| Quality of labour | 3 | High level of skill, local and immigrant |
| General unrest | 3 | Low |
| **Politics** | | |
| History | 3 | English based, stable |
| Stability | 6 | Probable constitutional change soon |
| External dangers | 3 | Low, strongly pro-Western |
| **Location** | | |
| Hostile borders | 5 | Practically indefensible but no foreseeable dangers |
| Critical sea routes | 2 | None locally |
| **Mineral Resources** | | |
| Adequacy of reserves | 3 | Excellent and expanding |
| Costs of production | 7 | Rising abnormally in the labour sector |
| Development appeal | 6 | Good in the long run but difficult near-term |
| **Financial and Economic** | | |
| Currency performance | 5 | Subjected to marginal pressures only |
| Vulnerability to manipulation | 5 | Not a major constraint |
| External indebtedness | 5 | Containable |
| Productivity | 7 | Vulnerable in spite of excellent geology |
| Foreign ownerships | 8 | High but now subject to restriction |
| **Energy and Other Natural Resources** | | |
| Domestic energy sources | 6 | Still weak in oil and gas but improving |
| Domestic energy production | 7 | Expanding but vulnerable in oil |
| Sources of outside supplies | 2 | Not necessary but long term oil reserves are in doubt |
| **Dependence on Foreign Natural Resources** | 2 | Low |

*These components are assessed on a scale from 1 (lowest) to 10 (highest) risk. Grouped together, they form the basis for the country risk factor assessment opposite.*

## OUTPUT AND RATINGS OF PRINCIPAL MINERALS

| *Mineral* | *Output (Tonnes)* *1981* | *(1971)* | *Percentage of Western World Output* | *Western World Rating* | *Mine (M)/ Smelter (S)* | *Major Producers* |
|---|---|---|---|---|---|---|
| Aluminium | 379,000 | (246,000) | 3.2 | 8 | S | Alcoa, Comalco, Alcan, Nabalco |
| Antimony | 1,130 | 1,149 | 2.7 | 8 | M | New England Antimony Mines |
| Asbestos | 80,000 | (1,000) | 4.2 | 6 | M | Woodsreef Mines Ltd |
| Bauxite | 25,541,000 | (12,732,000) | 33.7 | 1 | M | Alcoa, Comalco, Alcan, CRA, Nabalco |
| Bismuth | 900 | – | 31.5 | 1 | S | Peko-Wallsend |
| Cadmium | 1,027 | (558) | 7.7 | 6 | S | E Z Industries |
| Copper | 223,000 | (177,000) | 3.4 | 8 | M | MIM, Renison Goldfields, CRA |
| Gold | 16 | (21) | 1.7 | 8 | M | MIM, Western Mining, Central Norseman, Peko-Wallsend, Telfer (BHP/ Newmont) |
| Iron Ore | 86,000,000 | (55,000,000) | 16.0 | 2 | M | Hamersley, Mount Newman, Cliffs Western, Goldsworthy |
| Lead | 393,000 | (403,000) | 15.9 | 2 | M | |
| Manganese | 2,200,000 | (1,000,000) | 13.9 | 3 | M | BHP |
| Nickel | 74,500 | (31,100) | 15.2 | Joint 2 | M | Western Mining, Agnew (Seltrust/MIM), Greenvale (Metals Exploration/Freeport) |
| Silver | 777,000 | (672,000) | 9.3 | 5 | M | |
| Tin | 12,100 | (9,400) | 6.0 | 5 | M | Renison Goldfields, Cleveland Aberfoyle |
| Titanium (Sands) | 1,625 | (1,130) | 35.5 | 1 | M | Renison Goldfields, Cons. Rutile, Allied Eneabba, Westralian Sands |
| Tungsten | 3,500 | (1,350) | 12.7 | Joint 1 | M | Peko-Wallsend |
| Uranium | 2,861 | – | 6.7 | 5 | M | Mary-Kathleen (WMC/BP) |
| Zinc | 504,000 | (453,000) | 11.4 | Joint 2 | M | MIM, EZ, CRA |

Other mineral production: diamonds, cobalt

# BOLIVIA

## COUNTRY RISK FACTOR — 6.6

*This factor is assessed from 1.0 (lowest) to 10.0 (highest) risk. For components of this assessment see later table in this section.*

The relatively high risk rating of 6.6 would be even higher were it not for good energy resources and a low risk of external hostilities. Domestically there are serious political, economic and social problems and a high degree of dependence on mineral exports. Because of the difficult terrain and lack of infrastructure, mining costs are high. Tin, in which Bolivia ranks fourth in world production terms, is the mainstay of both the mining industry and the economy, with supporting output of silver, tungsten, lead, zinc and copper. As a mining country it faces severe problems associated with a high level of external debt, a glaring lack of infrastructure, an uncompetitive level of operating costs and a weakened market. Foreign involvement in the mining industry is being encouraged but the lack of a stable political base remains a major disincentive.

### MINERAL PRODUCTION

A mixture of state and private interests control the industry, the major state company, COMIBOL, producing about two-thirds of the total output. The private sector includes local companies, overseas interests and minor cooperatives. Broad planning guidelines flow down from the Ministry of Mines and Metallurgy but the above mentioned problems preclude any immediate hopes for changes in direction or expansion of the industry.

### THE MINING INDUSTRY

COMIBOL controls over 90 mines covering the full range of produced minerals. With the exception of tungsten all show a declining trend. The state organisation ENAF operates a large tin smelter, purchases the majority of Bolivia's tin and markets the refined metal worldwide. Medium sized mining companies in the private sector control 20-25 operations employing about 8,000 people in total. Small mines account for about 20,000 people but are, on the whole, badly organised and lacking operating capital.

### INVESTMENT AND DEVELOPMENT

The USA shows great financial concern and interest in Bolivia, as a strategically placed country and a source of critical imports. The country's political instability and high external debt level is manifest in frequent changes in mining policy, particularly with regard to taxation. Exploration incentives are not particularly attractive but a presence is maintained by US, West German and Belgian interests. Bolivia has only a limited road infrastructure and general facilities.

## OUTPUT AND RATINGS OF PRINCIPAL MINERALS

| *Mineral* | *Output (Tonnes) 1981* | *(1971)* | *Percentage of Western World Output* | *Western World Rating* | *Mine (M)/ Smelter (S)* | *Major Producers* |
|---|---|---|---|---|---|---|
| Antimony | 15,296 | (12,000) | 36.4 | 1 | M | COMIBOL, EMUSA |
| Silver | 193 | (211) | 2.3 | 8 | M | COMIBOL, COMSUR |
| Tin | 29,800 | (30,300) | 14.8 | 4 | M | COMIBOL, COMSUR |
| Tungsten | 3,200 | (1,900) | 11.4 | Joint 1 | M | COMIBOL, COMSUR |
| Zinc | 47,000 | (46,000) | 1.1 | – | M | COMIBOL, COMSUR |

Other mineral production: bismuth, cadmium, copper, lead

COMIBOL (Corporacion Minera de Bolivia)
COMSUR (Cia. Minera del Sur): grouping of medium sized mining companies

## COMPONENTS OF RISK ASSESSMENT

| *Factor* | *Rating* | *Comment* |
|---|---|---|
| **Labour** | | |
| Strike incidence | 7 | Medium-high |
| Quality of labour | 5 | Adequate for type of industry |
| General unrest | 8 | High |
| **Politics** | | |
| History | 9 | Troubled |
| Stability | 9 | Extremely volatile |
| External dangers | 4 | Not high |
| **Location** | | |
| Hostile borders | 4 | Not high |
| Critical land/sea routes | 8 | Very poor infrastructure, single smelting area |
| **Mineral Resources** | | |
| Adequacy | 6 | Medium |
| Costs of production | 9 | High |
| Development appeal | 9 | Low |
| **Financial and Economic** | | |
| Currency performance | 8 | Poor |
| Vulnerability to manipulation | 8 | High |
| External indebtedness | 9 | High |
| Productivity | 8 | Low, aggravated by geological and geographic conditions |
| Foreign ownerships | 3 | Low |
| **Energy and Other Natural Resources** | | |
| Domestic energy sources | 4 | Not rich in indigenous energy sources |
| Domestic energy production | 4 | |
| Sources of outside supplies | 4 | Low per capita usage; so little need to import energy |
| **Dependence on Foreign Natural Resources** | 5 | Few used domestically |

***These components are assessed on a scale from 1 (lowest) to 10 (highest) risk. Grouped together, they form the basis for the country risk factor assessment opposite.***

# BOTSWANA

**COUNTRY RISK FACTOR** **5.9**

*This factor is assessed from 1.0 (lowest) to 10.0 (highest) risk. For components of this assessment see later table in this section.*

Since independence in 1966 Botswana's performance has outstripped most other African states, but from a very low base. GNP rose from $69 to $1,000 per capita between 1966 and 1981 largely as a result of the development of the mining industry. There are three sectors: coal, copper-nickel-cobalt and diamonds, of which diamonds represents about 60 per cent of total exports and has the potential to retain a place in the top three producers worldwide. The recession in diamonds has had a marked and adverse effect on the country's performance, about which little can be done in the short term although a corrective budget has been introduced.

The large landmass and small population present a major problem. The country is half that of South Africa in size but the population only 800,000. Infrastructure costs are high and development most suited to small volume, high value minerals such as diamonds. Transport is a particular problem whilst in both access and financial stability Botswana has a long term dependence on South Africa. This is heightened by the political problems of neighbouring Angola and Namibia, both of which act as a deterrent to any move away from the close links with South Africa.

### MINERAL PRODUCTION

Although the development of the copper-nickel-cobalt mine at Selebi-Pikwe in the late 1960s was a financial disaster it provided infrastructure and laid the basis of a mining policy. There is evidence of sufficient ore reserves in the immediate area to maintain output, which has given Botswana a world rating in both nickel and cobalt production. The associated development of Morupule Colliery has tapped only a fringe of the estimated 20,000 million tonnes of coal reserves, providing around 400,000 tonnes a year to supply the nickel facility and local electricity generation. Diamonds are the most immediately promising mineral. In partnership with De Beers, two of a swarm of kimberlite pipes in the Orapa-Jwaneng areas have been developed to bring capacity to five million carats a year, putting Botswana into fourth position in the world rating. A high proportion is gemstone and open-cast working is also possible. But given its reliance on a single mineral the country must be classed in the same difficult category as Chile and Zambia although its domestic industry is well placed for recovery.

### THE MINING INDUSTRY

South African influence is overriding in the presence of De Beers in diamonds and Anglo American Corporation in

## OUTPUT AND RATINGS OF PRINCIPAL MINERALS

| *Mineral* | *Output (Tonnes) 1981* | *(1971)* | *Percentage of Western World Output* | *Western World Rating* | *Mine (M)/ Smelter (S)* | *Major Producers* |
|---|---|---|---|---|---|---|
| Cobalt | 280 | Nil | 1.2 | – | M | Botswana RST |
| Diamonds (million carats) | 5,101 | 1.0 | 16.2 | 3 | M | De Beers Botswana Mining Co |
| Nickel | 18,300 | Nil | 3.7 | 8 | M | Botswana RST |

Other mineral production: copper, coal

## COMPONENTS OF RISK ASSESSMENT

| *Factor* | *Rating* | *Comment* |
|---|---|---|
| **Labour** | | |
| Strike incidence | 3 | Low |
| Quality of labour | 5 | Good supply of unskilled, attractive to expatriates |
| General unrest | 5 | Low, but small population base gives rise to factions |
| **Politics** | | |
| History | 4 | Strong, family-style leadership since independence |
| Stability | 5 | Dependence on South Africa adds to stability |
| External dangers | 7 | Proximity to Angola and Namibia is a threat |
| **Mineral Resources** | | |
| Adequacy of reserves | 5 | Low, excepting diamonds |
| Costs of production | 5 | Competitive |
| Development appeal | 7 | Government rules hardening, location remote, infrastructure poor |
| **Location** | | |
| Hostile borders | 7 | Namibia/Angola could abuse neutrality |
| Critical sea/land routes | 8 | Dependent on South Africa |
| **Financial and Economic** | | |
| Currency performance | 7 | Strong links to the South African rand |
| Vulnerability to manipulation | 7 | Dependent on the rand |
| External indebtedness | 7 | Growing as diamond sector stays in recession |
| Productivity | 4 | Excellent in diamond sector |
| Foreign ownerships | 6 | High, but government code, if held, could be beneficial |
| **Energy and Other Natural Resources** | | |
| Domestic energy sources | 5 | Only coal known, but very large reserves |
| Domestic energy production | 4 | Self sufficient in coal with scope for exports |
| Sources of outside supplies | 7 | Routes through South Africa could present political embargoes |
| **Dependence on Foreign Natural Resources** | 9 | High |

***These components are assessed on a scale from 1 (lowest) to 10 (highest) risk. Grouped together, they form the basis for the country risk factor assessment opposite.***

coal and copper-nickel. The Sebebi-Pikwe venture also includes, as a founding partner, Amax of the USA through the holding company, Botswana RST. From inception the copper-nickel matte was purchased by Amax for refining in its Louisville, USA, smelter, but the contract hit major snags which resulted in some 25 per cent being shipped to Zimbabwe for treatment. Coal represents a big export potential. Shell Coal is in partnership with the government on a mine planned near Morupule in the mid 1980's.

### INVESTMENT AND DEVELOPMENT

The investment climate is better than for many other African states, the major problems being external in the form of Namibia's state of hostilities, the economic dependence on South Africa and the lack of infrastructure. These will preclude rapid further development but a return to health in the potentially very profitable diamond sector would enable the labour generating mass industries such as coal and copper-nickel to be expanded.

# BRAZIL

**COUNTRY RISK FACTOR** **6.0**

*This factor is assessed from 1.0 (lowest) to 10.0 (highest) risk. For components of this assessment see later table in this section.*

Whilst the minerals industry in all forms accounts for less than 20 per cent of the country's GNP it is of growing economic importance to Brazil and of strategic interest to the world markets. Brazil's political system remains relatively stable whilst a liberal foreign investment regime has encouraged major expansion in the mining sector. If the hydro-potential from the major river systems is realised, then the large-scale development of a major minerals industry through to refining and fabrication could be assured. As an investment haven, Brazil has promised much in the past decade but has not yet delivered. Domestic inflation and infrastructure costs are daunting and there is a wide imbalance in domestic living standards. Retaining political stability should ensure a high level of Western investment, irrespective of the less certain mineral potential of some regions.

## MINERAL PRODUCTION

Brazil has eight minerals in the top ten world production tables with vast, high grade surface deposits of iron ore, bauxite, manganese and nickel. The country has very large gold resources and is strong in tin, tantalum and asbestos. Output growth since 1971 has been exceptional.

Firm plans exist to increase output within market limits in all existing areas but with a major thrust into gold, alumina, aluminium and energy. These plans will hinge not on the availability of ore reserves but on the ability of the financial systems to cope with an increasing foreign debt burden and to continue to attract foreign capital.

## THE MINING INDUSTRY

The industry is characterised by a mix of government control, private local ownership, foreign company interests and joint ventures as the result of several major investment policy changes since 1960. The scale of development now favours joint-venture operations, particularly in the bauxite, alumina and aluminium sectors. Several of the

## OUTPUT AND RATINGS OF PRINCIPAL MINERALS

| *Mineral* | *Output (Tonnes)* 1981 | (1971) | *Percentage of Western World Output* | *Western World Rating* | *Mine (M)/ Smelter (S)* | *Major Producers* |
|---|---|---|---|---|---|---|
| Aluminium | 256,000 | (81,000) | 2.1 | — | S | ALBRAS, ALCAN, CVRD |
| Asbestos | 150,000 | Nil | 7.9 | 4 | M | Eternit |
| Bauxite | 4,662,000 | (566,000) | 6.1 | 4 | M | ALBRAS, ALCAN, CVRD |
| Beryllium | 35 | (42) | 10.0 | 2 | M | Various |
| Columbium ore | 14,500 | (5,858) | 86.9 | 1 | M | CBMM |
| Gold | 35 | (9) | 3.6 | 4 | M | Mineracao Morro Velho (Anglo American) |
| Iron Ore | 102,000,000 | (32,500,000) | 19.2 | 1 | M | CVRD, MBR |
| Manganese ore | 2,300,000 | (1,950,000) | 14.5 | 2 | M | CVRD |
| Nickel | 2,300 | (2,800) | <1.0 | — | M | CODEMIN/BRASIMET |
| Silver | 24 | (19) | <1.0 | — | M | Various |
| Tantalum | 114 | (383) | 25.5 | joint 1 | M | Various |
| Tin | 7,300 | (3,500) | 3.6 | 6 | M | Paranapanema, Brascan/Cesbra, Brumadinko |
| Zinc | 67,000 | (17,000) | 1.6 | — | M | Various |

CBMM (Companhia Brasiliera de Metallurgia e Mineracao)
CVRD (Companhia Vale do Rio Doce)
Cesbra (Companhia Estanifera do Brasil)

## COMPONENTS OF RISK ASSESSMENT

| *Factor* | *Rating* | *Comment* |
|---|---|---|
| **Labour** | | |
| Strike incidence | 5 | Average performance to date |
| Quality of labour | 7 | Improving with industrial expansion |
| General unrest | 5 | Medium |
| **Politics** | | |
| History | 7 | Troubled but improved in recent years |
| Stability | 4 | Good with recent improvements |
| External dangers | 4 | Low |
| **Location** | | |
| Hostile borders | 3 | No foreseeable problems |
| Critical land/sea routes | 7 | Relies totally on Atlantic sea routes |
| **Mineral Resources** | | |
| Adequacy of reserves | 3 | Very high in most minerals |
| Cost of production | 6 | Economies of scale but high in infrastructure |
| Development appeal | 6 | Currently excellent but changeable |
| **Financial and Economic** | | |
| Currency performance | 8 | Poor history but slowly improving |
| Vulnerability to manipulation | 8 | High |
| External indebtedness | 9 | Very high |
| Productivity | 5 | Good |
| Foreign ownerships | 7 | Growing dependence |
| **Energy and Other Natural Resources** | | |
| Domestic energy sources | 7 | Inadequate but improving. Vulnerable in oil |
| Domestic energy production | 5 | Growing |
| Sources of outside supply | 5 | Politically safe |
| **Dependence on Foreign Natural Resources** | 5 | Medium and improving |

***These components are assessed on a scale from 1 (lowest) to 10 (highest) risk. Grouped together, they form the basis for the country risk factor assessment opposite.***

largest Western companies are deeply involved in the Brazilian minerals industry, including Alcan, Alcoa, Shell, BP, Billiton, VAW, and Reynolds.

### INVESTMENT AND DEVELOPMENT

Brazil is attempting to maintain a very high rate of mining expansion, coupled with establishing daunting infrastructure requirements whilst struggling to control a rising foreign debt burden which now has a service ratio above 55 per cent. Investment incentives are particularly favourable in the north and north-east and for those minerals in which Brazil still has a production deficit e.g. copper, lead, nickel, zinc and energy. The remittance-of-profits policy allows wide scope but encourages reinvestment in Brazil.

# CANADA

---

**COUNTRY RISK FACTOR** **4.4**

*This factor is assessed from 1.0 (lowest) to 10.0 (highest) risk. For components of this assessment see later table in this section.*

---

Canada is the world's leading exporter of minerals in value terms and is the foremost producer of nickel, zinc, and asbestos. It is also a major supplier of aluminium, lead, silver and copper together with by-products including cadmium, cobalt, selenium and platinum. The country is deficient in chromium, bauxite, manganese and tin. Gold output is second only to South Africa and production is currently being expanded from just under 50 to over 60 tonnes a year.

Its unusual natural advantages of long term reserves of a wide range of economic minerals are assisted by an abundance of energy resources. Offsetting these are historic political problems, a recent history of poor labour relations and a current economic crisis in which high unemployment, inflation, poor industrial demand and the 'Canadianisation' of natural resource projects are dampening investment appeal. If the politico-economic climate can be improved the risk-rating could drop towards 3.0-3.5.

## MINERAL PRODUCTION

Canada has retained its leading position in the production of most major minerals over the last ten years remaining first in nickel and zinc, second in aluminium and third in copper and lead. It lost ground in its share of world output as a result of new production coming onstream elsewhere in aluminium and nickel and through falling domestic output in lead and zinc. It moved ahead in copper through the large scale development of resources in British Columbia.

Proven, untapped reserves of practically the full range of produced minerals could be developed to correct this trend. Emphasis on the production of high-technology minerals and precious metals is encouraging. The expansion of primary aluminium output through the greater application of hydro-electric power in British Columbia and Quebec is envisaged but similar plans are also being made in other parts of the world.

## THE MINING INDUSTRY

The present structure of the mining industry, dominated by about a dozen groups, dates back to the mergers and acquisitions at the turn of the century. Fully-owned Canadian concerns, such as Noranda and Inco, and foreign dominated companies, such as Rio Algom are both to be found in this group.

Left wing successes in the provincial elections of the early 1970s were soon followed by punitive tax regimes and the expropriation of mining assets in British Columbia, Saskatchewan, Newfoundland and Quebec. The Canadian Development Corporation was formed in 1972 by government initiative to inject major investment into mining operations with the intention that ownership would eventually pass to Canadian institutions and individuals. The venture has been relatively successful with investments including 34.9 per cent of Texasgulf Inc and several wholly owned ventures in oil and gas.

## INVESTMENT AND DEVELOPMENT

Small company participation is encouraged and mining stocks are amongst the most actively traded on the four exchanges of Montreal, Winnipeg, Toronto and Vancouver. Capital development costs have been a major inhibiting factor and the investment climate has deteriorated seriously since 1975.

With inflation remaining stubbornly in double figures the Canadian dollar has slipped below the rate of 80 cents to the US dollar. Unemployment, even in the mining sector is high.

The volume of known, but unexploited, mineral resources is great, complementing long life reserves at most of the major mines. The outlook is unsettled with competition intense in many minerals from other world suppliers and transport problems, already considerable through climate and distance, worsened by frequent labour disputes.

Nevertheless Canada remains the only assured non-South African source of a wide range of key minerals such as asbestos and the platinum group metals.

## COMPONENTS OF RISK ASSESSMENT

| *Factor* | *Rating* | *Comment* |
|---|---|---|
| **Labour** | | |
| Strike incidence | 7 | High by Western standards |
| Quality of labour | 3 | High proportion of skilled workers |
| General unrest | 6 | Some interprovincial, largely ethnic differences |
| **Politics** | | |
| History | 6 | Some French/English conflict |
| Stability | 6 | Danger of provincial fragmentation |
| External dangers | 2 | Low, close to US border |
| **Location** | | |
| Hostile borders | 2 | Low, bordering USA |
| Critical land/sea routes | 1 | None |
| **Mineral Resources** | | |
| Adequacy of reserves | 3 | Excellent |
| Cost of production | 7 | High by Western standards |
| Development appeal | 5 | Medium, some history of political interference |
| **Financial and Economic** | | |
| Currency performance | 7 | Very poor against the US dollar |
| Vulnerability to manipulation | 7 | High |
| External indebtedness | 6 | Medium |
| Productivity | 6 | Medium |
| Foreign ownerships | 8 | High |
| **Energy and Other Natural Resources** | | |
| Domestic energy sources | 1 | High in all sectors |
| Domestic energy production | 2 | Good, although recession has affected the oil-dependent Western provinces |
| Sources of outside supply | 1 | Not necessary |
| **Dependence on Foreign Natural Resources** | 2 | Low |

***These components are assessed on a scale from 1 (lowest) to 10 (highest) risk. Grouped together, they form the basis for the country risk factor assessment opposite.***

**OUTPUT AND RATINGS OF PRINCIPAL MINERALS**

| *Mineral* | *Output (Tonnes) 1981* | *(1971)* | *Percentage of Western World Output* | *Western World Rating* | *Mine (M)/ Smelter (S)* | *Major Producers* |
|---|---|---|---|---|---|---|
| Aluminium | 1,125,000 | (102,000) | 9.0 | 2 | S | Aluminium Co of Canada (ALCAN), Canadian Reynolds |
| Antimony | 1,670 | (150) | 4.4 | 5 | M | Consolidated Durham Mines |
| Asbestos | 1,335,000 | (1,489,000) | | | M | Asbestos Corp, Bell Asbestos Carey, Johns-Manville |
| Bismuth | 180 | (121) | 7.4 | 5 | S | Cominco Ltd, Hudson Bay M&S Ltd, Noranda, Texasgulf |
| Cadmium | 1,293 | (1,875) | 9.7 | 2 | S | Noranda |
| Cobalt | 1,361 | (2,265) | 5.5 | Joint 5 | M/S | Inco, Falconbridge, Sherritt Gordon |
| Columbium (pentoxide) | 2,000 | (988) | 11.4 | 2 | M | Molycorp/Kennecott |
| Copper | 693,000 | (648,000) | 10.7 | 3 | M | Inco, Noranda, Texasgulf, Rio Algom |
| Gold | 49.5 | (70) | 5.1 | 2 | M/S | Dome Mines, Pamour, Kerr-Addison |
| Iron Ore | 47,000,000 | (43,000,000) | 8.9 | 4 | M | IOC, Quebec Cartier, Wabush |
| Lead | 332,000 | (369,000) | 13.5 | 3 | M | Cominco, Brunswick M&S |
| Molybdenum | 10,900 | (12,085) | 12.1 | 3 | M | Bethlehem, Brenda, Gibraltar, Lornex, Noranda, Placer, Teck Utah |
| Nickel | 155,000 | (266,800) | 31.6 | 1 | M | Inco Ltd, Falconbridge Ltd, Sherritt Gordon Ltd |
| Platinum Group | 10.9 | (14.6) | 9.9 | 2 | M | Inco Ltd, Falconbridge Ltd, Sherritt Gordon Ltd |
| Selenium | 363 | (313) | 25.4 | 2 | S | Noranda |
| Silver | 1,204 | (1,398) | 14.7 | 3 | M | Noranda, Cominco, Inco |
| Tellurium | 45 | (18.6) | 15.3 | Joint 3 | S | Noranda |
| Tin | 264 | (133) | 0.1 | — | M/S | Various |
| Titanium (ilmenite) | 960,000 | N/A | 20.0 | 2 | M/S | Quebec Iron and Titanium |
| Tungsten | 2,725 | N/A | 9.9 | Joint 4 | M | Canada Tungsten |
| Uranium ($U_3O_8$) | 6,400 | (3,639) | 16.0 | 3 | M | Rio Algom Ltd, Denison Ltd, Eldorado Nuclear Ltd |
| Zinc | 1,096,000 | (1,114,000) | 24.7 | 1 | M | Canadian EZ, Cominco, Texasgulf, Hudson Bay M&S |

Also produces tantalum, silicon

# CHILE

**COUNTRY RISK FACTOR** **5.7**

*This factor is assessed from 1.0 (lowest) to 10.0 (highest) risk. For components of this assessment see later table in this section.*

Chile holds a pre-eminent position in the world copper market, having the largest level of known reserves and being the second largest producer in annual output terms in 1981 and overtaking the USA in 1982. Its high dependence on this mineral, however, underlines its vulnerability although it does hold the number two position in molybdenum and measureable gold output.

The country has undergone a series of political and economic transitions since 1970 and now enjoys one of the few recent success stories amongst South American countries. After a brief flirtation with communism under Allende in the early 1970s, a period which saw the US multinationals dismissed and the copper industry nationalised, Chile is once more firmly under the control of a right-wing military regime. The country's internal politics have earned it enemies in both international political camps but its economic recovery since 1979 has been impressive. The peso is now tied to the US dollar, the annual inflation rate was reduced from a high of 200 per cent to single figures although by late 1982 it was again climbing. The copper industry has not only been maintained in excellent repair but is prepared for expansion.

New incentives for foreign investment have attracted back multinational companies. The emphasis remains on copper but precious metals and fuels are also being actively pursued. The objective of a doubling of copper output to two million tonnes a year by the 1990s demands massive foreign, private sector capital investment which, if achieved, will be to the detriment of competitive sources of copper supply elsewhere.

## MINERAL PRODUCTION

Mass orebodies of surface, low grade porphyry copper occur throughout the Rocky Mountains-Andes formation from Alaska to the tip of South America but it is thought that the majority of total tonnage lies within the Chilean boundaries. The deposits are accompanied by high by-product values of molybdenum with gold and silver. Copper was first developed in Chile by US entrepreneurs such as Braden in the 1880s and, although losing first place in the world production league to the USA by 1900, has never fallen below the second position.

On recovery from the arbitrary nationalisations of the early 1970s, which saw major operators Anaconda and Kennecott thrown out, the administration first corrected the typically South American-style inflation rate and then opened up the return path for foreign investment. By 1982, the state-controlled output of 900,000 tonnes a year copper from Codelco had been supplemented by over 100,000 tonnes from the small mines sector. To this was being added the potential of foreign developments by St Joe Minerals, Getty/Utah, Anaconda and Foote Minerals of the USA, Preussag of Germany and a coal prospect in which Burnett and Hallamshire of the UK

## OUTPUT AND RATINGS OF PRINCIPAL MINERALS

| *Mineral* | *Output (Tonnes) 1981* | *(1971)* | *Percentage of Western World Output* | *Western World Rating* | *Mine (M)/ Smelter (S)* | *Major Producers* |
|---|---|---|---|---|---|---|
| Copper | 1,081,000 | (708,000) | 16.7 | 2 | M | Codelco, Exxon, St Joe Minerals, ENAMI |
| Gold | 11.4 | (1.7) | 1.2 | 12 | M | ENAMI, St Joe |
| Iron Ore | 7,800,000 | (11,000,000) | 1.0 | – | M | CAP |
| Molybdenum | 15,420 | (6,300) | 16.0 | 2 | M | Codelco, Exxon |
| Rhenium metal | 3.6 | N/A | 31.4 | Joint 1 | M/S | Codelco |
| Selenium | 9.0 | N/A | 1.0 | – | S | |

Other mineral production: coal, selenium
CODELCO (Corporacion Nacional del Cobre de Chile) state owned
ENAMI (Empresa Nacional de Minera) state owned

## COMPONENTS OF RISK ASSESSMENT

| *Factor* | *Rating* | *Comment* |
|---|---|---|
| **Labour** | | |
| Strike incidence | 6 | Recurrent but strict labour laws pertain |
| Quality of labour | 7 | Adequate in mining sector |
| General unrest | 6 | Long history of problems but currently steady |
| **Politics** | | |
| History | 8 | Troubled, predominantly right-wing |
| Stability | 6 | Currently well controlled but history of overturn |
| External dangers | 5 | Difficulties with Argentina |
| **Location** | | |
| Hostile borders | 5 | Disputes with Argentina, particularly long borders |
| Critical land/sea routes | 6 | Cape Horn and Panama canal for most exports |
| **Mineral Resources** | | |
| Adequacy of reserves | 4 | Very large in narrow range of minerals |
| Cost of production | 6 | Containable |
| Development appeal | 5 | Growing |
| **Financial and Economic** | | |
| Currency performance | 4 | Massive improvements since 1980 |
| Vulnerability to manipulation | 8 | High with copper exports priced in $US or £ sterling |
| External indebtedness | 4 | Reducing, exchange rate fixed to $US |
| Productivity | 5 | Good |
| Foreign ownerships | 4 | Encouraged again with reasonable formulae |
| **Energy and Other Natural Resources** | | |
| Domestic energy sources | 8 | Lacking oil, gas |
| Domestic energy production | 7 | Coal now being stepped up |
| Sources of outside supply | 4 | Wide choice, few restrictions of embargoes |
| **Dependence on Foreign Natural Resources** | 6 | The range of key minerals produced is small |

***These components are assessed on a scale from 1 (lowest) to 10 (highest) risk. Grouped together, they form the basis for the country risk factor assessment opposite.***

have an interest. Foote is investigating a major lithium project which it is thought could contribute 20 per cent of world output after 1984.

### THE MINING INDUSTRY

The state organisation, Codelco, controls the large mines taken over at nationalisation. In 1981 they produced 893,611 tonnes of copper and 15,360 tonnes of molybdenum. The balance, a recorded 188,000 tonnes of copper, was accounted for by medium and small mines part state and part privately owned. The new mining laws, aimed at attracting the foreign expertise and capital necessary to double the size of the industry in the next 10 to 15 years, guarantees financial and legal protection to concessionaires even in the event of expropriation. Critics of the South American scene may treat this lightly but it represents, as in the case of countries such as Indonesia, an attempt to accommodate the reality of the world market place and its increasing competitiveness.

**Major Prospecting and Development Targets in Chile 1982**

| *Name* | *Investor* | *Metal* | *Ore Reserves (Million Tonnes)* | *Possible Output (Tonnes p.a.)* |
|---|---|---|---|---|
| El Indio (on stream in 1982) | St Joe Minerals (USA) | Cu, Au, Ag | 3.4 | 50,000 Copper |
| Los Bronces | Exxon (USA) | Cu | 1,000 | 100,000 Copper |
| La Escondida | Getty/Utah (USA) | Cu | 200 | 100,000 Copper |
| Quebrada Blanca | Superior Oil (Canada) | Cu | N/A | N/A |
| Pelambres | Anaconda (USA) | Cu | 428 | 100,000 Copper |
| Atacama | Foote Minerals (USA) | Lithium | – | 1,000 Lithium |
| Pecket | Burnett (UK)+Chileans | Coal | 50 | 500,000 Coal |

## INVESTMENT AND DEVELOPMENT

The turnaround in Chile's economic performance and investment outlook caught many observers by surprise. US companies moved back as soon as possible after the end of the Allende regime in 1973 and their enthusiasm must be of concern to Chile's competing neighbours Brazil, Peru and Bolivia. Regulations now allow for the remittance of net profits, based on a 50 per cent tax rate after allowances, and compensation for loss of earnings and investment capital in the case of expropriation. Major prospects and developments are listed in the table.

Perhaps the most significant feature of Chile's mining plans, aided by its recent performance, is that if fulfilled they will seriously affect the outlook for such copper dependent nations as Peru, Zambia, Zaire and Philippines. This will particularly be the case if world economic growth rates continue to fall below expectation.

Events have shown that Central and South American countries have difficulty in maintaining this sort of performance and this is reflected in the risk rating of 5.7 compared to Canada estimated at 4.4 despite its major economic problems.

Chile's ability to build upon its recovery looked less positive by 1983 when its external indebtedness, domestic inflation, unemployment and productivity all deteriorated. Fortunately the problems have been recognised but the major question is whether an ostensibly free-enterprise system can flourish within politico-military boundaries.

# CHINA

## COUNTRY RISK FACTOR

Conditions render risk factor assessments impossible in Communist countries

With a population of around 1,000 million and a landmass roughly equal to that of Canada, China urgently needs to expand its industry in all sectors from mineral exploration through to fabrication. It has long been known to have excellent minerals potential and since the death of long-standing leader Mao-Tse Tung in 1976 has pursued a more outgoing investment policy. Following the 'Gang of Four' purge in 1979-81 and the coincident realisation that the first flush of foreign-investment euphoria had been overdone, the pace of development has slowed.

In reappraisal, long term enthusiasm has not waned and several major Western companies are engaged in both exploration and construction. China's use of the market is also changing, with an increased presence in London, which may detract from the importance of the traditional twice-yearly Canton Fair as a shop-front for China's exports of strategic minerals.

### MINERAL PRODUCTION

Per capita, China's minerals output is still small. It rests on a large coal and iron ore industry but steel production lags behind domestic demand. Many of the steel additives which would be absorbed locally are therefore available for export and include manganese, tungsten and vanadium. Other critical exports are antimony, bauxite, cadmium, mercury, tin, tungsten and titanium. There is a growing deficiency of aluminium, copper, lead, nickel, silver and zinc. A high level of geological exploration has produced claims of some major orebodies in almost the full range of minerals. China has adopted a policy of securing self-sufficiency in energy and basic steel-production. It must also pursue a high level of strategic mineral exports to maintain a foreign exchange balance.

## OUTPUT AND RATINGS OF PRINCIPAL MINERALS

| | *Output (Tonnes)* | | *Percentage of Total World Output* | *Mine (M)/ Smelter (S)* |
|---|---|---|---|---|
| | *1981* | *(1971)* | | |
| Aluminium | 358,000 | (145,000) | 2.3 | S |
| Antimony | 10,000 | (12,000) | 16.5 | M |
| Asbestos | 130,000 | (160,000) | 2.6 | M |
| Bauxite | 1,700,000 | (550,000) | 1.0 | M |
| Cadmium | 130,000 | (145,000) | 1.0 | S |
| Copper | 177,000 | (130,000) | 2.2 | M |
| Iron Ore | 110,000,000 | (44,000,000) | 12.2 | M |
| Lead | 160,000 | (120,000) | 4.4 | – |
| Manganese | 1,700,000 | N/A | 5.9 | M |
| Mercury* | 1,206 | N/A | 18.0 | M |
| Molybdenum | 11,000 | N/A | 1.1 | M |
| Silver | 59,000 | (25,000) | 1.0 | M |
| Tin | 16,000 | (23,000) | 6.8 | M |
| Tungsten | 13,060 | (8,000) | 26.0 | M |
| Vanadium | 5,000 | N/A | 14.0 | M |
| Zinc | 160,000 | (110,000) | 2.6 | M |

*China has long been a supplier of mercury but with a low level of exports.

Other minerals production: chromite, coal, cobalt, tantalum, columbium, gold, platinum.

## THE MINING INDUSTRY

The Ministry of Coal Industry and the Ministry of Metallurgical Industry oversee mining activities and control a large geological field force in parallel with the considerable resources of the Ministry of Geology. Many of the working mines are small but reserves could support world-sized operations. Opencast mining is practised on a large scale in both the ferrous and non-ferrous sectors. Foreign interest following the death of Mao produced an embarrassing rush of Western capital and technology into the minerals industry. The economic base was simply not strong enough to support it and the abrupt halt called to this sudden high level of expansion was considered by some sources to have been overdone. The most advanced projects covered the coal, iron ore, bauxite/aluminium and copper industries.

## INVESTMENT AND DEVELOPMENT

A serious Western multi-national presence remains, particularly through US and Japanese companies but major problems exist. Capital costs have generally been greater than at first assumed, the entire infrastructure is ill-prepared to handle massive expansion and there is a poor base upon which to attract foreign borrowings. The long term availability of Chinese minerals will be essential to world industry but there will be some serious distortions in supply as output is being expanded, the rate of domestic growth varies and economic cycles abroad lead to fluctuating demand.

# CUBA

## COUNTRY RISK FACTOR

Conditions render risk factor assessments impossible in Communist countries

The political U-turn performed by Cuba in 1959, transforming it from a pro-American into a Soviet-Socialist type state under the leadership of Fidel Castro was a severe blow to the USA. The large island lies only 145 kilometres (90 miles) south of key West Florida. It has a strong agricultural base but also holds large reserves of nickel, cobalt, chromium and manganese. Under its present political banner it does not present investment opportunities but sells its minerals on the open market as well as filling two critical Soviet import needs: cobalt and nickel. Some oil is produced, but not enough to cover domestic needs. The economy is massively over-reliant on sugar and depends heavily on development aid from the Soviet Union.

### MINERAL PRODUCTION

The conversion of Cuba to communism created a geographically strategic choke point and a source of critical minerals outside the control of the Western market system. Nickel output at around 40,000 tonnes a year is the sixth largest in the world but has not grown measurably since the early 1970s. Reserves are considered to be amongst the largest in the world. Cobalt is a valued by-product. Accurate figures are not released but in addition to the above metals commercial quantities of manganese and chromite are produced. The Comecon Bloc, particularly the Soviet Union and Poland, are the major customers but Cuban material, notably nickel, still finds its way onto the open market.

### THE MINING INDUSTRY

The mines, all openpit operations in Oriente Province, are the Marti and Pinceres, feeding the Nicaro Smelter and the Yamaniquey and Atlantic which supply the Moa Smelter. The lateritic ores contain on average 1.3 per cent nickel, and the extraction process is heavily energy intensive.

### INVESTMENT AND DEVELOPMENT

Earlier expansion plans saw Cuban nickel output rising to beyond 100,000 tonnes a year on the assumption that Comecon countries, mostly deficient in the ore themselves, would provide an increasing demand. The world downturn, presence of a growing free market surplus and the Soviet need to maintain a balance of trade with the West have tempered this projection. Serious attention is being paid to the establishment of a domestic iron and steel industry, based on the availability of local ores. As with the Soviets, the thrust of overall policy now has to be export trade to maintain growth but the market in both major products—sugar and nickel—is in long term oversupply. In 1983, a series of expansion programmes in the nickel sector were put on ice for several years.

## OUTPUT AND RATINGS OF PRINCIPAL MINERALS

| | *Output (Tonnes) 1981* | *(1971)* | *Percentage of World Output Output* | *Mine (M)/ Smelter (S)* | *Major Producers* |
|---|---|---|---|---|---|
| Cobalt | 1,750* | N/A | 5.9 | M/S | Cubaniquel |
| Nickel | 39,000 | (40,000) | 5.5 | M/S | Cubaniquel |

*Unconfirmed estimate

Other minerals production: copper, chromite, manganese

# FINLAND

**COUNTRY RISK FACTOR** **5.9**

*This factor is assessed from 1.0 (lowest) to 10.0 (highest) risk. For components of this assessment see later table in this section.*

The heavy emphasis on state control and the links with Russia make Finland a difficult country to assess. Attention is currently focused in the minerals field on looking abroad for sources of ore supply to replace decreasing domestic reserves.

## MINERAL PRODUCTION

A useful mix of metals is mined including copper, nickel, cobalt, vanadium, chromite and zinc but a progressive reduction in output levels has been noted in recent years in copper and zinc. Prospecting activity is on the increase to correct a rapidly depleting ore reserve base and nickel appears to have the best prospects.

## THE MINING INDUSTRY

The three major mining companies, Outokumpu Oy, Rautaruukki Oy, and Kemira Oy are state controlled. The one private company is Myllykoski Oy, producing copper, zinc and cobalt. Outokumpu has agreements for copper ore mining in Norway and Canada, and zinc in Ecuador. Co-operation with the Soviet Union has increased.

## INVESTMENT AND DEVELOPMENT

Whilst there is widespread activity in minerals for such a small country it does not amount to much in world terms other than that the spread includes some of the more strategic metals: cobalt, chromite, nickel and vanadium. Because of the paucity of ore supplies, even for domestic consumption, it is Finnish companies which seek to invest abroad rather than vice-versa. Recent ventures have included equity shares in prospects in Norway & Mexico.

## OUTPUT AND RATINGS OF PRINCIPAL MINERALS

| *Mineral* | *Output (Tonnes)* | | *Percentage of Western World Output* | *Western World Rating* | *Mine (M)/ Smelter (S)* | *Major Producers* |
|---|---|---|---|---|---|---|
| | *1981* | *(1971)* | | | | |
| Cadmium | 618 | (120) | 4.6 | 9 | S | Outokumpu Oy |
| Chromite | 165,000 | (139,400) | 2.9 | 5 | M | Outokumpu Oy |
| Cobalt | 1,250 | (900) | 4.9 | 5 | M | Outokumpu Oy |
| Nickel | 6,900 | (3,600) | 1.4 | — | M | Outokumpu Oy |
| Vanadium | 2,400 | (1,500) | 11.4 | 3 | S | Rautaruukki Oy |
| Zinc | 53,600 | (63,700) | 1.3 | — | M | Outokumpu Oy |

Other minerals production: copper, gold, iron ore, lead, silver

## COMPONENTS OF RISK ASSESSMENT

| *Factor* | *Rating* | *Comment* |
|---|---|---|
| **Labour** | | |
| Strike incidence | 3 | Low |
| Quality of labour | 5 | Medium-high |
| General unrest | 3 | Low |
| **Politics** | | |
| History | 7 | Forced political changes from 1939 onwards |
| Stability | 7 | Medium-low but unlikely to change |
| External dangers | 7 | Little scope to change; proximity to Russia |
| **Location** | | |
| Hostile borders | 7 | Proximity to Russia |
| Critical land/sea routes | 7 | High |
| **Mineral Resources** | | |
| Adequacy of reserves | 8 | Low |
| Cost of production | 7 | High |
| Development appeal | 8 | Low, due to effective state ownership of mining industry |
| **Financial and Economic** | | |
| Currency performance | 5 | Relatively stable |
| Vulnerability to manipulation | 5 | Hardly material |
| External indebtedness | 5 | Medium |
| Productivity | 5 | Medium |
| Foreign ownerships | 2 | Low |
| **Energy and Other Natural Resources** | | |
| Domestic energy sources | 8 | Low |
| Domestic energy production | 7 | Low |
| Sources of outside supply | 5 | Safe |
| **Dependence on Foreign Natural Resources** | 7 | Growing |

***These components are assessed on a scale from 1 (lowest) to 10 (highest) risk. Grouped together, they form the basis for the country risk factor assessment opposite.***

# FRANCE

## COUNTRY RISK FACTOR 6.3

*This factor is assessed from 1.0 (lowest) to 10.0 (highest) risk. For components of this assessment see later table in this section.*

The country's gloomy outlook for the remainder of the 1980s contrasts strongly with its progress in the previous two decades. The turning point appears in retrospect to have been 1981 when France swung decisively to the left, ending 20 years of Conservative government. The expected spate of nationalisations has hit the banking and commercial sector reducing investment enthusiasm considerably. The currency has come under severe attack, particularly against the neighbouring D-Mark, and the added stringencies of recession contribute to gloomy assessment.

### MINERAL PRODUCTION

France felt energy reality when the OPEC crisis threatened its oil dependency, highlighted the run down of its coal industry and caught the nuclear power industry in its infancy. It is still very vulnerable. It has a useful spread of minerals, with bauxite, aluminium, cadmium, cobalt and iron ore all in the top ten, but reserves are unexceptional and the range is rather narrow. In the European context the presence of bauxite, uranium and iron ore is particularly useful but the country does not score high geologically.

### THE MINING INDUSTRY

The pattern of development is typical of a well established, industrial country, where a raw materials production base has been outstripped by a refining and fabrication capacity demanding imported ores and concentrates. As a result several large companies operate with a small mining presence, large refining/fabrication capacity and complex foreign interests. Larger companies include Vieille Montaigne, Royal Asturienne des Mines, the PUK Group (nationalised in 1982), Penarroya, Compagnie de Mokta (the last two both subsidiaries of IMETAL) and foreign-owned companies including Alcan and Alusuisse.

### INVESTMENT AND DEVELOPMENT

Investors are now much more wary and most new projects are domestically and government inspired. It is directed towards the energy sector and nuclear power in particular given the limited geological scope available.

The French government, via state-owned Bureau de Recherches Geologiques et Minieres (BRGM), aims to promote at home and abroad the exploitation of mineral resources. BRGM has exploration subsidiaries in many countries including Senegal, Volta, New Caledonia, Zaire and Gabon and has played a major part in many ventures.

## OUTPUT AND RATINGS OF PRINCIPAL MINERALS

| *Mineral* | *Output (Tonnes)* | | *Percentage of Western World Output* | *Western World Rating* | *Mine (M)/ Smelter (S)* | *Major Producers* |
|---|---|---|---|---|---|---|
| | *1981* | *(1971)* | | | | |
| Aluminium | 436,000 | (375,000) | 3.5 | 6 | S | PUK, Alcan |
| Bauxite | 1,828,000 | (3,184,000) | 2.4 | 10 | M | PUK, BAP, Alusuisse |
| Cadmium | 660 | (579) | 5.0 | 8 | S | Penarroya, Vieille Montagne |
| Cobalt | 600 | N/A | 2.8 | 6 | S | PUK |
| Iron Ore | 21,600,000 | (56,420,000) | 2.5 | 8 | M | Arbed, Mokta |
| Lead | 19,000 | (37,000) | 1.0 | – | M | Penarroya, (CFMU) |
| Nickel | 10.1 | (9.9) | 2.1 | – | S | Societe Le Niquel |
| Silver | 53 | (110) | 1.0 | – | M | Vieille Montagne, Penarroya |
| Tungsten | 744 | (405) | 2.7 | – | M | |
| Uranium (metal content) | 2,555 | (2,155) | 6.0 | 7 | M | Cogema |
| Zinc | 37,400 | (25,000) | 1.0 | – | M | Vieille Montagne, Penarroya |

Other minerals production: copper, gold
PUK: (Pechiney Ugine Kuhlmann)
BAP (Bauxite et Alumines de Provence SA)

## COMPONENTS OF RISK ASSESSMENT

| *Factor* | *Rating* | *Comment* |
|---|---|---|
| **Labour** | | |
| Strike incidence | 6 | Medium—rising |
| Quality of labour | 4 | Good |
| General unrest | 6 | Medium—rising |
| **Politics** | | |
| History | 7 | Several major upheavals since 1945 |
| Stability | 7 | Questionable with left wing majority and large Communist minority |
| External dangers | 7 | Communist influence could affect Western affiliations |
| **Location** | | |
| Hostile borders | 4 | No serious problems |
| Critical land/sea routes | 6 | Average European seaport vulnerability |
| **Mineral Resources** | | |
| Adequacy of reserves | 6 | Average |
| Cost of production | 7 | Rising and exposed on energy supply |
| Development appeal | 8 | Low with present government |
| **Financial and Economic** | | |
| Currency performance | 7 | Poor in relation to international pool |
| Vulnerability to manipulation | 7 | High, but regulated by EMS controls |
| External indebtedness | 6 | High, but containable |
| Productivity | 4 | Good |
| Foreign ownerships | 4 | Reducing |
| **Energy and Other Natural Resources** | | |
| Domestic energy sources | 8 | Low and growth of nuclear power industry affected by environmentalist pressure groups |
| Domestic energy production | 8 | Low |
| Sources of outside supply | 7 | OPEC biased, particularly Iran |
| **Dependence on Foreign Natural Resources** | 7 | Medium-high |

*These components are assessed on a scale from 1 (lowest) to 10 (highest) risk. Grouped together, they form the basis for the country risk factor assessment opposite.*

# GABON

## COUNTRY RISK FACTOR 5.2

*This factor is assessed from 1.0 (lowest) to 10.0 (highest) risk. For components of this assessment see later table in this section.*

Gabon is favoured in both African and world terms. It has a low country risk assessment, reasonable domestic and export oil supplies, a per capita income of over $3,000, major uranium and manganese production and apparently harmonious foreign participation agreements. It has retained strong French interests and appears to have settled for a formula of 25 per cent government to 75 per cent outside equity in its mineral projects. Infrastructure is lacking and the choice of known minerals is small but the background, the transition to independence from France in 1960 and the recent stability all assist a healthier than average investment rating.

### MINERAL PRODUCTION

Manganese will probably remain the major product, despite the world steel slowdown, retaining fourth place jointly with India at almost two million tonnes annual sales. Reserves are massive and only transport problems hinder expansion. Oil is a critical foreign exchange earner but reserves, currently supporting a 150,000 bbd industry, are limited on current leases. Uranium output, sold to French, Japanese and Belgian interests, is being expanded from 1,000 to 1,500 tonnes a year capacity. Useful iron ore reserves have been delineated at Belinga in the north east but their development is tied to the expansion of the railway network. Lead, zinc, barite and gold occurrences have all warranted attention but whilst the political climate may attract, infrastructure costs are still a major hurdle.

### THE MINING INDUSTRY

Government partnership with foreign interests is the hallmark of Gabonese development, French companies being particularly favoured. Elf and Shell are the main oil operators. COMILOG is a 25 per cent government, plus US Steel, BRGM and Mokta partnership. It controls the manganese industry whilst COMUF (25 per cent government plus PUK, BRGM, Mokta and Parisbas) dominates uranium output although new capacity will involve Korean and other French interests.

### INVESTMENT AND DEVELOPMENT

The whole of West Africa, mostly independent since 1960 and having weathered a succession of civil wars, coups and insurrections, is regarded with some suspicion by the investment sector. Gabon currently enjoys a good rating, has a lengthy Atlantic sea coast and shares borders with countries whose own stability has improved in the recent past. As long as manganese and oil revenues can sustain an acceptable standard of living, there will be no urgent pressure to tighten prevailing investment formulae and thus foreign capital will continue to be attracted. The necessary road and rail transportation systems will need to be developed since this is the most immediate barrier to rapid growth in mineral exploitation from a broad base of domestic reserves.

## OUTPUT AND RATINGS OF PRINCIPAL MINERALS

| *Mineral* | *Output (Tonnes) 1981* | *(1971)* | *Percentage of Western World Output* | *Western World Rating* | *Mine (M)/ Smelter (S)* | *Major Producers* |
|---|---|---|---|---|---|---|
| Manganese | 1,800,000 | (1,868,000) | 11.4 | Joint 4 | M | COMILOG (Gabon Govt) |
| Uranium | 1,021 | (1,250) | 2.5 | – | M | COMUF (Gabon Govt) |

COMILOG (Compagnie Miniere de L'Ogooue SA)
COMUF (Compagnie des Mines d'Uranium de Franceville)

## COMPONENTS OF RISK ASSESSMENT

| *Factor* | *Rating* | *Comment* |
|---|---|---|
| **Labour** | | |
| Strike incidence | 5 | Reasonably controlled |
| Quality of labour | 6 | Lacks some foreign skills |
| General unrest | 5 | Fairly low after the first few years of independence |
| **Politics** | | |
| History | 7 | Troubled in the early years of independence |
| Stability | 5 | Excellent, particularly in relation to its location. One Party state |
| External dangers | 5 | Containable, with neighbours the Congo, Cameroon and Equitocial |
| **Location** | | |
| Hostile borders | 4 | Not an obvious problem, with a unifying French influence |
| Critical land/sea routes | 4 | Long coastline, good river transport |
| **Mineral Resources** | | |
| Adequacy of reserves | 4 | Excellent |
| Cost of production | 6 | Rising with the rate of economic advancement |
| Development appeal | 6 | Good but with high infrastructure costs |
| **Financial and Economic** | | |
| Currency performance | 5 | Improving with growth of export earnings |
| Vulnerability to manipulation | 7 | High as the rate of capital borrowing grows |
| External indebtedness | 7 | Vulnerable to oil, manganese and uranium prices |
| Productivity | 5 | Good |
| Foreign ownerships | 5 | Containable through a government participation formula |
| **Energy and Other Natural Resources** | | |
| Domestic energy sources | 4 | Good |
| Domestic energy production | 4 | Good although oil production may have peaked at 160,000 b/d |
| Sources of outside supply | 4 | Safe |
| **Dependence on Foreign Natural Resources** | 5 | Medium |

*These components are assessed on a scale from 1 (lowest) to 10 (highest) risk. Grouped together, they form the basis for the country risk factor assessment opposite.*

# GERMANY (FR)

## COUNTRY RISK FACTOR 5.5

*This factor is assessed from 1.0 (lowest) to 10.0 (highest) risk. For components of this assessment see later table in this section.*

Housing the Western world's third most productive economy, West Germany is highly dependent on the import of raw materials and the export of finished goods. It has a vulnerable energy profile. Its coal reserves are only marginally economic, its oil import dependency high and its nuclear power plans vigorously opposed by a strong domestic lobby.

The country's industrial performance since post-World War Two reconstruction has been impressive in all respects: overall growth, productivity, labour relations, employment and financial husbandry. Politically the country remains right-wing but there is no clear party majority. World recession is now biting, unemployment rising and the outlook less optimistic than for many years.

### MINERAL PRODUCTION

Other than coal, the only mineral produced on a large scale is primary aluminium. West German output ranks third in the world behind the USA and Canada but is produced from imported ores and falls 30 per cent short of consumption. Home produced lead falls about 90 per cent short and zinc is over 60 per cent deficient. West Germany is totally dependent on imports for tin and most other minerals. Efforts are being made to reduce import dependency on energy by modernising the coal industry, maintaining the small oil base, reducing oil consumption, diversifying import sources and developing nuclear power. This policy has taken precedence over any serious attempt to expand the domestic metallic minerals base.

### THE MINING INDUSTRY

Coal mining is concentrated on the two major coalfields from whose names the major mining companies, Ruhrkohle AG and Saarbergwerke AG derive. Synthetic oil-from-coal technology is also being developed again. Lead and zinc mining have declined since the 1965-75 period. Interest in domestic ore production is slight.

### INVESTMENT AND DEVELOPMENT

The pre-occupation with energy channels investment into long-term developments such as coal, oil-from-coal and nuclear power projects where security of supply takes precedence over return on investment. The world surplus of raw materials, indicating a low price base over the next economic cycle, conflicts with the increasing costs of developing new capacity. This, together with the country's limited geological wealth, acts against the development of Germany's mining potential in spite of the presence of a large captive market.

## OUTPUT AND RATINGS OF PRINCIPAL MINERALS

| *Mineral* | *Output (Tonnes)* | | *Percentage of Western World Output* | *Western World Rating* | *Mine (M)/ Smelter (S)* | *Major Producers* |
|---|---|---|---|---|---|---|
| | *1981* | *(1971)* | | | | |
| Aluminium | 729,000 | (428,000) | 5.8 | 4 | S | A-HR, VAW, HAW, Kaiser, Leichtmet-Oil |
| Cadmium | 1,073 | (982) | 8.1 | 4 | S | Metallgesellschaft, Preussag |
| Iron Ore | 1,600,000 | (1,800,000) | 1.0 | – | M | Krupp Stahl, Bochum, E-H-W |
| Lead | 29,000 | (50,000) | 1.2 | – | M | Berzelius, PBB, Preussag, Norddeutsche |
| Silver | 31 | 30 | <1.0 | – | M | |
| Zinc | 111,000 | (165,000) | 2.5 | 10 | M | Metallgesellschaft, Preussag |

VAW (Vereinigte Aluminium Werke GmbH)
EHW (Estel-Hoesch-Werke Ag)

## COMPONENTS OF RISK ASSESSMENT

| *Factor* | *Rating* | *Comment* |
|---|---|---|
| **Labour** | | |
| Strike incidence | 3 | Low |
| Quality of labour | 3 | Excellent |
| General unrest | 7 | Rising as unemployment increases |
| **Politics** | | |
| History | 4 | Stable since 1949 |
| Stability | 7 | No clear ruling majority, inhibits policy making |
| External dangers | 3 | Low, as focus of both power-blocs |
| **Location** | | |
| Hostile borders | 6 | Proximity to Eastern Bloc |
| Critical land/sea routes | 7 | Coastal access is into North Sea-English Channel |
| **Mineral Resources** | | |
| Adequacy of reserves | 8 | Low |
| Cost of production | 7 | High on geological grounds |
| Development appeal | 6 | Medium |
| **Financial and Economic** | | |
| Currency performance | 3 | Excellent historically |
| Vulnerability to manipulation | 6 | Growing as economic pressures rise |
| External indebtedness | 4 | Low |
| Productivity | 3 | High |
| Foreign ownerships | 3 | Low |
| **Energy and Other Natural Resources** | | |
| Domestic energy sources | 7 | Low in oil, marginal in economic coal |
| Domestic energy production | 7 | Rising in coal but on a high cost base |
| Sources of outside supply | 7 | Vulnerable to OPEC embargoes |
| **Dependence on Foreign Natural Resources** | 8 | High |

*These components are assessed on a scale from 1 (lowest) to 10 (highest) risk. Grouped together, they form the basis for the country risk factor assessment opposite.*

# GUYANA

## COUNTRY RISK FACTOR 7.3

*This factor is assessed from 1.0 (lowest) to 10.0 (highest) risk. For components of this assessment see later table in this section.*

Guyana's critical economic problems are squarely represented in its mineral rating. It has only the one major product, bauxite, whose tonnage has halved since 1971, the year in which the government nationalised the mines and relieved the operator, Alcan, of all managerial and technical responsibility. Output has also fallen in the other staple products including rice and sugar. There is negative economic growth, rising unemployment and inflation, a dearth of foreign exchange and severe political strictures upon what is, de facto, a democracy. With a low population density and lack of infrastructure it will be difficult to develop the promising indications of minerals countrywide, ignoring present government disincentives. The parliamentary opposition is Marxist oriented whilst the two main parties also divide ethnically into historically African and Asian descent. A major territorial dispute lodges between Guyana and its larger neighbour, Venezuela, the outcome of which will affect sovereignty of half Guyana's land mass. A forthright foreign investment policy might turn the economic situation round but after nearly 20 years in office the government is unlikely to succeed now.

### MINERAL PRODUCTION

Minor alluvial dredging of gold and diamonds has continued for many years and a small manganese operation persisted until about 1970. Bauxite has been the keystone of the economy for many years and was profitable, in spite of a high overburden ratio, until the massive deposits of Guinea and Australia were exploited on a large scale in the 1965-70 period. The reserves of refractory grade bauxite are the world's largest but competition from the Chinese is an increasing problem. The industry has deteriorated greatly since nationalisation, partly because of a weakening market. Outside help has had to be brought in to overcome stripping problems in the mine and a general run-down of alumina production facilities. It may already be too late to recover the competitive edge in an oversupplied market, which bodes ill for an economy reliant on the single product. Irregular prospecting has indicated more serious gold deposits, uranium and oil whilst the hydro-electric potential of the country, with its high annual rainfall, massive river systems and mountainous terrain is obvious. Theoretically attractive, the country is nevertheless assessed as a poor risk and is unlikely to find backers in the current world market, particularly whilst present government policies are enforced.

### THE MINING INDUSTRY

The arbitrary nationalisations in 1971 of the bauxite mines operated by the US company Reynolds and the larger bauxite and alumina facilities of Alcan the Canadian company were central to the Guyanese government's self-determination policy. They resulted in an under-capitalised industry, starved of essential skills and, combining with a weakening market and rising oil price, resulted in a deterioration which is critical to the entire economy.

## OUTPUT AND RATINGS OF PRINCIPAL MINERALS

| *Mineral* | *Output (Tonnes) 1981* | *(1971)* | *Percentage of Western World Output* | *Western World Rating* | *Mine (M)/ Smelter (S)* | *Major Producers* |
|---|---|---|---|---|---|---|
| Bauxite | 1,907,000 | (4,234,000) | 2.4 | Joint 7 | M | Guyana Mining Enterprise Ltd |

## COMPONENTS OF RISK ASSESSMENT

| *Factor* | *Rating* | *Comment* |
|---|---|---|
| **Labour** | | |
| Strike incidence | 8 | High |
| Quality of labour | 7 | Difficulty in retaining expatriate and local skilled workers |
| General unrest | 8 | High |
| **Politics** | | |
| History | 7 | Troubled since independence |
| Stability | 8 | Endangered if present government fails |
| External dangers | 7 | Communism a real threat if opposition party gains ground |
| **Location** | | |
| Hostile borders | 7 | Large territorial claims by neighbouring Venezuela |
| Critical land/sea routes | 5 | Open access to Atlantic |
| **Mineral Resources** | | |
| Adequacy of reserves | 6 | Moderate in bauxite, good but unexplored promise in others |
| Cost of production | 7 | Relatively high |
| Development appeal | 9 | Low |
| **Financial and Economic** | | |
| Currency performance | 8 | Poor |
| Vulnerability to manipulation | 8 | High |
| External indebtedness | 8 | High |
| Productivity | 8 | Low |
| Foreign ownerships | 4 | Very restricted, but help now sought |
| **Energy and Other Natural Resources** | | |
| Domestic energy sources | 8 | Very low, no oil or coal known |
| Domestic energy production | 8 | Hydro potential very expensive |
| Sources of outside supply | 7 | Not politically constrained but restricted by lack of hard currency |
| **Dependence on Foreign Natural Resources** | 7 | High |

*These components are assessed on a scale from 1 (lowest) to 10 (highest) risk. Grouped together, they form the basis for the country risk factor assessment opposite.*

## INVESTMENT AND DEVELOPMENT

Minor attempts have been made to draw in foreign investors but these are wide-ranging, embryonic and often involved the use of world aid funds. It is likely that any attempt to set up new mining ventures involving foreign participation will be hamstrung with protection clauses and under the present regime will fail to recognise the demands of competition in an oversupplied market.

If the hydro electric potential of the Mazaruni River could be harnessed to drive a planned 150,000 tonnes a year aluminium smelter, the bauxite and alumina industries brought back up to capacity and fringe industries such as gold and diamonds revived, then minerals could provide the platform upon which the economy could be rebuilt. Barring a declared change of direction this is unlikely to be achieved in the next few years.

# INDIA

## COUNTRY RISK FACTOR 7.0

*This factor is assessed from 1.0 (lowest) to 10.0 (highest) risk. For components of this assessment see later table in this section.*

In spite of a large landmass India is poor in minerals. It supports the basics of a steel industry and is concentrating on energy, particularly the search for oil. Industrial production has been rising but oil imports are proving a crippling burden. The current 15 million tonnes a year of domestic oil production will need to rise towards the 20-year 100 million tonnes a year plan, whilst at the same time domestic demand must only be inflated moderately from the current 31 million tonnes a year level. India maintains a delicate balance between the Soviets and the West in its minerals development and trading. It invites the supply of technology from both camps. With the exception of export quantities of iron ore, manganese, bauxite and mica, India's medium term aim must be towards self sufficiency with an urgent underlying need to provide employment in the high-volume, labour intensive sectors.

### MINERAL PRODUCTION

Coal reserves are very large and oil reserves are promising. After coal, iron ore is the largest resource, the 40.5 million tonnes a year output allowing over one-third export capacity, with exports planned to grow towards a target of 40 million tonnes a year by 1984-85. Japan and Romania are large customers. Manganese output of 1.65 million tonnes a year puts India in joint fourth position with Gabon, but dwindling reserves are a cause for concern. Large bauxite reserves are located in the east coast states of Orissa, Bihar and Andhra Pradesh. Output of about two million tonnes a year is well in excess of local aluminium conversion needs and plans exist for an alumina industry in which Russia has expressed interest. Lead, zinc and copper production are all below domestic requirements and, whilst prospecting has shown that current levels might be trebled, this is still a low figure and will demand a high level of capital expenditure and infrastructure.

### THE MINING INDUSTRY

Understandably, the most developed resource companies are in the coal, iron and steel sectors. The National Mineral Development Corporation (NMDC) mines the

### OUTPUT AND RATINGS OF PRINCIPAL MINERALS

| *Mineral* | *Output (Tonnes) 1981* | *(1971)* | *Percentage of Western World Output* | *Western World Rating* | *Mine (M)/ Smelter (S)* | *Major Producers* |
|---|---|---|---|---|---|---|
| Aluminium | 213,000 | (178,000) | 1.7 | 10 | S | Indian Aluminium Corp<br>Hindustan Aluminium Corp<br>Baharat Aluminium Corp |
| Bauxite | 1,912,000 | (1,487,000) | 2.5 | Joint 7 | M | Indian Aluminium Corp<br>Hindustan Aluminium Corp<br>Baharat Aluminium Corp |
| Cadmium | 111 | N/A | 0.8 | – | S | Hindustan Zinc |
| Copper | 25,000 | (9,500) | 0.4 | – | M | Hindustan Copper |
| Gold | 2.6 | (3.7) | <1.0 | – | M | Bharat Gold (state) |
| Iron Ore | 40,500,000 | (30,000,000) | 4.7 | 5 | M | NMDC (state) |
| Lead | 15,200 | N/A | 0.6 | – | M | Hindustan Zinc |
| Manganese | 1,650,000 | (1,700,000) | 11.5 | Joint 4 | M | Manganese Ore (India) Ltd |
| Silver | 12 | (4) | 1.0 | – | | |
| Zinc | 34,200 | (13,000) | 0.8 | – | M | Hindustan Zinc |

Other minerals production: coal, gemstones

## COMPONENTS OF RISK ASSESSMENT

| *Factor* | *Rating* | *Comment* |
|---|---|---|
| **Labour** | | |
| Strike incidence | 5 | Medium |
| Quality of labour | 4 | Medium-low |
| General unrest | 6 | Limited and masked by a high unemployment rate |
| **Politics** | | |
| History | 7 | Recurrent constitutional problems since independence in 1948 |
| Stability | 7 | Medium-low profile |
| External dangers | 7 | Recent hostilities with Pakistan |
| **Location** | | |
| Hostile borders | 6 | Not presently but note above |
| Critical land/sea routes | 7 | Access to West involves Suez or Cape of Good Hope |
| **Mineral Resources** | | |
| Adequacy of reserves | 7 | Low except in coal and iron |
| Cost of production | 7 | High on infrastructure, low on labour |
| Development appeal | 8 | Low |
| **Financial and Economic** | | |
| Currency performance | | Rupee is non negotiable |
| Vulnerability to manipulation | | Rupee is non negotiable |
| External indebtedness | 8 | High |
| Productivity | 8 | Medium-low |
| Foreign ownerships | 3 | Low |
| **Energy and Other Natural Resources** | | |
| Domestic energy sources | 7 | Good in coal, improving in oil |
| Domestic energy production | 6 | Good in coal, improving in oil |
| Sources of outside supply | 7 | Not restricted politically, but financially |
| **Dependence on Foreign Natural Resources** | 9 | High |

*These components are assessed on a scale from 1 (lowest) to 10 (highest) risk. Grouped together, they form the basis for the country risk factor assessment opposite.*

ore exported by the Minerals and Minerals Trading Corporation (MMTC). Manganese is similarly handled, the largest producer being Manganese Ore (India) Ltd and major exporters MMTC, EMCO and Mysore Minerals Ltd. Although there is an increasing demand for aluminium in the electrical cable industry, India still has an export surplus which finds its way onto the free market. Five smelters comprise the 350,000 tonnes a year capacity, three being in the 100,000 tonnes range. Depressed conditions make it unlikely that any immediate expansion will come on stream. Power shortages are a major mining industry restriction.

### INVESTMENT AND DEVELOPMENT

A high level of government involvement at both state and federal level characterises Indian mining in all its forms. The country has a high development-aid profile and strong relationships with Soviet countries such as the USSR, Romania and Mozambique, making it less than secure as an investment haven.

# INDONESIA

## COUNTRY RISK FACTOR 5.7

*This factor is assessed from 1.0 (lowest) to 10.0 (highest) risk. For components of this assessment see later table in this section.*

Indonesia is strategic on all important considerations: its position in South East Asia, where capitalist and communist ideologies come face to face; its obvious mineral wealth including oil; its large landmass and population. The most serious setback in the impressive minerals-based development of the country came with the redefining of foreign investment parameters in the mid 1970s (when several large projects were already committed) and it is debatable whether present rules are sufficiently attractive in the more depressed economic climate. Growth remains impressive but care must be taken to ensure that it does not fall into the Mexican trap. Major political and ethnic problems lie close to the surface. Recent prosperity plus the individual leadership style of President Suharto are instrumental in containing them.

### MINERAL PRODUCTION

Both Indonesia's leading hard minerals have a high strategic profile. The country is second only to Malaysia in tin mining, having almost doubled its output since 1971. Fourth place in nickel is an improvement of two places in the same period but output, at 45,500 tonnes a year, has quadrupled. Oil, coal and gas apart, other minerals have proved less exciting and costs are now squeezing the major products. Because of export controls (see Tin, Section Two) annual output, at 35,000 tonnes a year, will probably fall by over 10 per cent in the 1982-84 period. There are nevertheless ample reserves to retain a rising output base. Nickel is suffering more, as a result of the pull-out of the PT Pacific Nickel Consortium and the remaining operating companies will probably cut back drastically to match dwindling market demand. Copper output from the only operator, Freeport Indonesia, is also constrained but grades are better than the world average so that the current 63,000 tonnes a year could be maintained. Bauxite is also suffering from a depressed market (Japan being the major customer) but the first aluminium smelter, due for completion at 75,000 tonnes a year capacity by the end of 1982, will ultimately aid the vertical integration of this sector.

### THE MINING INDUSTRY

The high-profile oil and gas sector is operated by production-sharing or contract-of-work agreement between the state company Pertamina and foreign oil majors. Coal is largely produced by foreign joint venture companies operating within a defined framework. In the other sectors there are two major foreign companies involved: in nickel, PT INCO Indonesia operates a 60,000 tonnes a year capacity mine on Sulawesi but output may be dropped to no more than 10,000 tonnes a year if market conditions do not improve. The second, government owned company PT Aneka Tambang, operates two smaller plants. Copper is won from the Ertsburg complex of Freeport Indonesia at Irian Jaya where there is also a little by-product gold and silver. High grades will probably sustain this output but significant expansion is hardly a starter. Bauxite production is also under state control and there is a distinct lack of enthusiasm by foreign aluminium companies to become involved in the integrated aluminium project on other than a technical basis. Tin is won by a mixture of state, local privately owned and foreign joint-venture projects, such as Koba. An increasing amount is won from offshore.

## OUTPUT AND RATINGS OF PRINCIPAL MINERALS

| *Mineral* | *Output (Tonnes)* 1981 | (1971) | *Percentage of Western World Output* | *Western World Rating* | *Mine (M)/ Smelter (S)* | *Major Producers* |
|---|---|---|---|---|---|---|
| Bauxite | 1,203,000 | (1,238,000) | 1.6 | – | M | PT Aneka Tambang |
| Copper | 63,000 | Nil | 1.0 | – | M | Freeport Indonesia |
| Gold | 1.0 | (0.3) | <1.0 | – | M | Freeport Indonesia |
| Nickel | 45,500 | (12,000) | 9.3 | 4 | M | PT Inco, PT Aneka Tambang |
| Tin | 35,000 | (20,000) | 17.5 | 2 | M | PT Koba Tin, PT Riau, PT Tambang Timah |

Other mineral production: silver, coal, oil

## COMPONENTS OF RISK ASSESSMENT

| *Factor* | *Rating* | *Comment* |
|---|---|---|
| **Labour** | | |
| Strike incidence | 4 | Low |
| Quality of labour | 6 | Good |
| General unrest | 4 | Currently low |
| **Politics** | | |
| History | 8 | Troubled since post World War Two |
| Stability | 7 | Currently sound but could change |
| External dangers | 7 | Always a possibility in SE Asia |
| **Location** | | |
| Hostile borders | 7 | An area of established East-West tension |
| Critical land/sea routes | 8 | Capable of disruption at sea |
| **Mineral Resources** | | |
| Adequacy of reserves | 4 | Good |
| Cost of production | 7 | Rising |
| Development appeal | 6 | Medium |
| **Financial and Economic** | | |
| Currency performance | 4 | Good |
| Vulnerability to manipulation | 5 | Low on exports of energy, but could change |
| External indebtedness | 4 | Low-medium |
| Productivity | 5 | Average |
| Foreign ownerships | 8 | Tightly controlled but critically placed |
| **Energy and Other Natural Resources** | | |
| Domestic energy sources | 3 | Excellent |
| Domestic energy production | 3 | Excellent |
| Sources of outside supply | 1 | Not necessary |
| **Dependence on Foreign Natural Resources** | 8 | High in non-energy sector |

*These components are assessed on a scale from 1 (lowest) to 10 (highest) risk. Grouped together, they form the basis for the country risk factor assessment opposite.*

## INVESTMENT AND DEVELOPMENT

Indonesia's foreign investment rules are tight but in some circumstances workable. They attracted INCO, possibly as a means of maintaining its world lead in nickel production, but tipped the scales against the PT Pacific Consortium (US Steel, Hoogovens, Amoco) and possibly decided Kaiser against an involvement in the plans for a fully integrated aluminium industry. The regulations effectively place a ceiling on remittable profits and require an eventual surrender of control to local interests. Whilst the demand for energy minerals remains high (and the proximity to Japan must weigh heavily) the economic climate will probably remain reasonably favourable. Yet in the non-energy sector, the most abundant minerals—copper, tin, nickel and bauxite—face strong competition from other countries.

# IRELAND

---

**COUNTRY RISK FACTOR** **6.3**

*This factor is assessed from 1.0 (lowest) to 10.0 (highest) risk. For components of this assessment see later table in this section.*

---

Ireland merits inclusion amongst the world mineral countries because a single producer, Tara Mines, constitutes Europe's largest source of zinc concentrates and is a major supplier of lead. Supporting operations at Silvermines, Tipperary and Avoca, Wicklow, are highly marginal and have poor reserves whilst the Tynagh Mine was finally closed in 1980. Scope for minerals expansion countrywide has been indicated by active prospecting since government tax incentives in the 1960s, now rescinded, prompted enthusiasm.

Ireland has severe economic problems. A member of the EEC since 1973, it broke its ties with the British pound in 1980 and has since seen its own currency, the punt, slip from par to below 80 per cent of sterling values. Unemployment and inflation are at historic highs and much of the development appeal expected from EEC membership has evaporated. The country is also politically split, largely on economic issues and, since 1979-80, has been difficult to govern. Offshore oil development has generated interest since the early 1970s but prospecting results have not yet remotely matched expectations. Eventual operating conditions may prove arduous.

## MINERAL PRODUCTION

Whilst there seems little prospect of onshore hydrocarbon finds the past 20 years have shown sufficient copper, lead, zinc and silver mineralisation to continue to warrant attention. Most deposits, such as the Silvermines, Tynagh and Avoca, are typically 20-year operations whilst the Tara lead-zinc orebody is of world-ranking proportions, estimated at more than 60 million tonnes of 9.9 per cent zinc and 2.7 per cent lead. The operation has been plagued by strikes, for which reason 1980 has been selected for inclusion in the tables as a more representative output year than 1981. It is capable of expansion in the Bula lease on the northern, outcrop end so that this source could produce in total 250,000 tonnes a year of zinc and 70,000 tonnes a year of lead. Barytes production is still above 200,000 tonnes a year although falling whilst the sporadic coal reserves are being given new consideration. The major alumina undertaking by Alcan Ltd in the River Shannon estuary should come on stream in 1983 with the first phase of its 800,000 tonnes a year capacity. The timing may be unfortunate but shipping costs to Europe should be much more attractive than for Caribbean or Australian alumina.

---

## OUTPUT AND RATINGS OF PRINCIPAL MINERALS

| *Mineral* | *Output (Tonnes) 1980* | *(1971)* | *Percentage of Western World Output (1980)* | *Western World Rating* | *Mine (M)/ Smelter (S)* | *Major Producers* |
|---|---|---|---|---|---|---|
| Lead | 59,000 | (54,000) | 2.3 | – | | Tara Mines, (Northgate, Charter Cons) Silvermines |
| Silver | 24 | (14.5)est | 1.0 | – | | Silvermines |
| Zinc | 229,000 | (88,000) | 5.0 | 9 | | Tara Mines, (Northgate, Charter Cons) Silvermines, Mogul of Ireland Ltd |

Other mineral production: barytes, copper

## COMPONENTS OF RISK ASSESSMENT

| *Factor* | *Rating* | *Comment* |
|---|---|---|
| **Labour** | | |
| Strike incidence | 7 | Fairly high |
| Quality of labour | 6 | Medium |
| General unrest | 7 | Underlying religious divisions |
| **Politics** | | |
| History | 7 | Externally cohesive, domestically split |
| Stability | 6 | Medium |
| External dangers | 4 | Low |
| **Location** | | |
| Hostile borders | 1 | None, (excluding Northern Ireland dispute) |
| Critical land/sea routes | 2 | Low |
| **Mineral Resources** | | |
| Adequacy of reserves | 7 | Medium/low |
| Cost of production | 7 | High by Western standards |
| Development appeal | 8 | Low |
| **Financial and Economic** | | |
| Currency performance | 8 | Poor, punt continues to fall against other EEC currencies |
| Vulnerability to manipulation | 8 | High |
| External indebtedness | 7 | High |
| Productivity | 6 | Medium |
| Foreign ownerships | 8 | High |
| **Energy and Other Natural Resources** | | |
| Domestic energy sources | 8 | Low. Widespread expectations of major oil find dashed in 1981 |
| Domestic energy production | 8 | Low |
| Sources of outside supply | 3 | Safe |
| **Dependence on Foreign Natural Resources** | 8 | High |

*These components are assessed on a scale from 1 (lowest) to 10 (highest) risk. Grouped together, they form the basis for the country risk factor assessment opposite.*

### THE MINING INDUSTRY

The Irish industry has strong links with Canada through Northgate Exploration, Noranda and some junior companies. Charter Consolidated of the UK is a partner in the Tara venture where a massive debt burden has had to be restructured. Offshore oil exploration continues to dominate the energy materials scene but, whilst several of the major companies such as Shell, BP, Chevron and Elf are represented there is more hope than promise from early results.

### INVESTMENT AND DEVELOPMENT

The investment climate is cool to frosty. The major adverse turnabout in government policy at the time of negotiating a mining lease for Tara in 1968-70 was a major blow to prospective developers. The Bula prospect has effectively been on ice with similar constraints for almost ten years. The associated problems of a weak currency, high inflation and a marked deterioration in labour attitudes do not mark Eire as a high priority minerals investment area.

# ITALY

| COUNTRY RISK FACTOR | 6.0 |
|---|---|

*This factor is assessed from 1.0 (lowest) to 10.0 (highest) risk. For components of this assessment see later table in this section.*

Although Italy appears in only two mineral ratings—asbestos and antimony—the country has a very broad spread of produced minerals and known occurrences. This gives it some strategic importance both domestically and as a member of the EEC. The country's political problems have delayed the formulation of a good mining policy upon which the industry could develop. Equally, there is a history of serious economic difficulties. But the determination to tackle the energy problem by local exploitation of oil and gas and, to a lesser extent, lignite and geothermal sources is a very positive step.

## MINERAL PRODUCTION

Asbestos production is important although demand is declining. Aluminium output has doubled in a decade to keep pace with local demand. Mercury has been significant and may again feature if the market holds although major cutbacks were in force in the 1976-81 period at the Monte Amiata mine. Although the range of minerals is impressive Italy is heavily dependent on imports for an industrialised country. Moreover incentives remain lacking for serious expansion despite the country's interesting prospects.

## THE MINING INDUSTRY

The SAMIM Group, part of the major state-controlled ENI company, is the leading producer in most sectors, along with its sister company SOLMINE, mining operations being integrated through to smelting and refining.

## INVESTMENT AND DEVELOPMENT

Italy is taking advantage of the EEC 'Crest Committee' raw materials programme to investigate new development possibilities but outside the oil and gas sector the approach is of a sporadic nature. The government has legislation in hand which will rationalise and hopefully stimulate mining across the spectrum but its application is subject to customary delays. With a high import-dependence and serious balance of payments problems it is reasonable to assume Italy might give increased impetus to its mining industry.

## OUTPUT AND RATINGS OF PRINCIPAL MINERALS

| *Mineral* | *Output (Tonnes) 1981* | *(1971)* | *Percentage of Western World Output* | *Western World Rating* | *Mine (M)/ Smelter (S)* | *Major Producers* |
|---|---|---|---|---|---|---|
| Aluminium | 274,000 | (136,000) | 2.2 | – | S | SAVA, Aluminio Italia |
| Antimony | 696 | (1,275) | 1.7 | 10 | M | SAMIM |
| Asbestos | 137,000 | (130,000) | 7.2 | 5 | M | Balangero Asbestos |
| Bismuth | 19 | N/A | 1.0 | – | S | SAMIM |
| Cadmium | 489 | (350) | 3.7 | – | S | SAMIM |
| Lead | 21,300 | (32,000) | 0.9 | – | M | SAMIM, Silius Mining Co |
| Mercury | 25 | (1,550) | 1.0 | – | M | SAMIM |
| Silver | 55 | N/A | 1.0 | – | M | SAMIM, Penarroya |
| Zinc | 42,400 | (106,000) | 1.0 | – | M | SAMIM |

Other minerals production: copper, iron ore, manganese, bauxite, germanium

SAMIM (Societa Azionaria Minero-Metallurgica SpA)

## COMPONENTS OF RISK ASSESSMENT

| *Factor* | *Rating* | *Comment* |
|---|---|---|
| **Labour** | | |
| Strike incidence | 6 | Medium-high |
| Quality of labour | 4 | Good |
| General unrest | 5 | Medium |
| **Politics** | | |
| History | 8 | One of acute change |
| Stability | 7 | Proving very difficult to adopt a stable government |
| External dangers | 7 | Possibly high on strong communist presence |
| **Location** | | |
| Hostile borders | 1 | None |
| Critical land/sea routes | 7 | Mediterranean area presents obvious dangers |
| **Mineral Resources** | | |
| Adequacy of reserves | 7 | Rather low in currently mined minerals |
| Cost of production | 7 | Medium-high |
| Development appeal | 7 | Medium-low on economic and incentive grounds |
| **Financial and Economic** | | |
| Currency performance | 7 | Poor |
| Vulnerability to manipulation | 7 | High but mitigated by membership of European Monetary System |
| External indebtedness | 7 | High |
| Productivity | 6 | Medium-high |
| Foreign ownerships | 3 | Low |
| **Energy and Other Natural Resources** | | |
| Domestic energy sources | 5 | Reasonable in oil but low in coal |
| Domestic energy production | 6 | Growing with reserve limits |
| Sources of outside supply | 7 | OPEC-biased |
| **Dependence on Foreign Natural Resources** | 7 | Medium-high |

*These components are assessed on a scale from 1 (lowest) to 10 (highest) risk. Grouped together, they form the basis for the country risk factor assessment opposite.*

# JAMAICA

## COUNTRY RISK FACTOR 7.0

*This factor is assessed from 1.0 (lowest) to 10.0 (highest) risk. For components of this assessment see later table in this section.*

Since independence Jamaica has had a troubled political history and a growing left wing tendency which, had the outcome of the 1980 elections favoured the Labour Party, might have encouraged some alignment with the Soviet bloc. The surprise outcome from the election has prompted the USA to offer positive support in both financial aid and the purchase of the country's main exports—bauxite and alumina. The fiscal situation is extremely serious, unemployment probably above 25 per cent and the second industry, tourism, far from recovery. The island remains politically insecure not helped by being surrounded by similarly volatile Caribbean influences. The USA is the main destination for the island's bauxite production but industrial cutbacks brought a sharp fall in demand in 1982-83. The partnership ventures into which the North American operating companies were obliged to enter with the government in the 1970s still cause serious problems, largely related to taxation and are a deterrent to further investment.

### MINERAL PRODUCTION

Other than local cement, the only mineral production is bauxite with large conversion capacity into alumina. Until the Australian deposits were opened up in the 1970s, Jamaica was the leading producer and its capacity of over 12 million tonnes a year still leaves it in third place just behind Guinea. Its alumina conversion capacity of over three million tonnes a year is an important source for the US market.

### THE MINING INDUSTRY

The bauxite mines were developed by North American companies Alcan, Alcoa, Kaiser, Reynolds as fully owned subsidiaries but, in common with most Third World countries in the 1970s, Jamaica sought, and obtained, changes leading to major participation including sales and tax formulae which greatly enhanced the share of revenue retained locally. Levels of tax and other levies have remained in contention since then and are claimed by the companies as a barrier to expansion. But in reality the world over-supply of bauxite is a greater problem, caused in large part by the investment programmes in Australia and Guinea. Production levels are certain to fall below the 1981 output of 11.6 million tonnes in the next few years unless the US government directs its buying as a political support operation. The USA has purchased 1.6m tonnes bauxite from Jamaica for its strategic stockpile and a recent 1982-83 innovation has been the barter of alumina for vehicles with General Motors and with Chrysler Corporation.

### INVESTMENT AND DEVELOPMENT

Having identified oil imports as its most serious financial burden, Jamaica is actively reducing consumption and searching for domestic supplies. This will lessen, but not seriously improve the critical imbalance created by falling bauxite sales. The prospects for further investment in the industry are remote in the near term, barring supply disruptions or changes in the political climate in other producing countries.

## OUTPUT AND RATINGS OF PRINCIPAL MINERALS

| *Mineral* | *Output (Tonnes) 1981* | *(1971)* | *Percentage of Western World Output* | *Western World Rating* | *Mine (M)/ Smelter (S)* | *Major Producers* |
|---|---|---|---|---|---|---|
| Bauxite | 11,606,000 | (12,000,000)est | 15.4 | 3 | M | Alcan Jamaica Ltd<br>Reynolds, Jamalco (ALCOA), ALPART (Anaconda, Kaiser, Reynolds) |

Other minerals production: conversion of bauxite to alumina

## COMPONENTS OF RISK ASSESSMENT

| *Factor* | *Rating* | *Comment* |
|---|---|---|
| **Labour** | | |
| Strike incidence | 7 | Fairly high but mitigated by high unemployment |
| Quality of labour | 6 | Adequate following long history of bauxite mining |
| General unrest | 8 | High |
| **Politics** | | |
| History | 8 | Difficult since independence |
| Stability | 6 | Improving |
| External dangers | 7 | Proximity to Cuba |
| **Location** | | |
| Hostile borders | 6 | None, but located in Caribbean hot-spot |
| Critical land/sea routes | 4 | Probably secure in proximity to USA |
| **Mineral Resources** | | |
| Adequacy of reserves | 7 | High in bauxite but not in other minerals |
| Cost of production | 7 | Fairly high |
| Development appeal | 7 | Low but signs of improvement |
| **Financial and Economic** | | |
| Currency performance | 8 | Very poor against US dollar, fallen 80 per cent in last 5 years |
| Vulnerability to manipulation | 8 | High |
| External indebtedness | 8 | High |
| Productivity | 5 | Adequate |
| Foreign ownerships | 5 | Contained |
| **Energy and Other Natural Resources** | | |
| Domestic energy sources | 9 | Nil in fossil fuels as yet, exploratory petroleum drilling underway |
| Domestic energy production | 9 | Nil |
| Sources of outside supply | 5 | No political constraints, only currency availability and huge costs with current oil bill of US$425 million a year |
| **Dependence on Foreign Natural Resources** | 9 | High |

***These components are assessed on a scale from 1 (lowest) to 10 (highest) risk. Grouped together, they form the basis for the country risk factor assessment opposite.***

# JAPAN

**COUNTRY RISK FACTOR** **5.0**

*This factor is assessed from 1.0 (lowest) to 10.0 (highest) risk. For components of this assessment see later table in this section.*

Japan profited more than any other country, including West Germany, from the consumer-fed economic growth period of 1960-1980. But even before this it could not sustain its raw materials appetite from domestic sources. The policy adopted, with notable success, was to import semi-processed minerals from a wide choice of sources and to refine them in the country. This developed to produce a refining industry which is second only to the USA in the output of aluminium and copper and Canada in nickel, whilst Japan now holds first place in cadmium, zinc and a range of minor metals. The dependence on imported fuel is overwhelming and the sudden OPEC price increases of 1973-78 greatly upset Japan's oil-based refining industry. The country can probably be considered the technological leader in refining techniques but has even so become uncompetitive with areas offering lower fuel costs. The decision to favour imported coal made Japan's trade links with Australia even more important. Another move was to form joint-venture raw materials projects abroad, securing supplies of metal in return for capital and technology. Links with the Philippines are particularly strong, and those with South American resource-rich countries are being strengthened.

The quality and price competitiveness of Japanese goods have proved formidable to the Western producers, enabling the country to achieve virtual supremacy in automobile and shipbuilding and seriously challenge the USA in electronics. The few industries Japanese companies have not seriously attempted to conquer are aerospace and, since 1945, armaments. There are now signs that this growth may have peaked. Emerging nations such as Taiwan, Korea and Hong Kong can provide cheaper labour whilst others can better Japan's raw material costs. The joint-venture policy abroad, extended into manufacturing, may prove the saving factor but there is a growing Western lobby demanding freer access to Japanese markets.

## OUTPUT AND RATINGS OF PRINCIPAL MINERALS

| *Mineral* | *Output (Tonnes) 1981* | *(1971)* | *Percentage of Western World Output* | *Western World Rating* | *Mine (M)/ Smelter (S)* | *Major Producers* |
|---|---|---|---|---|---|---|
| Aluminium | 771,000 | (887,000) | 6.2 | 3 | S | Mitsubishi Light Metal, NLM Sumitomo Aluminium Smelting |
| Bismuth | 315 | (655) | 12.9 | 3 | M | Mitsui Mining & Smelting Co |
| Cadmium | 2,036 | (2,675) | 15.3 | 1 | S | Mitsui |
| Cobalt | 2,400 | N/A | 11.1 | 4 | S | Nippon Mining, Sumitomo |
| Copper (mined) | 52,000 | (121,000) | <1.0 | — | M | Dowa Mining, Mitsubishi |
| Copper (refined) | 1,050,000 | — | 14.5 | 2 | S | Mitsui, Mitsubishi, Nippon |
| Indium | 5 | (9) | 10.0 | Joint 2 | S | Sumitomo |
| Lead | 47,000 | (71,000) | 1.9 | — | M | Mitsubishi |
| Selenium | 453 | (238) | 31.7 | 1 | S | Mitsui, Sumitomo |
| Silver (domestic ores) | 279 | (351) | 3.3 | 7 | S | Mitsubishi |
| Tellurium | 80 | (36) | 27.1 | Joint 1 | S | Mitsui |
| Zinc | 242,000 | (294,000) | 5.5 | 5 | M | Mitsui, Toho Zinc Co, Dowa Mining |

Other minerals produced and refined: coal, gold, tungsten, titanium, germanium, zirconium, ferro alloys

NLM (Nippon Light Metal)

## COMPONENTS OF RISK ASSESSMENT

| *Factor* | *Rating* | *Comment* |
|---|---|---|
| **Labour** | | |
| Strike incidence | 2 | Low |
| Quality of labour | 3 | High |
| General unrest | 3 | Low, due to paternalistic attitudes of large mining and refining companies |
| **Politics** | | |
| History | 2 | Long established and developed system |
| Stability | 3 | Impressive since World War Two |
| External dangers | 3 | Low, particularly on strength of US connections |
| **Location** | | |
| Hostile borders | 3 | One minor dispute with USSR |
| Critical land/sea routes | 8 | Long distances from sources of raw materials supply, particularly Australia |
| **Mineral Resources** | | |
| Adequacy of reserves | 8 | Low |
| Cost of production | 7 | High in mineral sector |
| Development appeal | 8 | Low on geological considerations |
| **Financial and Economic** | | |
| Currency performance | 4 | Excellent in past 20 years but export bias is now a danger. Yen kept artificially low in 1982 to aid exports. |
| Vulnerability to manipulation | 7 | Medium-high |
| External indebtedness | 2 | Low |
| Productivity | 4 | Low on geological considerations |
| Foreign ownerships | 2 | Not encouraged |
| **Energy and Other Natural Resources** | | |
| Domestic energy sources | 8 | Low |
| Domestic energy production | 8 | Low |
| Sources of outside supply | 7 | OPEC biased but improving |
| **Dependence on Foreign Natural Resources** | 7 | High |

*These components are assessed on a scale from 1 (lowest) to 10 (highest) risk. Grouped together, they form the basis for the country risk factor assessment opposite.*

### MINERAL PRODUCTION

Japan has minor outputs of lead, copper, coal, precious metals and industrial minerals but it is only in zinc that it registers on the mine-production worldscale. The outside view has to be that whilst domestic exploration and development are encouraged, there is such little hope of denting the reliance on foreign sources that Japanese production is not to be taken seriously. But by-product strategic metals output from imported concentrates rates highly against the possibility of supply disruption of semi-processed raw materials into Japan.

The energy plan, although frequently changed, is the focus of attention. The seventh policy, introduced in 1982, is to maintain domestic coal output at about

20 million tonnes a year, about one-fifth of planned consumption levels.

## THE MINING INDUSTRY

The Metal Mining Agency of Japan conducts surveys and feasibility studies into mineral deposits at home and abroad. In addition to a detailed survey in Japan its work has covered China, Canada, Thailand, Peru, Philippines, Brazil and Niger. Individual companies are also active. Many of Japan's largest companies are either integrated from mining and smelting through to manufacturing or include a major metals commitment. These include Mitsui, Mitsubishi, Dowa, Nippon, Marubeni, Furukawa and Sumitomo. With the exception of aluminium, where power costs even using coal-based electricity are uncompetitive, there is an underlying growth base in the refinery output of most metals, keyed into strengthening links with overseas raw materials sources.

## INVESTMENT AND DEVELOPMENT

Whatever new investment is possible in the domestic mining industry can easily be accommodated by local sources. Japanese development elsewhere is of much greater importance. Australia provides the greatest volume of materials in the form of bauxite, coal, copper and zinc concentrates but Japan has been wary in its investment. It is now markedly cooler in its planned bauxite involvements (Gladstone and Worseley) since domestic demand has been falling and has cautioned Australia that persistent strikes and rising costs are damaging its trading relationships. Investment in the Philippines copper industry is increasing, although there may be competition from China with whom oil-barters have been reported. Japan imports about seven per cent of Canada's mineral output. The option has been for guaranteed offtake in coal and copper particularly, rather than equity financing. There are already direct investments in Brazil and Zaire, prepayment purchasing agreements with Zambia and others and related activities on a world scale. There is no discernible directed policy and it may be that Japan sees the surplus in supplies as being of a long term nature, allowing it to be selective and to concentrate on conversion technology and end-uses. Japan is also to begin purchases for an official metal stockpile in October 1983. The metals will include cobalt, nickel, chromium, tungsten, molybdenum, manganese and vanadium and the purchasing agency is aiming for five days of consumption of each.

# MALAYSIA

| COUNTRY RISK FACTOR | 5.9 |
|---|---|

*This factor is assessed from 1.0 (lowest) to 10.0 (highest) risk. For components of this assessment see later table in this section.*

**Beneath the surface of a successful commodities-based economy, Malaysia has both ethnic and political problems. As long as growth can be maintained there is little fear of a return to the Communist insurgence problems put down by the British in 1948, but as the pacesetter in a vulnerable corner of South East Asia, Malaysia must never be overlooked. Four products—oil, rubber, tin and palm oil—dominate in that order of importance both the domestic economy and export earnings. Government policy has become increasingly nationalistic since 1975, but is at pains to prove that it is buying back control on a considerate and economic basis.**

## MINERAL PRODUCTION

Malaysia, the size of Japan but with little more than one tenth of its population, offers only tin as a world-ranking product. It remains the clear output leader but both the tonnage and share of the world market have declined steadily since 1971. Most observers cite state control of mining land, a high tax profile and foreign investment disincentives as the major problems. Production fell from 73,800 tonnes in 1971 to 60,000 tonnes in 1981. Exports declined from 93 per cent to less than 70 per cent and accessible reserves were depleted. Political considerations are denying access to reserves by the gravel-pump miners, who supply over half total production. Bauxite output fell sharply in 1981 to 0.71 million tonnes from a peak of 0.92 million tonnes in 1980. The one important copper mine, Malmut, appears capable of maintaining output at between 25,000 and 30,000 tonnes a year of contained copper with useful gold and silver by-products. A proposed smelter to handle the output may see realisation in 1983-84.

## THE MINING INDUSTRY

The dismantling of the 'colonial' structure of the tin mining industry is largely complete. The federal-controlled Malaysia Mining Corporation (MMC) dominates the

## OUTPUT AND RATINGS OF PRINCIPAL MINERALS

| *Mineral* | *Output (Tonnes) 1981* | *(1971)* | *Percentage of Western World* | *Western World Output* | *Mine (M)/ Smelter (S) Rating* | *Major Producers* |
|---|---|---|---|---|---|---|
| Bauxite | 700,000 | (1.12m) | 1.0 | — | M | Southeast Asia Bauxite Ltd (Alcan) |
| Copper | 28,600 | (1,000) | 0.4 | — | M | *Overseas Mineral Resources Development |
| Gold | 2.43 | (0.15) | 8.0 | — | M | Overseas Mineral Resources Development |
| Ilmenite | 155,000 | (189,400) | 3.2 | 7 | M | |
| Iron Ore | 532,000 | (4.42m) | 0.1 | — | M | |
| Monazite | 225 | (2,000) | 0.7 | 5 | M | Tronoh Mines, Kinta Kellas Tin Dredging |
| Silver | 14.70 | N/A | 0.2 | — | M | Overseas Mineral Resources Development |
| Tin | 59,938 | (73,800) | 30.6 | 1 | M/S | MMC |
| Tantalum (from concentrates) | 4.5 | N/A | 1.0 | 8 | M | MMC |
| Zirconium | 400 est | (1,400) | 0.1 | — | M | |

*Full name—Overseas Mineral Resources Development Sabah Bhd

## COMPONENTS OF RISK ASSESSMENT

| *Factor* | *Rating* | *Comment* |
|---|---|---|
| **Labour** | | |
| Strike incidence | 7 | Medium-high |
| Quality of labour | 5 | Good |
| General unrest | 6 | Under control but with a history of unrest |
| **Politics** | | |
| History | 7 | Troubled |
| Stability | 5 | Currently under control |
| External dangers | 7 | Currently containable, but always a threat in South East Asia |
| **Location** | | |
| Hostile borders | 7 | None currently but qualified as in 'Politics' |
| Critical land/sea routes | 7 | Easily manipulated in South East Asia |
| **Mineral Resources** | | |
| Adequacy of reserves | 4 | Excellent in tin and oil |
| Cost of production | 7 | Rising, due to high taxes charged |
| Development appeal | 8 | Low |
| **Financial and Economic** | | |
| Currency performance | 5 | Reasonable, in line with high export earnings |
| Vulnerability to manipulation | 6 | High on dollar bias and dependency on commodity market |
| External indebtedness | 5 | Medium-low |
| Productivity | 4 | Medium-high |
| Foreign ownerships | 8 | Disincentives growing |
| **Energy and Other Natural Resources** | | |
| Domestic energy sources | 3 | Excellent in oil, with output at 250,000 b/d |
| Domestic energy production | 5 | Good |
| Sources of outside supply | 5 | No political constraints |
| **Dependence on Foreign Natural Resources** | 7 | High |

***These components are assessed on a scale from 1 (lowest) to 10 (highest) risk. Grouped together, they form the basis for the country risk factor assessment opposite.***

---

sector, having subjugated the interests of London Tin and Charter Consoldiated, merged with the Malayan Tin Dredging Group and restructured the smelting activities of Straits Trading through the formation of the Malaysia Smelting Corporation (MSC). MMC is the Western world's largest single tin producing group and has extra interests in Nigeria and Thailand.

### INVESTMENT AND DEVELOPMENT

There is little doubt about Malaysia's ambition to exert a major influence on the world tin industry through to marketing. There is serious opposition, not just internationally, but within the federated states, to the manner in which the mining law is currently administered.

# MEXICO

## COUNTRY RISK FACTOR 5.1

*This factor is assessed from 1.0 (lowest) to 10.0 (highest) risk. For components of this assessment see later table in this section.*

Ranking in the top ten in the output of eight key minerals, third in Western World oil production and sharing a border with the USA, Mexico has many advantages. The rapid advance of its minerals output, particularly oil, prompted a programme of domestic expansion in which foreign borrowings have played a vital role. The recession of 1979-82 has eroded profitability in the minerals sector to such an extent that by mid-1982 the country was technically bankrupt, unable to meet its interest payment obligations and forced to negotiate with its creditors. This will have a serious impact on the level of investment in both Mexico and other developing countries. The risk assessment of only 5.1 reflects the many natural advantages Mexico offers but the rapid deterioration of its fiscal base will prove at least a short term deterrent.

### MINERAL PRODUCTION

Copper and silver have proved the best growth areas since 1971 apart from oil. Mexico has moved into the copper rankings to eighth place behind the Philippines and has the capacity to go higher. Its leading position in silver is assured with the prospects of moving from 1,400 to over 2,000 tonnes if world conditions permit. The slowdown in lead and zinc reflects falling export demand—with a similar cutback in by-product bismuth and cadmium—whilst iron ore and manganese production have been stepped up to meet the demands of the domestic steel industry and the US market.

### THE MINING INDUSTRY

Government incentives to the industry have increased in the form of lower taxes on previous metals and in tax credits for expansion generally. Labour, electricity and

### OUTPUT AND RATINGS OF PRINCIPAL MINERALS

| *Mineral* | *Output (Tonnes) 1981* | *(1971)* | *Percentage of Western World Output* | *Western World Rating* | *Mine (M)/ Smelter (S)* | *Major Producers* |
|---|---|---|---|---|---|---|
| Aluminium | 43,200 | Nil | 0.3 | — | S | Alumino SA de CV |
| Antimony | 1,800 | (3,200) | 4.7 | 3 | M | |
| Bismuth | 725 | N/A | 23.4 | 2 | S | *Penoles |
| Cadmium | 600 | (192) | 4.5 | 7 | S | IMMSA, *Penoles, MR de A, Zincamex |
| Copper | 230,000 | (75,000) | 3.6 | 8 | M | IMMSA, Mexicana de Cobre, Frisco SA de CV |
| Gold | 5.0 | (4.7) | 1.0 | — | M | IMMSA |
| Iron Ore | 5,350,000 | (2,800,000) | 1.0 | — | M | Sidermex |
| Lead | 142,000 | (160,000) | 5.7 | 5 | M | *Penoles, MR de A |
| Manganese | 200,000 | (95,000) | 1.0 | — | M | Cia Minera Autlan SA de CV |
| Mercury | 52 | (1,200) | 1.4 | — | M | Almex SA, Amescua y Asociados |
| Silver | 1,410 | (1,140) | 17.1 | 1 | M | *Penoles, Fresnillo, IMMSA, MR de A |
| Zinc | 204,000 | (261,000) | 4.6 | 6 | M | *Penoles, MR de A |

Other minerals produced: oil, molybdenum, barytes, coal, fluorite, uranium
IMMSA (Industrial Minera Mexico SA)
*Full name: Industrias Penoles SA
MR de A: (Minera Real de Angeles SA de CV)—partowned by government

## COMPONENTS OF RISK ASSESSMENT

| *Factor* | *Rating* | *Comment* |
|---|---|---|
| **Labour** | | |
| Strike incidence | 6 | Low but unrest is growing |
| Quality of labour | 4 | Good |
| General unrest | 7 | Rising as economic conditions deteriorate |
| **Politics** | | |
| History | 5 | Good |
| Stability | 5 | Good |
| External dangers | 4 | Low, although close to Caribbean influence |
| **Location** | | |
| Hostile borders | 3 | Low |
| Critical land/sea routes | 3 | Low |
| **Mineral Resources** | | |
| Adequacy of reserves | 3 | Excellent |
| Cost of production | 7 | Rising |
| Development appeal | 7 | Dented by economic problems |
| **Financial and Economic** | | |
| Currency performance | 8 | Low with a forced devaluation of the peso in the wake of its debt crisis |
| Vulnerability to manipulation | 8 | High |
| External indebtedness | 9 | Untenable |
| Productivity | 4 | Good |
| Foreign ownerships | 7 | Government formula not conducive to major foreign presence |
| **Energy and Other Natural Resources** | | |
| Domestic energy sources | 3 | High in oil |
| Domestic energy production | 3 | Growing in both oil and coal |
| Sources of outside supply | 1 | Not necessary |
| **Dependence on Foreign Natural Resources** | 4 | Not critical |

***These components are assessed on a scale from 1 (lowest) to 10 (highest) risk. Grouped together, they form the basis for the country risk factor assessment opposite.***

equipment prices however have all moved up sharply. Private enterprise thrives although there is State participation in several operations, particularly uranium. Foreign companies involved in Mexican raw-materials at either the prospecting or development stage include BP, CRA, Outokumpu (Finland) and Billiton. Foreign investment rules are likely to be relaxed as a result of recent financial problems.

### INVESTMENT AND DEVELOPMENT

In August 1982 the Mexican government admitted to a public sector foreign debt of $US 60 billion, twice that of Poland, and froze all foreign currency transactions. The rescue package negotiated amongst Western banks will prove a serious deterrent to any further expansion in industry where copper, silver, oil and coal would have received priority.

# NAMIBIA

**COUNTRY RISK FACTOR** **6.7**

*This factor is assessed from 1.0 (lowest) to 10.0 (highest) risk. For components of this assessment see later table in this section.*

Namibia, formerly South West Africa, is administered by South Africa, under a trusteeship which dates back to the League of Nations in 1919. For many years there has been intense international pressure for this to be relinquished in favour of independence for the region's one million inhabitants. South Africa has resolutely refused on the grounds that none of the constitutions so far proposed would protect either ethnic interest or the integrity of the long common border. The issue is complicated by the political problems of Angola, Namibia's immediate northern neighbour. An effective state of guerilla warfare exists in Namibia, the anti-South African SWAPO group using Angolan territory as a springboard. The country is rich in diamonds, uranium and base metals, to which coal finds have been recently added. It is geographically unattractive and greatly lacking infrastructure but could attract large scale foreign investment given a permanent political solution.

## MINERAL PRODUCTION

Namibia rates fourth in Western world uranium output primarily because of RTZ's Rossing Mine, the largest of its type. The country is second to South Africa in gem diamonds and has measureable, although decreasing, production of base metals. Silver output, at 50 tonnes a year, is increasing. The weakness of world markets is cushioned by the special circumstances surrounding the two major minerals. Uranium production is mostly forward sold on long-term contract and the diamond output has a high proportion of gemstones. Both are obtained from large, open cast facilities. A large belt of metallic mineralisation running roughly east-west supports copper, lead, zinc and silver operations although these are marginal in current terms with output stagnant to decreasing. Much of the country remains unexplored in detail. The diamond operations, controlled by De Beers, work a rich strip of coastal sand at the mouth of the Orange River. Production is limited by market demands rather than reserves. The discovery of an area of power-station quality coal near the Botswana border could prove an economic turning point.

## THE MINING INDUSTRY

The industry is foreign controlled, the major participants being De Beers/CDM in diamonds, RTZ and partners in the Rossing Uranium mine and a number of North American companies in base metals. Tsumeb (Amax, Newmont) is the largest of these, operating four mines,

## OUTPUT AND RATINGS OF PRINCIPAL MINERALS

| *Mineral* | *Output (Tonnes)* 1981 | (1971) | *Percentage of Western World Output* | *Western World Rating* | *Mine (M)/ Smelter (S)* | *Major Producers* |
|---|---|---|---|---|---|---|
| Cadmium | 70 est | (266) | 0.5 | – | M | Tsumeb |
| Copper | 44,300 | (35,000) est | 0.7 | – | M | Tsumeb, Oamites |
| Diamonds (gem) | 1.25* | (1.56) | 5.0 | 4 | M | CDM |
| Germanium (in ore) | 7.3 | N/A | 5-7 | 4 | M | Tsumeb |
| Lead | 47,000 | (73,000) | 1.9 | – | M | Tsumeb, Iscor |
| Silver | 50 | (54) | <1.0 | – | M | Tsumeb, Oamites |
| Tin | 800 | (300) est | <1.0 | – | M | Iscor |
| Uranium | 4,645 | Nil | 9.0 | 5 | M | Rossing (RTZ Ltd) |
| Zinc | 36,000 | (49,000) | 0.8 | – | M | Tsumeb |

*million carats
CDM: Consolidated Diamond Mines (De Beers subsidiary)

## COMPONENTS OF RISK ASSESSMENT

| *Factor* | *Rating* | *Comment* |
|---|---|---|
| **Labour** | | |
| Strike incidence | 7 | Growing |
| Quality of labour | 7 | Low |
| General unrest | 9 | Very high |
| **Politics** | | |
| History | 7 | Uneasy |
| Stability | 9 | Highly unstable |
| External dangers | 8 | High |
| **Location** | | |
| Hostile borders | 7 | Angola, South Africa could become hostile |
| Critical land/sea routes | 8 | High profile on both |
| **Mineral Resources** | | |
| Adequacy of reserves | 3 | Good and capable of expansion |
| Cost of production | 7 | Low on labour, high on infrastructure |
| Development appeal | 7 | Very low now, could easily improve |
| **Financial and Economic** | | |
| Currency performance | 5 | Tied to South Africa |
| Vulnerability to manipulation | 5 | Tied to South Africa |
| External indebtedness | 4 | Low |
| Productivity | 4 | High through diamond sector |
| Foreign ownerships | 8 | High |
| **Energy and Other Natural Resources** | | |
| Domestic energy sources | 8 | Low, as uranium cannot be directly employed but improving on a large coal discovery |
| Domestic energy production | 7 | Very low but coal could start up shortly |
| Sources of outside supply | 8 | Capable of manipulation in oil via South Africa |
| **Dependence on Foreign Natural Resources** | 6 | High but mostly coming from South Africa |

***These components are assessed on a scale from 1 (lowest) to 10 (highest) risk. Grouped together, they form the basis for the country risk factor assessment opposite.***

four concentrators and an integrated smelter. Falconbridge Nickel operates the Oamites copper/silver mine whilst the South African government controlled ISCOR runs a lead-zinc operation plus a tin mine.

## INVESTMENT AND DEVELOPMENT

Geologically, Namibia represents rich pickings. Its economic potential is attractive. But it has a low population base—at 1.2 persons per square kilometre—confined mainly to the coastal outlets and infrastructure costs will be heavy. In spite of the political uncertainties exploration activity continues quite actively. Consolidated Diamond Mines (CDM) has an on-going programme and a subsidiary of Utah International is also active. West Germany paid particular attention to the region some years ago, given its colonial influence remaining from pre-World War One days. If the country is to emerge as an effective economic unit its relationship with South Africa will be critical to its success. As in the case of Botswana, this will be regardless of whatever political facade covers its constitution.

# PAPUA NEW GUINEA

## COUNTRY RISK FACTOR 5.5

*This factor is assessed from 1.0 (lowest) to 10.0 (highest) risk. For components of this assessment see later table in this section.*

Peaceful since full independence from Australia in 1975, Papua New Guinea nevertheless has some ethnic problems. Its relative stability has encouraged mineral exploration and in addition to supporting one of the world's largest copper-gold mines on Bougainville Island, has shown exceptional mineral potential in other areas. The country is financially sound, has a ready labour supply and offers proximity to both Australian and Japanese markets.

### MINERAL PRODUCTION

Bougainville Copper Limited ranks as the world's fourth largest copper and sixth largest gold producer. It is the largest single gold supplier outside South Africa. It will be joined in the late 1980s by a similar development at Ok Tedi in Western Province. Ore grades are slightly better than at Bougainville and in the early stages gold production will be important, perhaps reaching 12 tonnes a year. Another major copper find has been located at Frieda River on the West Irian border whilst gold-silver, nickel-cobalt, chrome and natural gas have also been delineated. Provided market conditions allow and government incentives remain, Papua New Guinea could enter the 1990s as the Western World's fourth largest producer of gold, sixth in copper and significant also in silver, cobalt and nickel.

### THE MINING INDUSTRY

The majority of the Bougainville Copper Company is owned by Australia's CRA, part of the RTZ Group, with 20 per cent government participation and the balance held by the public. A similar pattern is envisaged for the Ok Tedi project where BHP, Amoco and three West German companies are in joint venture, and Frieda River where the partners are MIM, CRA, Norddeutsche Affinerie and the Japanese ORMD Group. Placer and Renison GoldFields are also actively involved in exploration.

### INVESTMENT AND DEVELOPMENT

The political climate is currently attractive to minerals development but the most obvious problems are the massive infrastructure costs in rugged terrain and the increasing dependence of the economy upon the metals market. This might cause a tightening of investment rules. Both climate and language provide problems but the proximity to Australia and to the Japanese market are major plus factors.

## OUTPUT AND RATINGS OF PRINCIPAL MINERALS

| *Mineral* | *Output (Tonnes)* *1981* | *(1971)* | *Percentage of Western World Output* | *Western World Rating* | *Mine (M)/ Smelter (S)* | *Major Producers* |
|---|---|---|---|---|---|---|
| Copper | 165,000 | Nil | 2.6 | 10 | M | Bougainville Copper Ltd |
| Gold | 17.2 | (0.7) | 1.8 | 6 | M | Bougainville Copper Ltd |
| Silver | 37 | Nil | 1.0 | – | M | Bougainville Copper Ltd |

## COMPONENTS OF RISK ASSESSMENT

| *Factor* | *Rating* | *Comment* |
|---|---|---|
| **Labour** | | |
| Strike incidence | 6 | Spasmodic problems |
| Quality of labour | 7 | Raw but capable of advancement |
| General unrest | 6 | Spills over from ethnic differences |
| **Politics** | | |
| History | 7 | Turbulent over several decades |
| Stability | 5 | Improving since 1975 independence |
| External dangers | 4 | Low |
| **Location** | | |
| Hostile borders | 5 | None so long as Indonesia remains stable |
| Critical land/sea routes | 6 | Ever present in South East Asia |
| **Mineral Resources** | | |
| Adequacy of reserves | 3 | Excellent |
| Cost of production | 6 | Medium-high |
| Development appeal | 5 | Tempered by high infrastructure costs |
| **Financial and Economic** | | |
| Currency performance | 4 | Good |
| Vulnerability to manipulation | 6 | Affected by metal prices |
| External indebtedness | 4 | Containable |
| Productivity | 5 | Good |
| Foreign ownerships | 7 | High |
| **Energy and Other Natural Resources** | | |
| Domestic energy sources | 4 | Potentially excellent in gas and hydro, but distance of natural gas reserves from the market has made development, as yet, uneconomic |
| Domestic energy production | 8 | Low |
| Sources of outside supply | 3 | Good relationships with Australia |
| **Dependence on Foreign Natural Resources** | 7 | High but overall demand is low |

***These components are assessed on a scale from 1 (lowest) to 10 (highest) risk. Grouped together, they form the basis for the country risk factor assessment opposite.***

# PERU

**COUNTRY RISK FACTOR** **5.7**

*This factor is assessed from 1.0 (lowest) to 10.0 (highest) risk. For components of this assessment see later table in this section.*

The return to democracy in Peru in 1980 after 12 years of military rule was an important step in the policy of rebuilding the economy and encouraging foreign participation. The country is rich in minerals, the exports of which account for about half of foreign exchange earnings. The new mining code, adopted in 1981, is progressive but Peru remains plagued by inflation and severe labour unrest.

## MINERAL PRODUCTION

Copper is the mainstay of the mining industry, the country now ranking fifth with about 400,000 tonnes a year capacity, almost double that of 1971. Silver production ranks a close second to Mexico but is about half the value of copper output whilst gold, at over seven tonnes a year is becoming increasingly important. Zinc and lead, in which Peru rates third and fourth respectively have also increased measurably. Antimony, bismuth, selenium and tellurium are exported in sizeable quantities.

## THE MINING INDUSTRY

Two large producers, the state-owned Centromin and Minero Peru, plus the privately owned Southern Peru Copper Corporation (SPCC) make up a large-mining sector. SPCC provides almost three-quarters of total copper output from two large mines, Toquepala and Cuajone for which further major expansions are planned. Similarly ambitious programmes are detailed for the state companies, particularly in copper and gold but lack of adequate finance and poor world market conditions

## OUTPUT AND RATINGS OF PRINCIPAL MINERALS

| *Mineral* | *Output (Tonnes) 1981* | *(1971)* | *Percentage of Western World Output* | *Western World Rating* | *Mine (M)/ Smelter (S)* | *Major Producers* |
|---|---|---|---|---|---|---|
| Antimony | 235 | (686) | 0.6 | – | M | |
| Bismuth | 545 | (651) | 17.6 | 3 | S | Centromin |
| Cadmium | 312 | (171) | 2.3 | – | S | |
| Copper | 328,000 | (213,000) | 5.0 | 6 | M | SPCC, Centromin, Minero Peru |
| Gold | 7.2 | N/A | 0.7 | – | M | Centromin, Ocona, Castrovirreyna |
| Indium | 150 * | (35) | 19.5 | Joint 2 | S | Centromin |
| Iron Ore | 6,000,000 | N/A | 1.0 | – | M | Hierroperu |
| Lead | 187,000 | (147,000) | 7.6 | 4 | M | Centromin, Atacocha, Milpo |
| Molybdenum | 900 | (700) | 1.0 | 4 | M | SPCC |
| Selenium | 25 | N/A | 1.8 | – | S | Centromin |
| Silver | 1,244 | (1,243) | 15.2 | 2 | M | Centromin, Arcata, (Hochschild) |
| Tellurium | 21 | (25) | 7.1 | 6 | S | Centromin |
| Tin | 1,380 | N/A | <1.0 | – | M | |
| Zinc | 497,000 | (311,000) | 11.2 | 3 | M | Centromin, San Ignacio, Minero Peru |

*000 ounces

Other minerals produced: tungsten, coal, oil

## COMPONENTS OF RISK ASSESSMENT

| *Factor* | *Rating* | *Comment* |
|---|---|---|
| **Labour** | | |
| Strike incidence | 8 | High |
| Quality of labour | 6 | Medium |
| General unrest | 8 | High |
| **Politics** | | |
| History | 7 | Disturbed |
| Stability | 7 | Questionable |
| External dangers | 3 | Low |
| **Location** | | |
| Hostile borders | 3 | None currently |
| Critical land/sea routes | 7 | Access only to Pacific |
| **Mineral Resources** | | |
| Adequacy of reserves | 3 | High |
| Cost of production | 7 | High |
| Development appeal | 4 | Better than average for a developing region |
| **Financial and Economic** | | |
| Currency performance | 7 | Medium-low |
| Vulnerability to manipulation | 7 | Export minerals dependency |
| External indebtedness | 7 | High |
| Productivity | 8 | Low due to labour unrest |
| Foreign ownerships | 6 | Growing, state monopoly on marketing held by Minpeco partially removed in 1981 |
| **Energy and Other Natural Resources** | | |
| Domestic energy sources | 4 | Oil reserves available |
| Domestic energy production | 4 | Crude oil output at 193,000 b/d in 1981, 70 per cent consumed internally |
| Sources of outside supply | 3 | Not politically constrained |
| **Dependence on Foreign Natural Resources** | 5 | Medium |

***These components are assessed on a scale from 1 (lowest) to 10 (highest) risk. Grouped together, they form the basis for the country risk factor assessment opposite.***

are inhibiting factors. Most of Peru's metal is sold on a free market price basis through the state marketing organisation, Minpeco. The political undertones of Peru's persistent labour disputes are a serious cause for concern.

### INVESTMENT AND DEVELOPMENT

The mining law changes of 1981 were generally favourable to foreign investment but the rapid growth in output in recent years has actually been reversed by recurrent labour problems. Overseas companies entrenched in Peru include Asarco, which owns 52.3 per cent of SPCC, Homestake Mining Co and St Joe Minerals Co whilst several of the majors support active exploration offices. Government missions to Peru include those of France, Germany, Japan, Spain and the UK. If the labour situation can be improved, Peru should maintain 1981 levels of output with perhaps minor additions but could benefit greatly from any increases in world demand towards the end of the 1980s.

# PHILIPPINES

| COUNTRY RISK FACTOR | 6.0 |
|---|---|

*This factor is assessed from 1.0 (lowest) to 10.0 (highest) risk. For components of this assessment see later table in this section.*

Increasingly reliant on the export of minerals and poor in oil, the Philippines has a growing debt service ratio. 1981 was described as bleak but 1982 saw even lower levels of prices. Copper and gold are roughly equal revenue earners but about 60 per cent of gold is won as a by-product of copper operations. This has prompted a government subsidy package for the copper industry which has generated some international ill-feeling. Strong US links, typified in company stockholding and long term loan facilities, are being complemented by a Japanese presence in both minerals purchases and industrial investment. Politically the country appears to be holding together well but minerals expansion has led to rising expectations which may prove difficult to sustain.

### MINERAL PRODUCTION

Copper production rose from below 200,000 to over 300,000 tonnes a year between 1971 and 1981 with the development of relatively high grade mines with good by-product values. The Philippines now stands sixth in the world table of copper producers. Gold production rose, through copper by-product recovery and the opening of new gold mines, to almost 25 tonnes, and the Philippines ranks fifth in this mineral. Recession in the steel industry has hit both nickel and chromite demand but each remains important to the country and the Western world. In chromite the major alternative source is South Africa and expansion in ferro chrome seems a logical progression. Nickel output is well down at 29,000 tonnes a year from a peak 38,300 tonnes in 1980 but reserves and capacity are available to meet any improvement in demand. Coal production is being expanded towards a target of 12 per cent of total energy demand from the present one per cent, whilst indigenous oil is moving towards 17 per cent of demand. By-product cobalt production now approaching 1,400 tonnes a year is also becoming important strategically.

### THE MINING INDUSTRY

Private ownership, through large, publicly quoted companies, dominates the mining industry, beneath a framework of the Ministry of Natural Resources and the Chamber of Mines. Foreign cooperation exists through technical assistance, loans, some joint venture equity, stock holdings and marketing agreements. The government decision in 1982 to grant a subsidy to copper mines, based on the shortfall between the market price and the cost of production, was viewed internationally as interfering with market forces and delaying the industry correction. It had little effect on the long term position of the country as a mining investment area.

15 companies produce copper, the leaders being Atlas Consolidated (140,000 tonnes a year), Marcopper (36,000), Marinduque (27,000) and Benguet (28,000). In

### OUTPUT AND RATINGS OF PRINCIPAL MINERALS

| *Mineral* | *Output (Tonnes) 1981* | *(1971)* | *Percentage of Western World Output* | *Western World Rating* | *Mine (M)/ Smelter (S)* | *Major Producers* |
|---|---|---|---|---|---|---|
| Chromite | 545,000 | (422,000) | 9.7 | 2 | M | Benguet, Acoje |
| Cobalt | 1,360 | N/A | 5.5 | 6 | M | Marinduque |
| Copper | 302,000 | (200,000) | 4.6 | 6 | M | Atlas Consolidated, Western Minolco Benguet, Lepanto |
| Gold | 24.9 | (19.7) | 2.5 | 5 | M | Atlas Consolidated, Lepanto, Western Minolco |
| Nickel | 29,000 | (200) | 6.0 | 5 | M | Marinduque |
| Silver | 63.0 | (61.0) | <1.0 | — | M | Atlas, Marinduque |

Other minerals produced: lead, zinc, manganese, molybdenum, coal, oil

## COMPONENTS OF RISK ASSESSMENT

| *Factor* | *Rating* | *Comment* |
|---|---|---|
| **Labour** | | |
| Strike incidence | 7 | Medium-high |
| Quality of labour | 5 | Medium-low |
| General unrest | 6 | Currently containable |
| **Politics** | | |
| History | 7 | Changeable |
| Stability | 6 | Recently sound |
| External dangers | 5 | None imminent |
| **Location** | | |
| Hostile borders | 7 | None immediate but general area is high profile |
| Critical land/sea routes | 7 | Transport routes easily disrupted |
| **Mineral Resources** | | |
| Adequacy of reserves | 3 | High |
| Cost of production | 7 | Medium-high |
| Development appeal | 6 | Reasonable |
| **Financial and Economic** | | |
| Currency performance | 7 | Medium-low |
| Vulnerability to manipulation | 7 | Medium-high |
| External indebtedness | 8 | High |
| Productivity | 6 | Medium |
| Foreign ownerships | 7 | Low but active through equity interests |
| **Energy and Other Natural Resources** | | |
| Domestic energy sources | 5 | Medium in coal, low in oil |
| Domestic energy production | 7 | Growing slowly |
| Sources of outside supply | 6 | Relatively safe |
| **Dependence on Foreign Natural Resources** | 7 | High |

***These components are assessed on a scale from 1 (lowest) to 10 (highest) risk. Grouped together, they form the basis for the country risk factor assessment opposite.***

addition Benguet produces 300,000 tonnes a year of chromite, and is the leading primary gold producer. Atlas produces coal and primary gold. The Surigao Mine of Marinduque produces most of the country's cobalt plus nickel and iron whilst the Sipalay copper mine produces by-product molybdenum, gold and silver.

### INVESTMENT AND DEVELOPMENT

The Philippines is rich in reserves of copper with associated by-products. At about £800 per tonne copper (65 cents per pound) and $300 per ounce gold, copper and gold reserves are equal but two-thirds of the 25 tonnes a year gold output is recovered from copper production. Market conditions may preclude further nearby copper expansion and will similarly affect the steel industry metals. Concentration on producing domestic fuels will also lessen the strategic importance of the Philippines whilst improving its credit rating.

The outward looking investment regime, the close affinity with the USA and proximity to Japan are all positive assessment factors.

# SOUTH AFRICA

**COUNTRY RISK FACTOR** **5.3**

*This factor is assessed from 1.0 (lowest) to 10.0 (highest) risk. For components of this assessment see later table in this section.*

South Africa, surrounded by unfriendly, passive or openly hostile neighbours and ostracised internationally for its racial policies might, on the face of it, deserve a higher risk rating than 5.3, which is marginally better than Chile. On examination the strength lies in the overwhelming dependence of the West upon its minerals and a reluctant acceptance that even its unpalatable domestic regime is more tolerable than a Soviet dominated or influenced region. Perhaps there is even a Soviet realisation that the fiscal and economic care and control exercised by the minority government might serve Russian foreign policy better than an untested and incohesive black administration thrust arbitrarily into the international ring. In spite of arms, trade and sporting boycotts there is an obvious international policy, galling though it may be to emergent African states, towards persuasion rather than force to bring about change in the last white-minority administration in the world.

The mineral statistics are impressive: first in both output and reserves of gold, platinum, chromite and manganese (all materials the USA badly lacks), first also in vanadium, second in antimony and diamonds and in the top ten in asbestos, titanium sands, copper, iron ore, lead, nickel and tin. It is rapidly moving towards first place in uranium, taking a large part of the export market in coal and combatting its lack of natural oil sources by synthetic production on a large scale. Self sufficient in almost every mineral, except tungsten and oil, it is a leading and growing exporter.

## OUTPUT AND RATINGS OF PRINCIPAL MINERALS

| *Mineral* | *Output (Tonnes)* | | *Percentage of Western World Output* | *Western World Rating* | *Mine (M)/ Smelter (S)* | *Major Producers* |
|---|---|---|---|---|---|---|
| | *1981* | *(1971)* | | | | |
| Antimony | 9,750 | (14,000) | 25.4 | 2 | M | Consolidated Murchison |
| Asbestos | 236,000 | (317,000) | 12.4 | 3 | M | GENCOR |
| Chromite | 3,090,000 | (1,542,000) | 54.8 | 1 | M | Barlow Group |
| Copper | 209,000 | (148,000) | 3.2 | 10 | M | Palabora Mining, JCI, Messina |
| Diamonds (cts) | 9.18* | (7.03)* | 29.2 | 1 | M | De Beers Ltd |
| Gold | 658 | (976) | 68.4 | 1 | M | AAC, GENCOR, GFSA, JCI, Anglovaal, Barlow-Rand Ltd |
| Iron Ore | 23,500,000 | (10,496,000) | 4.9 | 6 | M | ISCOR Ltd |
| Lead | 99,000 | Nil | 4.0 | 8 | M | GFSA/Phelps Dodge |
| Manganese | 6,000,000 | (3,418,000) | 41.7 | 1 | M | Samancor Ltd Associated Manganese Ltd |
| Nickel | 26,400 | (12,800) | 5.3 | 6 | M | JCI Ltd |
| Platinum Group | 96 | (54) | 87.8 | 1 | M | Rustenburg Ltd, Impala Ltd, Western Platinum |
| Silver | 224 | (105) | 2.7 | 8 | M | GFSA/Phelps Dodge |
| Tin | 2,400 | (2,000) | 1.2 | 10 | M | Rooiberg Tin, Union Tin |
| Uranium | 7,183 | (3,800) | 16.7 | 3 | M | AAC, GENCOR, JCI |
| Vanadium | 10,900 | (7,160) | 55.2 | 1 | M | Highveld Steel Ltd |
| Zinc | 86,700 | Nil | 1.9 | – | M | GFSA/Phelps Dodge, JCI |

Other minerals produced: aluminium, fluorite, coal, titanium ores
*million carats
AAC Anglo American Corporation
JCI Johannesburg Consolidated Investments
GFSA Gold Fields of South Africa Ltd
GENCOR General Mining Union Corporation Ltd
ISCOR South African Iron & Steel Corporation

## COMPONENTS OF RISK ASSESSMENT

| *Factor* | *Rating* | *Comment* |
|---|---|---|
| **Labour** | | |
| Strike incidence | 5 | Low in white but rising in black sector |
| Quality of labour | 4 | Skills restricted in black sector |
| General unrest | 7 | Growing |
| **Politics** | | |
| History | 7 | Polarisation of white divisions, English and Afrikaans |
| Stability | 7 | Under serious domestic pressure |
| External dangers | 8 | Pressure for domestic change from all international sources |
| **Location** | | |
| Hostile borders | 7 | Very large borders, all potentially hostile |
| Critical land/sea routes | 7 | Cape of Good Hope is critical alternative to Suez |
| **Mineral Resources** | | |
| Adequacy of reserves | 2 | Vast and developing |
| Cost of production | 3 | Low but rising in black sector |
| Development appeal | 5 | Lessened by political problems |
| **Financial and Economic** | | |
| Currency performance | 7 | Poor against US dollar, but reunification of rand in 1983 was an important forward step |
| Vulnerability to manipulation | 7 | Critically related to gold |
| External indebtedness | 4 | Medium, well backed by reserves/resources |
| Productivity | 4 | Growing |
| Foreign ownerships | 5 | Sizeable but contained |
| **Energy and Other Natural Resources** | | |
| Domestic energy sources | 3 | Excellent barring oil and gas |
| Domestic energy production | 3 | Export surpluses in coal and uranium |
| Sources of outside supply | 7 | Oil imports are politically suspect |
| **Dependence on Foreign Natural Resources** | 4 | Limited to oil and a few special raw minerals<br>Synthetic oil production well developed |

*These components are assessed on a scale from 1 (lowest) to 10 (highest) risk. Grouped together, they form the basis for the country risk factor assessment opposite.*

### MINERAL PRODUCTION

The development of minerals other than gold and diamonds, whose origins lie in the late nineteenth century and the Boer War struggle, is relatively recent. Platinum and nickel date from the 1950s, copper and uranium from the 1960s, the steel industry metals—iron ore, manganese and chromite—slowly over the past 30 years and lead, zinc and silver with the tapping of the deposits in the Northern Cape in the past decade. Coal expansion was a deliberate policy aimed at reducing dependence on oil post the 1973 OPEC crisis and the loss of friendly relationships with Iran following the fall of the Shah. Platinum was given a life injection with the environmentalist decision to fit motor vehicles with exhaust purifiers in 1975 and the mined by-products nickel and cobalt arrived without choice. With the broadening of the minerals base has come the exploration impetus in which international companies, whilst showing apparent diffidence, have played a significant role. There is no formally directed minerals policy. As deposits are found and evaluated, operating licences are granted largely on economic considerations. With the exception of gold, platinum and coal, most minerals deposits are of an individual nature, although some spectacular finds have been made. Reserves of the major minerals are of worldscale, particularly gold, platinum, manganese, chromite, iron ore, coal and diamonds. The reserve bases for copper, asbestos and tin give more cause for concern. In spite of a thriving and export-oriented agricultural sector, South Africa is a minerals based economy, likely to encourage all profitable sectors but placing particular emphasis on the dominant gold, the domestically important coal, and the strategically biased uranium, chrome and manganese sectors.

## THE MINING INDUSTRY

The Chamber of Mines, formed to regulate the activities of the growing gold mining industry, dominates South African mining, particularly the coal and gold sectors. It liaises with government, unions and owners, sets health and safety standards and acts as the reporting medium. Commercially, seven large mining finance houses own and operate the industry. All publicly quoted, they are: Anglo American Corporation, Anglovaal, Barlow Rand, De Beers, Gencor, Gold Fields of South Africa, and Johannesburg Consolidated Investments. Most have overseas interests and are in consortia within and without South Africa in minerals ventures with international groups and governments. Private enterprise is a feature of South African mining development and all the major companies are actively traded on the Johannesburg Stock Exchange and internationally. A particular feature of the industry is the granting of government mining leases to work minerals and a tax and lease system, particularly applicable to gold mining which allows for regulation of grades of ore worked, profitability and taxes paid to accommodate changes in the market price of the product. It is significant that in spite of its dominant position as a producer, South Africa has made little attempt to influence market prices artificially. It relies largely upon the prevailing market in most products, platinum being the exception where the two largest companies periodically adjust their selling or 'producer' price, although even in platinum, Rustenburg was forced in 1983 to officially take note of free market price levels.

## INVESTMENT AND DEVELOPMENT

Although it has felt it necessary to retain exchange control the government has maintained a commercial facility whereby investors can introduce development capital at an advantageous rate (the 'financial' Rand) and secure the commercial outflow of dividends. Since the Rand, formerly fixed to the US dollar, was allowed to float freely in 1980 it has fallen sharply, assisting the domestic performance of companies whose products are sold in dollars but worsening the balance of payments problem. Two successive austerity budgets have taken the edge off foreign investment but the growing pool of private investment funds, technically trapped inside the Republic, flows into major undertakings such as the Sasol synthetic oil programme and the construction of major new gold mines.

Informed financial sources feel that South Africa will remain a reasonable investment area into the foreseeable future, particularly in the gold mining sector and in the mining houses, whose spread of interests is such that little short of a complete overthrow could undermine their basic values. The closest investment parallel to South Africa is probably Australia. Key minerals in common, fighting for shares in the export market, include coal, manganese, iron ore and titanium beach sands. What Australia gains in stability it is now losing in dwindling investor confidence, labour disputes and fragmentation of investment policy at local level.

# SOVIET UNION

### COUNTRY RISK FACTOR

Conditions render risk factor assessments impossible in Communist countries

With the world's largest landmass, a population base greater than that of the USA and already large and proven reserves of most of the world's important minerals, the Soviet Union is the leading producer of oil, coal, asbestos, iron ore, lead, manganese, mercury, titanium and is second in chromite and silver. Its net import reliance is small and its export capability critical. Its domestic growth rate and dearth of foreign exchange are serious influences so that the impact of the Soviet minerals industry upon the world is major, yet changing. Oil export revenues swamp all others but diamonds, gold and platinum are also important. Import requirements in foodstuffs, manufactured goods and technology are vital to a domestic rate of growth which must remain high if the entire communist ideal is to be proved sustainable. A growing number of Comecon countries are now reliant on the Soviet Union for oil and other basics whilst the Soviet influence in Angola, Cuba and Mozambique also represents a financial drain. The Soviet use and pattern of use of Western commodity markets is a key factor in the world minerals balance as is the growth of barter and switch deals.

### OUTPUT AND RATINGS OF PRINCIPAL MINERALS

| *Mineral* | *1981 Output (Tonnes)* | *World Ranking* | *Percentage of Total World Output* | *Mine (M)/ Smelter (S)* |
|---|---|---|---|---|
| Aluminium | 2,400,000 | 2 | 15.4 | S |
| Antimony | 6,500 | 4 | 11.6 | M |
| Asbestos | 2,250,000 | 1 | 45.1 | M |
| Bauxite | 6,400,000 | 4 | 7.4 | M |
| Bismuth | 75 | — | 2.6 | S |
| Cadmium | 2,800 | 1 | 16.0 | S |
| Chromite | 2,400,000 | 2 | 26.3 | M |
| Cobalt | 2,250 | 3 | 8.4 | M |
| Copper | 1,150,000 | 2 | 13.7 | M |
| Diamond Industrial | 8.6* | 2 | 27.7 | M |
| Gem | 2.4* | 2 | 21.1 | M |
| Gold | 260 | 2 | 21.0 | M |
| Iron Ore | 241,000,000 | 1 | 28.3 | M |
| Lead | 570,000 | 1 | 16.5 | M |
| Manganese | 10,350,000 | 1 | 39.4 | M |
| Mercury | 2,137 | 1 | 32.6 | M |
| Molybdenum | 10,200 | 4 | 10.0 | M |
| Nickel | 145,000 | 2 | 20.6 | M |
| Platinum Group | 101 | 1 | 47.9 | M |
| Silver | 1,580 | 2 | 14.0 | M |
| Tin | 16,000 | Joint 5 | 6.8 | S |
| Titanium Sponge | 37,000 | 1 | 38(est) | M |
| Tungsten | 8,800 | 2 | 16.8 | M |
| Vanadium | 10,880 | 2 | 30.6 | S |
| Zinc | 1,010,000 | 2 | 16.5 | M |

*million carats

Note: Figures for selenium, tellurium, germanium, uranium are not reported.
A flask of mercury = 76 lbs or 35 kilos.

## MINERAL PRODUCTION

With central control of all aspects of industry from development to marketing, the succession of five-year plans provides the guideline. The 1981-85 plan is quite sober, recognising the need for improved productivity, self-sufficiency and the availability of key raw materials for export. The aims include: oil and gas to increase seven per cent to 620-645 million tonnes; coal up 7-12 per cent; steel products up 17 per cent. Base metal targets are ambitious: copper +20-25 per cent (240,000-300,000 tonnes a year), aluminium +15-20 per cent (360,000-480,000 tonnes a year), nickel +30 per cent (42,000 tonnes a year), and cobalt +30 per cent (675 tonnes). If this plan is followed through there could be repercussions worldwide, particularly if Soviet industry cannot absorb the expanded mining output.

## THE MINING INDUSTRY

The entire industry is nationalised and subjected to central policy and control in all respects. Infrastructure costs are particularly high and transport distances vast. The metals mining industry employs about three million workers and the coal industry about 2.2 millions. Self-sufficiency rather than cost is the keynote of the minerals industry which is administered by 12 or so separate ministries. Production goals are set at high political levels and are approved by the Supreme Soviet being encapsulated in the Five Year Plan.

Marketing represents a key sector of the industry both in disbursing the high volume of exports and in purchasing the narrow but critical range of imports: bauxite and alumina, tin, fluorine, antimony, tungsten, molybdenum and recently silver. The large tonnages moved through inter-Comecon, barter and switch trading are not assessable. There is increasing use of the Western Free Market system, particularly through London. Separate export and import agencies plus the overseeing Trade Delegation move the full range of LME metals (Cu, Al, Pb, Zn, Ag, Sn, Ni) plus many minor metals. Gold is released through the bullion markets and the banking system of London and Zurich. Part of the large cut diamond output is routed through a joint-venture Belgian company in Antwerp with uncut stones sold into the CSO network in London.

## INVESTMENT AND DEVELOPMENT

As there is no direct foreign investment in Soviet industry an assessment has to be made of officially released information and trading patterns. Western observers cite low utilisation, poor operating efficiencies and a general malaise of Soviet industry as long term problems. Even if Western costs are used to estimate the demands of the five year plan, the sums involved are large. Installing the planned 300,000 tonnes a year of copper capacity alone could be $3-3.50 billion, equal to the sale of 200 tons of gold at $400 per ounce.

Exploration costs are also very high, one source quoting the 1980 budget at 5,000 million roubles (US$ 6.7 billion). The export-oriented production of minerals must remain the focal point of the Soviet economy. Any serious changes in the world demand for minerals will create major distortions in the market for those which the Soviets must sell. Oil, gold, platinum and diamonds are the most critical.

# SPAIN

## COUNTRY RISK FACTOR 5.5

*This factor is assessed from 1.0 (lowest) to 10.0 (highest) risk. For components of this assessment see later table in this section.*

Spain returned to democratic government and a form of monarchy in 1975 on the death of General Franco. It is an uneasy situation compounded by a nationalist problem amongst the Basque minority but the difficulties are within Spain's borders. The country has good minerals output and potential but is suspect financially and has not built a full degree of confidence in the investment world. There is a lack of oil resources but coal deposits, although marginal, are being exploited. The proximity to the European market is an advantage and membership of the EEC is being sought.

### MINERAL PRODUCTION

Spain ranks first in Western world production of mercury, eighth in primary aluminium and zinc, tenth in lead and has a useful output of cadmium, copper, iron ore, silver, tin and tungsten. Coal output in 1981 was approximately 35 million tonnes in all forms, well ahead of 1980, but below domestic requirements. Production covers a useful mixture of power-generation lignite, general purpose bituminous and steel industry anthracite. Oil production is negligible. Uranium production, 210 tonnes in 1981, is planned to be expanded towards a target of 1,000 tonnes a year shortly. Reserves are thought to be in the region of 20,000 tonnes. Pyrites output is also important. Although mineral deposits occur throughout the country the heavy concentration of mining is in the south west. Deposits tend to be medium-small in size and the infrastructure of the industry has to be structured to accommodate this.

### THE MINING INDUSTRY

Private sector development is encouraged beneath the umbrella of a National Mineral Raw Materials Supply Plan and a National Energy Plan. Under this guidance the coal industry has expanded well but the iron and steel sector is in difficulties. The three aluminium producers, Aluminio Espanol, Endasa and Alugasa, have virtually doubled output to 400,000 tonnes a year since 1977 making substantial inroads into alumina imports. Exports are about half the total. Copper output is on the increase at over 50,000 tonnes a year to which will be

## OUTPUT AND RATINGS OF PRINCIPAL MINERALS

| *Mineral* | *Output (Tonnes)* | | *Percentage of Western World Output* | *Western World Rating* | *Mine (M)/ Smelter (S)* | *Major Producers* |
|---|---|---|---|---|---|---|
| | *1981* | *(1971)* | | | | |
| Aluminium | 397,000 | (126,000) | 3.2 | 8 | S | Aluminio Espanol, Endasa, Alugasa |
| Antimony | 660 | (111) | 1.7 | – | M | |
| Cadmium | 303 | (102) | 2.3 | – | S | Real Compania Asturiana SA |
| Iron Ore | 8,800,000 | (7,370,000) | <1.0 | – | M | |
| Lead | 83,000 | (70,000) | 3.4 | 10 | M | IRMM, RCA, Penarroya |
| Mercury | 1,150 | (1,551) | 32.0 | 1 | M | Cia Minas de Almaden |
| Zinc | 176,000 | (92,000) | 4.0 | 7 | M | Asturiana de Zinc, Exminesa |

Other minerals produced: copper, silver, tin, tungsten, coal
Exminesa: Exploracion Minera Internacional Espana SA

## COMPONENTS OF RISK ASSESSMENT

| *Factor* | *Rating* | *Comment* |
|---|---|---|
| **Labour** | | |
| Strike incidence | 4 | Fairly low |
| Quality of labour | 6 | Good |
| General unrest | 4 | Low |
| **Politics** | | |
| History | 7 | Uneasy since death of Franco in 1975, with several attempted military coups |
| Stability | 7 | Suspect |
| External dangers | 3 | Low |
| **Location** | | |
| Hostile borders | 1 | None |
| Critical land/sea routes | 4 | Reasonably placed |
| **Mineral Resources** | | |
| Adequacy of reserves | 6 | Medium |
| Cost of production | 4 | Fairly low |
| Development appeal | 4 | Good |
| **Financial and Economic** | | |
| Currency performance | 7 | Poor |
| Vulnerability to manipulation | 7 | High |
| External indebtedness | 7 | Medium-high |
| Productivity | 6 | Medium |
| Foreign ownerships | 6 | Being encouraged |
| **Energy and Other Natural Resources** | | |
| Domestic energy sources | 7 | Inadequate, but coal being developed |
| Domestic energy production | 8 | Inadequate, giving high volume/cost of fuel imports equating to $12,800 million in 1981 |
| Sources of outside supply | 4 | Safe |
| **Dependence on Foreign Natural Resources** | 6 | Medium |

*These components are assessed on a scale from 1 (lowest) to 10 (highest) risk. Grouped together, they form the basis for the country risk factor assessment opposite.*

added the expansion of the Cerro Colorado plant in Huelva. At the Sotiel (Huelva) mine of Minas de Almagrera, copper, lead and zinc will come on stream in 1983.

### INVESTMENT AND DEVELOPMENT

The medium, 5.5 risk rating of Spain reflects the positive domestic attitude to mining expansion and the proximity to the European market but also registers the relatively mediocre geological and infrastructure attractions. In the absence of measurable oil resources, coal appears quite attractive whilst iron and steel need expensive rationalisation in a depressed market. Aluminium too may have run its course for some years. The base metals, copper, lead and zinc present a medium investment risk whilst tin could be steadily built up although individual deposits are small. There is little room to manoeuvre in the mercury sector and it is doubtful that the oversupplied uranium market holds outside attractions.

# THAILAND

## COUNTRY RISK FACTOR 6.0

*This factor is assessed from 1.0 (lowest) to 10.0 (highest) risk. For components of this assessment see later table in this section.*

In spite of the severe problems, both economic and political, of its near neighbours, Thailand remains relatively stable whilst relying massively on tin as its main source of economic stability. Currency devaluation has helped stabilise exports whilst the commercial production of natural gas is also assisting. In spite of the latent instability of the entire South East Asia region, Thailand has managed to attract foreign interest and continues to build upon a mineral-based economy.

### MINERAL PRODUCTION

Tin remains the backbone. 1981 production at 31,500 tonnes was below the peaks of the previous two years but still represents nearly 16 per cent of Western world output. Small-boat suction dredge activity, associated with the serious problem of tin smuggling, showed a downturn in providing 33 per cent of output, but regular offshore and onshore dredges and gravel pump mines improved. Environmentalism is now restricting all-out expansion but gives no grounds for believing that output need fall substantially. Tungsten mining, in which the Doi Ngom field is the leading source, produces about 1,600 tonnes of metal in concentrate, almost six per cent of Western world output. Tantalum concentrates of 358 tonnes, containing about 59 tonnes metal content, are produced, representing over 13 per cent of the Western world total. Lead production is rising through the increased output of the Metallgesellschaft joint venture mine of KEM Limited and there is minor zinc, iron ore and a little gold. Antimony remains important, in spite of a prevailing market weakness, and accounts for over 4.5 per cent of Western supplies. Fuels are of increasing interest. Natural gas is now produced commercially from the Gulf of Thailand where reserves are estimated at 456,000 million cubic metres. Lignite production is moving towards the two million tonnes a year mark and there is also oil shale and natural oil potential.

### THE MINING INDUSTRY

A mixture of government owned, private Thai and foreign contract and joint venture operations runs the minerals industry, all these sectors being represented in tin. The major operators include the state owned Offshore Mining Organisation (OMO), Billiton, Tongkah Harbour Limited and Aokam. Foreign interests in Thai mining include Billiton, Metallgesellschaft, Vieille Montagne, Mechim, CRA Limited and AMAX.

### INVESTMENT AND DEVELOPMENT

The recent history and the location of the South East Asia Peninsula raise obvious question marks against permanent capital investment in Thailand but since 1980 the domestic signs have been positive. A responsible government approach to the economy, the encouragement of foreign participation and the active development of energy resources give the country a relatively healthy risk rating. A repeat of the 1970s successes of Indonesia is a distinct possibility but the ability to control economic growth will need to be closely monitored.

## OUTPUT AND RATINGS OF PRINCIPAL MINERALS

| *Mineral* | *Output (Tonnes) 1981* | *(1971)* | *Percentage of Western World Output* | *Western World Rating* | *Mine (M)/ Smelter (S)* | *Major Producers* |
|---|---|---|---|---|---|---|
| Antimony | 1,750 | (2,249) | 4.6 | 4 | M | |
| Lead | 16,700 | (5,500) | 0.7 | – | M | Kanchanaburi Exploration and Mining |
| Tantalum | 59 | N/A | 13.3 | 4 | S | Aokam Tin, Thaisarco |
| Tin | 31,500 | (21,700) | 15.6 | 3 | M | Aokam Tin, Tongkah Harbour |
| Tungsten | 1,590 | (4,865) | 5.8 | 7 | M | Southwest Consolidated Resources |

Other minerals: lignite, oil, gas, iron ore, zinc, gold

## COMPONENTS OF RISK ASSESSMENT

| *Factor* | *Rating* | *Comment* |
|---|---|---|
| **Labour** | | |
| Strike incidence | 5 | Average, many small operations each effectively run by one family reduces labour difficulties |
| Quality of labour | 5 | Reasonable and adaptable |
| General unrest | 6 | Fair in relation to the region |
| **Politics** | | |
| History | 7 | Troubled since World War Two |
| Stability | 7 | Improving although suspect |
| External dangers | 8 | Ever present |
| **Location** | | |
| Hostile borders | 8 | Borders Kampuchea (Cambodia) |
| Critical land/sea routes | 8 | In the heart of South East Asia |
| **Mineral Resources** | | |
| Adequacy of reserves | 5 | Good in tin, largely unproven in others |
| Cost of production | 5 | Sustainable |
| Development appeal | 6 | Improving |
| **Financial and Economic** | | |
| Currency performance | 7 | Improving with stronger controls. Baht devalued against US dollar in effort to promote exports in 1982 |
| Vulnerability to manipulation | 7 | Serious but improving with devaluation and gas revenues |
| External indebtedness | 6 | Probably containable |
| Productivity | 5 | Good |
| Foreign ownerships | 4 | A containable presence |
| **Energy and Other Natural Resources** | | |
| Domestic energy sources | 4 | Improving |
| Domestic energy production | 4 | Increasing rapidly, with Shell producing Thailand's first commercial crude oil at 4,600 b/d |
| Sources of outside supply | 4 | Safe and becoming less important |
| **Dependence on Foreign Natural Resources** | 8 | High |

*These components are assessed on a scale from 1 (lowest) to 10 (highest) risk. Grouped together, they form the basis for the country risk factor assessment opposite.*

# UNITED KINGDOM

## COUNTRY RISK ASSESSMENT 5.0

*This factor is assessed from 1.0 (lowest) to 10.0 (highest) risk. For components of this assessment see later table in this section.*

Britian's importance as a minerals producer waned by 1850 in all except coal although it has remained an important industrial user. It is also prominent in the refining and fabrication of specialist metals and is a world minerals trading and financial centre.

### MINERAL PRODUCTION

The legacy of Britain's mid nineteenth century position as the leading producer of most industrial metals remains in a large refining capacity in copper, lead, zinc and silver, a major, but shrinking, primary steel and tinplate industry and a high volume of recycling and secondary refining. Primary aluminium, tin and platinum group metals refining are also world ranking. By volume, coal is only about one half of its peak levels around the turn of the century but ranks third in the Western world and is the only truly viable coal industry in Western Europe. The discovery of gas followed by oil in the late 1960s has transformed the energy balance. Tin mining is enjoying a resurgence and there is a belief that the 1981 level of 3,900 tonnes could be at least doubled. Exploration is at its best level for some years and has yielded economic reserves of tungsten, barytes, fluorspar and isolated, but workable, deposits of lead, zinc and copper.

### THE MINING INDUSTRY

Coal remains the major product although the domestic base is shrinking whilst oil, in which the UK now ranks sixth in production terms, has transformed the balance of payments. The coal industry is over 95 per cent nationalised and, although the production of oil is largely in the hands of international operators, there are strict government controls over output and marketing. Three international mining groups—Consolidated Gold Fields, Charter Consolidated and RTZ Ltd—are UK based whilst British Petroleum and the Royal Dutch Shell Group operation have heavy mining commitments. The Cornish tin industry is in transition with RTZ Ltd showing determination to expand operations. Several major companies are funding projects of which the Hemerdon tin, tungsten deposit of Amax Inc in Devon is the most ambitious. On full production it could make the UK a net exporter of tungsten.

### INVESTMENT AND DEVELOPMENT

On paper, domestic minerals demand, the need to boost heavy industry and the level of taxation offer excellent incentives but a complex system of mineral ownership rights and tight environmental laws are serious deterrents. It is also ironic that a country which boasts long term reserves and an export surplus in coal, gas and oil plus one of the more advanced nuclear industries should be losing its refining capacity on energy cost considerations. There is excellent potential for expansion in British mining but clear lack of incentive.

## OUTPUT AND RATINGS OF PRINCIPAL MINERALS

| *Mineral* | *Output (Tonnes) 1981* | *(1971)* | *Percentage of Western World Output* | *Western World Rating* | *Mine (M)/ Smelter (S)* | *Major Producers* |
|---|---|---|---|---|---|---|
| Aluminium | 339,000 | (119,000) | 2.7 | 9 | S | ALCAN, Kaiser/RTZ, British Aluminium |
| Cadmium | 278 | (262) | 2.1 | — | S | RTZ |
| Tin | 3,869 | (1,800) | 1.9 | 8 | M | RTZ, Geevor, South Crofty |

Other minerals: copper, iron ore, lead, zinc, barytes, oil
Refined metal: copper, nickel, platinum

## COMPONENTS OF RISK ASSESSMENT

| *Factor* | *Rating* | *Comment* |
|---|---|---|
| **Labour** | | |
| Strike incidence | 4 | Major improvements since 1979-80 |
| Quality of labour | 3 | Highly skilled force |
| General unrest | 6 | Underlying political/class divisions |
| **Politics** | | |
| History | 3 | Long established democracy |
| Stability | 4 | Some minority problems, probably containable |
| External dangers | 3 | Low |
| **Location** | | |
| Hostile borders | 1 | None |
| Critical land/sea routes | 7 | English Channel, North Sea, proximity to Eastern Europe |
| **Mineral Resources** | | |
| Adequacy of reserves | 9 | Low in most minerals |
| Cost of production | 7 | High where domestic reserves exist |
| Development appeal | 6 | Improving slowly |
| **Financial and Economic** | | |
| Currency performance | 7 | Volatile since not linked to European Monetary System |
| Vulnerability to manipulation | 7 | Closely allied to economic health |
| External indebtedness | 6 | Improving with oil output |
| Productivity | 7 | Medium-low in developed terms |
| Foreign ownerships | 5 | Medium |
| **Energy and Other Natural Resources** | | |
| Domestic energy sources | 2 | Excellent barring uranium |
| Domestic energy production | 3 | Growing in non-coal sector, although North Sea oil output and development vulnerable to taxation levels |
| Sources of outside supply | 2 | Safe |
| **Dependence on Foreign Natural Resources** | 8 | High. In 1983, the UK Government began a limited strategic stockpiling programme to hedge against a disruption in supplies of chromium, manganese, vanadium and cobalt |

*These components are assessed on a scale from 1 (lowest) to 10 (highest) risk. Grouped together, they form the basis for the country risk factor assessment opposite.*

# UNITED STATES OF AMERICA

**COUNTRY RISK ASSESSMENT** **4.3**

*This factor is assessed from 1.0 (lowest) to 10.0 (highest) risk. For components of this assessment see later table in this section.*

The rise of the USA from underdeveloped status to first place as an industrial and military power took less than 60 years. It profited from the key elements of a young immigrant population, a large landmass, a compatible climate and a wealth of resource minerals. The outcome is an economy which, whilst now showing some premature ageing cracks, leads the world in minerals production and consumption and is the focal point of the international currency system. In broad terms the USA consumes 30 per cent of the world's raw materials which produce 35 per cent of its GNP. It ranks in the top ten in the production of no fewer than 20 of the most critical industrial minerals. With minor and temporary exceptions, the direction of the US economy governs that of the world generally.

## MINERAL PRODUCTION

The Californian gold discoveries in the mid 1800s increased world financial liquidity and offered opportunities for opening the North American sub-continent at a pace hitherto believed impossible. An understanding of the development which followed is critical to the appreciation of the current role of the USA in world minerals. When domestic demand began to outstrip domestic ability to provide, US interests moved abroad—into Chile, Africa and elsewhere—to secure supplies. Their continuity was threatened firstly by the two World Wars then by the successive conflicts in Korea (1950-53), the Middle East and Suez (1953-56) and Vietnam (1969-75). In the light of these experiences US minerals policy has been reappraised. The OPEC oil crisis has probably had the most significant effect, not only in reshaping minerals policy, but also in the corporate restructuring of industry.

The USA has known occurrences of almost all critical industrial minerals. In many cases, however, the country is deficient in current production, economic reserves or both. In the steel industry metals—manganese, nickel and chromium—it has a growing foreign dependency and in these, plus platinum and cobalt, Africa is the prime source. A marked deficiency in other minerals including antimony, asbestos, bauxite, chromium, columbium, diamond, fluorspar, titanium, and tantalum also has political connotations.

In addition to its strategic role, minerals production is vital economically. The 1981 value of processed raw materials, an estimated $240 billion, was about eight per cent of total GNP and the industry directly employs over one million people. Environmental restraints and increasing tax burdens have combined with market recession to dull the edge of mining investment but incentives in the domestic energy sector are growing, where several major producers have diversified into the production of non-fuel minerals. Other than in those materials mentioned there is no shortage of either developed or available reserves. Economic incentive is the only criterion.

## THE MINING INDUSTRY

Private enterprise in the shape of large, publicly quoted conglomerates, dominates the US minerals industry. Government involvement is limited to legislation, the export of critical minerals such as uranium and the withholding of certain statistics of strategic importance. As late as 1970, it was possible to identify each major company with the output of a single mineral or related group, e.g. copper, uranium, iron and steel, coal, oil. Since then considerable expansion, state involvement and merger activity has changed the format considerably. In particular, oil companies, awash with surplus revenues for some time after 1975, moved laterally into other fuels and into metals. The enthusiasm for investment abroad, encouraged by the belief that this would secure supplies to meet the domestic shortfall, has waned. A combination of foreign nationalisations, poor investment terms and general recession have contributed to this change of attitude.

Outside the domestic steel sector and the integrated oil companies, the largest metals producers in 1981 were: Alcoa (aluminium, $4,998 million); Amax (copper, molybdenum, coal, $2,799 million); Asarco (copper, silver, lead, zinc, $1,532 million); Cities Service (copper, oil $8,546 million); Fluor (includes St Joe Minerals, base metals); Kaiser Aluminium ($3,342 million); Kerr McGee (titanium, uranium, $3,826 million); Phelps Dodge (copper, $1,439 million); Reynolds (aluminium, $3,481 million); Utah International (copper, coal, $1,722 million). Figures in brackets refer to 1981 revenues.

## INVESTMENT AND DEVELOPMENT

The investment climate in the USA has always been considered relatively attractive but the 1974 OPEC shock gave great impetus to domestic energy production. Oil and gas prices were gradually decontrolled and a coal industry blueprint envisaged a doubling of production

## COMPONENTS OF RISK ASSESSMENT

| *Factor* | *Rating* | *Comment* |
|---|---|---|
| **Labour** | | |
| Strike incidence | 3 | Low, good union discipline |
| Quality of labour | 3 | Good, high level of mechanisation |
| General unrest | 3 | Low, some ethnic problems |
| **Politics** | | |
| History | 3 | Long established democratic system |
| Stability | 3 | Very good |
| External dangers | 2 | None foreseeable |
| **Location** | | |
| Hostile borders | 1 | None |
| Critical land/sea routes | 2 | Possible problems in Gulf of Mexico |
| **Mineral Resources** | | |
| Adequacy of reserves | 6 | Shortage or absence of several key minerals |
| Cost of production | 7 | Medium but rising |
| Development appeal | 6 | Excellent in selected industries but environmentalism a growing problem |
| **Financial and Economic** | | |
| Currency performance | 4 | Volatile but strengthening |
| Vulnerability to manipulation | 7 | High as the leading convertible currency |
| External indebtedness | 7 | High but balanced by industrial production capacity |
| Productivity | 7 | High but trailing Japan |
| Foreign ownerships | 3 | Fairly low |
| **Energy and Other Natural Resources** | | |
| Domestic energy sources | 4 | Excellent but oil imports still prevail |
| Domestic energy production | 3 | Growing rapidly |
| Sources of outside supply | 5 | Dependence on OPEC and Mexico |
| **Dependence on Foreign Natural Resources** | 7 | Serious in some key minerals, but these could mostly be developed domestically in emergency. The GSA (General Services Administration) has acted as agent to build up government stockpiles of selected minerals, with the aim of providing self-sufficiency in a traditional three year war scenario |

*These components are assessed on a scale from 1 (lowest) to 10 (highest) risk. Grouped together, they form the basis for the country risk factor assessment opposite.*

---

from 600 to 1,200 million tonnes a year by 1985. It has been successful to the degree allowed by market forces and has prompted a high level of both domestic and foreign investment. Other minerals have not fared so well. Environmental constraints, both in the production and end-use of certain metals, have proved dampening whilst the serious market surpluses resulting from the onset of recession in 1979 have reduced output in some base metals to below half capacity.

The long term trend is also not encouraging, serious reductions having registered in output since 1971 in lead, zinc and gold, whilst aluminium, copper and nickel have stagnated. If the USA can maintain its powerful industrial base, it may assist world recovery for it to encourage raw materials production abroad. Yet, as long as international capital is free to select its base, any serious economic upturn will probably see US mining projects once more well supported.

## OUTPUT AND RATINGS OF PRINCIPAL MINERALS

| *Mineral* | *Output (Tonnes) 1981* | *(1971)* | *Percentage of Western World Output* | *Western World Rating* | *Mine (M)/ Smelter (S)* | *Major Producers* |
|---|---|---|---|---|---|---|
| Aluminium | 4,491,000 | (3,561,000) | 36.0 | 1 | S | Howmet, Alumax, Anaconda, Conalco, Alcoa, Kaiser, Reynolds, Southwire |
| Antimony | 586 | (930) | 1.5 | – | M | |
| Asbestos | 81,000 | (118,000) | 4.3 | 6 | M | Manville Corporation |
| Bauxite | 1,510,000 | (2,020,000) | 2.0 | 11 | M | Alcoa, Reynolds |
| Beryllium* (beryl) | 2,000 | N/A | 68.9 | 1 | M/S | Brush-Wellman Inc |
| Cadmium | 1,871 | (3,597) | 14.1 | 2 | S | Amax, Asarco, Bunker Hill, National Zinc |
| Copper | 1,529,000 | (1,381,000) | 23.8 | 1 | M | Amax, Asarco, Phelps Dodge, Cities Service |
| Gallium* | 4.5 | N/A | 18.8 | 2 | S | Alcoa, Eagle Picher |
| Germanium* | 30 | N/A | N/A | 2 | S | Eagle Picher |
| Gold | 41 | (47) | 4.2 | 3 | M | Homestake, Kennecott, Newmont |
| Indium* | 10 | (8) | 41.5 | 1 | S | Indium Corp |
| Iron Ore | 74,000,000 | (83,600,000) | 14.0 | 3 | M | US Steel, Republic, National |
| Lead | 452,000 | (547,000) | 18.4 | 1 | M | Amax, Asarco, St Joe, Homestake |
| Lithium | 4,800 | (3,500) est | 90.0 est | 1 | M | Foote Minerals, Lithium Corp |
| Magnesium | 150,000 | N/A | 65.7 | 1 | S | Dow Chemical, Amax, Northwest Alloys |
| Mercury | 965 | (601) | 26.9 | 3 | M | Placer Amex |
| Molybdenum | 65,800 | (49,000) | 72.9 | 1 | M | Amax, Duval, Kennecott |
| Nickel | 11,000 | (16,000) | 2.2 | 11 | M | Amax, Hanna |
| Selenium | 270 | N/A | 19.0 | 3 | S | Phelps Dodge, Asarco |
| Silver | 1,168 | (1,293) | 13.6 | 4 | M | Asarco, Hecla, Phelps Dodge, Newmont |
| Tellurium* | 80 | N/A | 27.1 | Joint 1 | S | Phelps Dodge, Asarco |
| Tin | 96 | N/A | <0.1 | – | M/S | |
| Titanium | | | | | | |
| –Ilmenite | 550,000 | (650,000) est | 11.4 | 4 | M | Asarco |
| –Rutile | 30,000 | | 6.7 | 4 | M | Kerr-McGee, Gulf & Western |
| –Titanium Sponge | 27,000 | N/A | 49.0 est | 1 | S | Oremet, Timet |
| –Titanium Dioxide | 689,000 | N/A | 39.0 | 1 | S | Kerr McGee |
| Tungsten | 3,175 | (7,000) est | 11.5 | Joint 2 | M | Amax |
| Uranium | 14,115 | (11,247) | 32.9 | 1 | M | Kerr-McGee, Freeport, Phelps Dodge |
| Vanadium | 4,800 | (5,000) | 24.3 | 2 | M | Union Carbide |
| Zinc | 335,000 | (501,000) | 7.6 | 4 | M | Amax, Anaconda, Asarco, New Jersey Zinc |

Other minerals production: bismuth, platinum, tin, rare earth minerals, rhenium, silicon, zirconium
*Estimated as actual figures withheld

# ZAIRE

**COUNTRY RISK ASSESSMENT** **6.8**

*This factor is assessed from 1.0 (lowest) to 10.0 (highest) risk. For components of this assessment see later table in this section.*

One of the largest countries in Africa and one of the more industrialised, Zaire is important for its minerals output as well as its strategic location. It ranks first in cobalt and industrial diamonds and is listed in cadmium, tin and tantalum. By-product germanium is also important. The country's desperate finances, which deteriorated rapidly between 1975 and 1979, have received critical attention from the IMF and the Western banking system. Under strict guidelines, the performance has begun to improve, although low metal prices and a very poor diamond market have held it back. Although President Mobutu has retained power for over a decade, his grip has been challenged on both ethnic and economic fronts. Zaire received the largest sum ever recorded by an African nation from the IMF, US$ 1,050 million, in 1981 and there is said to be quiet optimism at early signs of improvement. Foreign exchange levels, external indebtedness and a dearth of vital manufactured supplies remain severe problems and the country is not equipped to weather a prolonged market recession.

## MINERAL PRODUCTION

Copper and industrial diamonds have been the mainstay of the economy for many years. At over 500,000 tonnes a year copper is once again approaching peak levels after several years of decline whilst cobalt, of which Zaire is the most important source, enjoys a higher basis-price level. The diamond industry has suffered on two counts: the industrial stones market has met severe competition from synthetics whilst gemstones have been caught in a severe deflationary spiral.

The tin industry, which registered output at over 6,000 tonnes in 1971, has halved output in spite of high price levels. Cobalt output may have registered as high as 15,000 tonnes in 1981 but, as sales were depressed and stocks high, the published figure will be nearer 13,600 tonnes.

There is an ambitious rehabilitation programme designed primarily to bring the mining facilities back up to standard so that further output increases will probably be marginal. A number of feasibility studies are in hand but these will be at least five years on the drawing board. The exception is the Tenke-Fungurume copper-cobalt project on which considerable development has already been done. This could add 40,000 tonnes a year of copper and 2,000 tonnes a year of cobalt. To counter weak market conditions, Zaire is planning temporary drastic cuts in cobalt output, perhaps down to 10,000 tonnes.

## THE MINING INDUSTRY

Private participation is again encouraged but the state-controlled Gecamines accounts for over 90 per cent of copper-cobalt output, mostly won from mines operating

## OUTPUT AND RATINGS OF PRINCIPAL MINERALS

| *Mineral* | *Output (Tonnes) 1981* | *(1971)* | *Percentage of Western World Output* | *Western World Rating* | *Mine (M)/ Smelter (S)* | *Major Producers* |
|---|---|---|---|---|---|---|
| Cadmium | 230 | (262) | 1.7 | – | S | Gecamines |
| Cobalt | 13,600 | (14,000) | 54.7 | 1 | M | Gecamines, Sodimiza |
| Copper | 505,000 | (406,000) | 7.8 | 5 | M | Gecamines, Sodimiza |
| Diamonds Gem | 0.42 * | >14.8 | 4.7 | Joint 5 | M | Bakwanga Mining |
| Ind | 9.90 * | | 43.5 | 1 | M | Bakwanga Mining |
| Germanium (in ore) | 22.5 | N/A | 15-20 | 3 | M | Gecamines |
| Gold | 3.2 | N/A | 0.3 | – | M | Sominki, Kilo Moto |
| Silver | 80 | (56) | 1.0 | – | M | Gecamines |
| Tantalum | 11,340 | (51,000) | 2.6 | 8 | M | Sominki |
| Tin | 2,550 | (6,500) | 1.2 | 9 | M | Sominki, Zairetain |
| Zinc | 76,000 | (109,000) | 1.7 | – | M | Gecamines |

*million carats

Other minerals: tungsten, columbite-tantalite, monazite, bauxite, coal

Gecamines: La Generale des Carrieres et des Mines du Zaire

## COMPONENTS OF RISK ASSESSMENT

| *Factor* | *Rating* | *Comment* |
|---|---|---|
| **Labour** | | |
| Strike incidence | 7 | Medium-high |
| Quality of labour | 6 | Medium for the region |
| General unrest | 8 | Recently high |
| **Politics** | | |
| History | 8 | Troubled |
| Stability | 6 | Well held, but resting on personalities |
| External dangers | 8 | Possible troubles via Congo and Angola |
| **Location** | | |
| Hostile borders | 7 | Proximity to Angola, Congo |
| Critical land/sea routes | 8 | High profile in east, west and southerly directions |
| **Mineral Resources** | | |
| Adequacy of reserves | 3 | Excellent |
| Cost of production | 8 | High |
| Development appeal | 7 | Moderate, but well qualified risks |
| **Financial and Economic** | | |
| Currency performance | 9 | Very poor |
| Vulnerability to manipulation | 8 | High |
| External indebtedness | 8 | High but accommodated by IMF etc |
| Productivity | 7 | Low but improving |
| Foreign ownerships | 3 | Low but encouraged |
| **Energy and Other Natural Resources** | | |
| Domestic energy sources | 4 | Some oil, good hydro potential |
| Domestic energy production | 5 | Low but improving |
| Sources of outside supply | 8 | Critical of foreign exchange and transport considerations |
| **Dependence on Foreign Natural Resources** | 7 | High but could be reduced by domestic expansion |

***These components are assessed on a scale from 1 (lowest) to 10 (highest) risk. Grouped together, they form the basis for the country risk factor assessment opposite.***

in South Shaba, close to the Zambian border. Gecamines estimated 1981 production is around 460,000 tonnes copper. The balance, exported in concentrate form, comes from the company Sodimiza, in which both Zairean and Japanese groups (including Nippon Mining Co) have an interest. Minerals marketing is conducted by the state organisation Sozacom, which has agents appointed world-wide.

Effectively landlocked, Zaire has major transport problems, even though hostilities in surrounding territories have largely ceased. Unrest in Angola leaves the westward link of the Benguela Railway inoperable. The alternative routes, each taking about 50 per cent of traffic, are east through Tanzania and south through Zambia. Both are costly and subject to major disruptions en route and at port. The Tenke-Fungurume project, started with major Western involvement in 1970 but essentially abandoned in 1975 has high value reserves running five per cent copper and 0.5 per cent cobalt. Shareholders include a subsidiary of the French Atomic Energy Commission, Britain's Charter Consolidated and Japan's Mitsui.

### INVESTMENT AND DEVELOPMENT

Western support for the Zairean economy, of which mining is the cornerstone, continues unstintingly. Mineral reserves are high and varied but a combination of a weak market in all the major products plus a poor financial track record do not enhance the region's rating. Two ambitious schemes now at the study stage are involved in the tin-lithium deposit of Manono, the largest of its type in the world, and an aluminium smelter plan to convert local bauxite by hydro-electricity from the Zaire River.

The decision to break away form the CSO diamond sales monopoly in 1981 may have damaged the foreign investment rating also. The reversal of the decision in March 1983 indicates that the 22 month experiment was not a success.

# ZAMBIA

## COUNTRY RISK FACTOR 6.8

*This factor is assessed from 1.0 (lowest) to 10.0 (highest) risk. For components of this assessment see later table in this section.*

Zambia's position as the non-communist world's third largest producer of copper and cobalt makes the established nature of its economic difficulties and its critical geographical location in Central Africa all the more serious. Its politically admirable severence of all links with Zimbabwe during the critical 1965-79 period hurt the economy deeply, a condition aggravated by the growing world surplus of copper, upon which Zambia relies for about 90 per cent of its foreign exchange. The industry is losing its competitive edge on a world basis, irrespective of the increased cost of oil upon which the country is totally import-dependent. During the independence period Zambia has swung from a modest exporter of foodstuffs into a serious importer and the entire economic structure relies heavily upon sympathetic treatment by the IMF and the Western banking system generally. The political regime led by the sole President since independence in 1964, Kenneth Kaunda, is now an official one-party system. This introduces a degree of stability but, with the attendent economic problems, an undercurrent of unrest is ever present. Unfortunately no immediate relief appears in sight for the market in either major mineral so that any improvement can only come via productivity and marketing.

### MINERAL PRODUCTION

The copperbelt of Zambia was developed by the British as a strategic colonial resource following World War One. At independence two large groups led by the USA conglomerate AMAX and the Anglo American Corporation owned and operated the industry. In successive moves between 1969 and 1981, the government took majority control of management and marketing. They finally merged the entire industry into one large concern (ZCCM), the government holding 60.3 per cent of the voting shares. Copper production is centred on eight established mines with reserves, at current rates, capable of sustaining output for between 25 and 40 years. By-product cobalt and lead/zinc values are in similar proportion. Production in 1981 of 587,000 tonnes mined, follows a pattern of consistent decrease since the early 1970s but still ranks third in the West. The rate of decrease has been arrested since 1979 but the country's ability to finance stocks in a weak market is limited. Much of the open-pit potential appears exhausted and, whilst underground both grades and volumes arithmetically support an expanding industry, domestic problems, foreign exchange and outside energy dependence are detrimental. Cobalt production is being expanded although the euphoria which followed the unusually active market conditions of 1978-79 has died down. Production capacity of 4,000 tonnes was well above output or demand and at about 12 per cent of Western requirements indicates the continuation of a weak market. Combined lead and zinc capacity of 76,000 tonnes is a useful revenue earner but of no strategic importance. A high proportion of Zambia's metal output is refined domestically and the brands produced enjoy premium values in the market.

### OUTPUT AND RATINGS OF PRINCIPAL MINERALS

| *Mineral* | *Output (Tonnes)* | | *Percentage of Western World Output* | *Western World Rating* | *Mine (M)/ Smelter (S)* | *Major Producers* |
|---|---|---|---|---|---|---|
| | *1981* | *(1971)* | | | | |
| Cobalt | 2,833 | (2,080) | 10.3 | 2 | M | ZCCM Ltd |
| Copper | 587,000 | (651,000) | 9.1 | 4 | M | ZCCM Ltd |
| Lead | 16,100 | (34,000) | 0.6 | – | M | ZCCM Ltd |
| Zinc | 39,700 | (69,000) | 0.9 | – | M | ZCCM Ltd |

Other minerals: gemstones, magnetite (iron ore)

## COMPONENTS OF RISK ASSESSMENT

| *Factor* | *Rating* | *Comment* |
|---|---|---|
| **Labour** | | |
| Strike incidence | 8 | High in recent years |
| Quality of labour | 7 | Slowly improving domestically but high expatriate turnover |
| General unrest | 8 | Disaffection brought on by poor living standards |
| **Politics** | | |
| History | 5 | Same leadership since 1964 |
| Stability | 7 | Strains of Kaunda leadership showing |
| External dangers | 4 | Lessened since end of Rhodesia's UDI (1979) |
| **Location** | | |
| Hostile borders | 4 | Long term unpredictability but currently calm |
| Critical land/sea routes | 8 | Landlocked, rail links prone to disruption, dependence on South Africa/Zimbabwe |
| **Mineral Resources** | | |
| Adequacy of reserves | 6 | Limited to copper/cobalt/lead/zinc, lowish grades |
| Cost of production | 7 | High, including import dependence |
| Development appeal | 6 | Lowish but well supported by Western bloc |
| **Financial and Economic** | | |
| Currency performance | 8 | Poor, consistent downward realignments |
| Vulnerability to manipulation | 7 | High |
| External indebtedness | 8 | High |
| Productivity | 7 | Medium-low |
| Foreign ownerships | 4 | Now greatly reduced |
| **Energy and Other Natural Resources** | | |
| Domestic energy sources | 8 | Very low, no oil, poor in coal |
| Domestic energy production | 7 | Low, some power from hydro at Kariba Dam |
| Sources of outside supply | 8 | Limited by transport and foreign exchange |
| **Dependence on Foreign Natural Resources** | 8 | High |

***These components are assessed on a scale from 1 (lowest) to 10 (highest) risk. Grouped together, they form the basis for the country risk factor assessment opposite.***

### THE MINING INDUSTRY

Zambia Consolidated Copper Mines Ltd (ZCCM), formed in April 1982 with retrospective effect from April 1981, is the world's second largest copper mining company after Chile's CODELCO, with a rated capacity of 700,000 tonnes a year. Its formation should greatly rationalise the running of the industry, hopefully with measurable productivity benefits. In spite of there being only one major employee union, the MUZ, labour disputes remain a serious problem. The difficulty in attracting and retaining skilled expatriates reflects overall economic conditions and the emphasis is strongly towards Zambianisation. Transport is a permanent problem. The distance to ports is long, the choice resting between the railroad through Tanzania to Dar; Zimbabwe/South Africa to the Indian Ocean ports or the reopened Beira railroad through Mozambique. With political restrictions lifted on all bar the Lobito railway through Angola, the economic and financial problems remain.

**INVESTMENT AND DEVELOPMENT**

Although foreign investment is still sought incentives offered do not reflect the competitive nature of the world industry. Having taken effectively 17 years (1964-81) to gain complete control of its major natural resource it would hardly be possible for Zambia to weaken its rules now.

The mines are losing money heavily and capital expenditures, financed by loans, show an unhealthy imbalance between new projects and replacements. Sporadic exploration ventures include an overview by the World Bank, uranium exploration by an international consortium and a Canadian-led oil search in the Zambesi basin.

There appear to be no short cuts to Zambia's return to economic health nor the lessening of its overriding dependence upon copper. Sadly, any prolonged upturn in the market would see fierce competition from Chile, Australia, the Philippines and elsewhere to bring on new capacity.

# ZIMBABWE

**COUNTRY RISK FACTOR** **6.3**

*This factor is assessed from 1.0 (lowest) to 10.0 (highest) risk. For components of this assessment see later table in this section.*

Zimbabwe is the non-communist world's second largest producer of white asbestos, third in chromite and ninth in nickel. It also produces useful amounts of antimony, gemstones, cobalt, copper, iron ore, lithium, silver and tin whilst coal is important domestically and the country is strategically placed for exports to surrounding countries.

Obvious internal political differences which emerged post 1979 independence are discouraging foreign investment as are rapid strides towards centralisation of the minerals industry. Whilst accepting a democratic basis of government as part of the settlement of independence from Britain, it has always been the desire of the leader, Robert Mugabe, to convert Zimbabwe into a one-party state on Marxist lines and this further inhibits the inflow of permanent foreign capital. Internal disputes centre on the long established tribal and, more recently developed, politico-military differences between the government and opposition. Tensions are likely to increase and the risk rating could move up accordingly.

## MINERAL PRODUCTION

Asbestos has remained a mainstay of the industry since its development pre-World War Two but whilst reserves are large, both output and sales are suspect; the former on domestic and the latter on world-environmental considerations. Gold is the largest export earner with the official figure of 11.6 tonnes probably understated. There are numerous small workings and several well-developed deep workings including the Coronation Syndicate(3), Falcon and Rio Tinto Zimbabwe. Nickel output grew rapidly in the 1960s and 1970s but has fallen as grades appear marginal and the world market slackens. Chief operators are the Bindura Group, incorporating Shangani Mining Corp, and the Blanket Mine, owned by Canada's Falconbridge Nickel. Copper output has fallen sharply since its mid-1970s peak as reserves dwindled at the mines of the main producer MTD (Mangula) Ltd, part of the South African Messina Group.

## OUTPUT AND RATINGS OF PRINCIPAL MINERALS

| *Mineral* | *Output (Tonnes)* *1981* | *(1971)* | *Percentage of Western World Output* | *Western World Rating* | *Mine (M)/ Smelter (S)* | *Major Producers* |
|---|---|---|---|---|---|---|
| Antimony | 263 | (200) | 0.7 | – | M | |
| Asbestos | 248,000 | (150,000) | 13.0 | 2 | M | Turner & Newall Group, Gencor |
| Chromite | 530,000 | (400,000) | 9.4 | 3 | M | Union Carbide |
| Coal | 3,000,000 | N/A | 1.0 | – | M | Wankie Colliery Ltd |
| Cobalt | 109 | N/A | 0.4 | – | M | By-product |
| Copper | 24,600 | (23,300) | 0.3 | – | M | MTD Mangula Ltd |
| Gold | 11.6 | (15.0) | 1.2 | – | M | Coronation Syndicate, Falcon Mines |
| Nickel | 15,000 | (11,600) | 3.1 | 9 | M | JCI Group, RTZ Group, Falconbridge |
| Silver | 27 | N/A | 0.3 | – | M | By-product |
| Tin | 1,100 | N/A | 0.5 | – | M | Kamativi Tin Ltd |

Other mineral production : gemstones, iron ore, lithium

Note: During the period of UDI (1965-79), mineral production figures were conjectural but based on recent results they appear to have been reasonably representative.

## COMPONENTS OF RISK ASSESSMENT

| *Factor* | *Rating* | *Comment* |
|---|---|---|
| **Labour** | | |
| Strike incidence | 7 | Growing as black/white differentials and politics emerge |
| Quality of labour | 7 | Exodus of skilled whites, shortage of artisans |
| General unrest | 8 | High as tribal/political differences persist |
| **Politics** | | |
| History | 4 | UK style democracy agreed at Independence 1979 |
| Stability | 8 | Dangerous level of internal feuding |
| External dangers | 5 | Now receded but could re-emerge if civil war occurs |
| **Location** | | |
| Hostile borders | 5 | Currently quiet but close to Namibia and Angola |
| Critical land/sea routes | 8 | Landlocked with large dependence on South Africa |
| **Mineral Resources** | | |
| Adequacy of reserves | 4 | Good domestically but lacking variety for exports |
| Cost of production | 5 | Reasonable but foreign exchange is an inhibition |
| Development appeal | 8 | Low because of government policies and political uncertainties |
| **Financial and Economic** | | |
| Currency performance | 7 | Under strain as export values fall |
| Vulnerability to manipulation | 7 | Growing |
| External indebtedness | 7 | Growing |
| Productivity | 6 | Medium |
| Foreign ownerships | 4 | Being reduced |
| **Energy and Other Natural Resources** | | |
| Domestic energy sources | 6 | Coal and hydroelectricity |
| Domestic energy production | 7 | Good coal and coke output with export potential |
| Sources of outside supply | 6 | No political constraints but transport routes vulnerable |
| **Dependence on Foreign Natural Resources** | 2 | Low |

*These components are assessed on a scale from 1 (lowest) to 10 (highest) risk. Grouped together, they form the basis for the country risk factor assessment opposite.*

Coal mining is centred on the only operating mine, Wankie Colliery, a largely open cast operation close to the Zambian border. Administered by the Anglo American Group, output is running in the order of three million tonnes a year. It is being increasingly utilised on electrical power generation and coke production for steelmaking. Chromite production has enjoyed favoured world status owing to the high-grade, metallurgical quality lump-ore produced. This is now under threat from recent technical advances in the use of lower-grade ores and from high labour costs associated with the mining of narrow vein structures. The lithium mineral (petalite) produced also has particular glass-making properties. Geologically there is scope for significant expansion of a broad range of mineral output but the constraints remain the world market, domestic difficulties and a dearth of foreign investment.

Without the attraction of foreign and white indigenous labour skills and capital it will suffer the problems seen in Zambia and Zaire. Its available minerals are becoming less critical to the importing world and can be replaced by those of South Africa almost in entirety. It is a position which calls for a policy similar to that of Indonesia but this may be incompatible with the increasing political discord.

## THE MINING INDUSTRY

British and then South African companies developed the, then, Rhodesian mining industry from colonisation in the late 1890s, expanded it significantly in the 1950-65 period and maintained it through UDI: 1965-79. The value of minerals produced peaked in 1980 at $Z415 million ($Z = $US 1.30) having increased successively for 17 years. The major British influences are the Turner and Newall Ltd asbestos mines at Shabani and Mashaba, the RTZ Group interests which include gold, nickel, emeralds, copper and platinum and the Coronation Syndicate (Lonrho) producing gold, silver and copper. South African interests are represented by the Messina Group (copper), Anglo American Corporation (coal), JCI (chrome, nickel) and Gencor (chrome, asbestos). Union Carbide of the USA is a major chrome producer.

## INVESTMENT AND DEVELOPMENT

In terms of its infrastructure and external links Zimbabwe is in a better position than its neighbours to maintain and develop a mining industry. The government is torn between the ideal of black national advancement and the practicalities of attracting foreign expertise and capital. Minerals export earnings are becoming increasingly important, as compared to agricultural earnings, and recent legislation is thus viewed seriously by both established companies and prospective investors. The most contentious acts have included:

—minimum wage stipulations and increases, the removal of depletion allowances and the reduction of capital investment allowances

—participation by government in strategic mining ventures and

—the decision to set up the government Minerals Marketing Corporation which, by definition, has wide powers. The marketing agency began operation in 1983 with sole control over all mineral sales, except gold.

The Zimbabwe mining industry is at a crossroads. Given the more overt political delicacy of white-land tenure it could rapidly outpace agriculture as the most important foreign exchange earner.

# SECTION FOUR:

# THE COMPANIES

# COMPANY/MINERAL MASTER TABLE

*This table highlights those minerals with which each of the main companies is involved either in terms of mining or refining. A two star grading has been adopted in the mining category to show those companies which have major involvements at this stage. Details on the great majority of the companies listed in the table appear in the Directory of Principal Companies which starts on page 259.*

| * Some involvement in mining<br>** Major involvement in mining<br>† Major involvement in refining | Agnico Eagle | Alcan Aluminium | Alcoa | Amax | Amoco | Anaconda | Anglo American | Asarco | Associated Manganese Mines of S Africa | Associated Minerals Consolidated | Atlas Consolidated Mining & Development | Barlow Rand | Benguet Corp | Beralt Tin & Wolfran |
|---|---|---|---|---|---|---|---|---|---|---|---|---|---|---|
| Aluminium | | **† | **† | * | | *† | | | | | | | | |
| Antimony | | | | | | | | * | | | | | | |
| Beryllium | | | | | | | | | | | | | | |
| Bismuth | | | | | | | | *† | | | | | | |
| Cadmium | | | | | | * | | *† | | | | | | |
| Chromium | | | | | | | * | | | | | ** | * | |
| Cobalt | | | | *† | | | | | | | | | | |
| Columbium | | | | | | | | | | | | | | |
| Copper | | | | * | * | ** | * | *† | | | * | | * | * |
| Diamond | | | | | | | | | | | | | | |
| Germanium | | | | | | | | | | | | | | |
| Gold | * | | | | * | * | **† | | | | * | ** | * | |
| Gallium | | | *† | | | | | | | | | | | |
| Indium | | | | | | | | * | | | | | | |
| Iron | | | | * | | | | | * | | | | | |
| Lead | | | | * | * | * | | **† | | | | | | |
| Lithium | | | | | | | | | | | | | | |
| Magnesium | | | | *† | | | | | | | | | | |
| Manganese | | | | | | | * | | **† | | | | | |
| Mercury | | | | | | | | | | | | | | |
| Molybdenum | | | | **† | | * | | * | | | * | | | |
| Nickel | | | | *† | * | | * | * | | | | | | |
| PGMs | | | | | | | | | | | | | | |
| Rare Earths | | | | | | | | | | | | | | |
| Rhenium | | | | | | | | | | | | | | |
| Selenium | | | | † | | | | *† | | | | | | |
| Silicon | | | | | | | | | | | | | | |
| Silver | * | | | * | | * | | * | | | * | | * | |
| Tantalum | | | | | | | | | | | | | | |
| Tellurium | | | | † | | | | *† | | | | | | |
| Tin | | | | | | | | | | | | | | * |
| Titanium | | | | | | | | | | ** | | | | |
| Tungsten | | | | * | | | | | | | | | | * |
| Uranium | | | | | * | * | * | | | | | *† | | |
| Vanadium | | | | | | | **† | | | | | | | |
| Zinc | | | | * | * | * | | **† | | | | | | |
| Zirconium | | | | | | | | | | ** | | | | |

| Bethlehem Copper Corp | Billiton (Royal Dutch Shell) | Boliden | Botswana RST | Bougainville (RTZ subsidiary) | Broken Hill Pty | Brunswick Mining & Smelting | Brush-Wellman | Canada Tungsten | Cominco | Consolidated Gold Fields | Codelco Chile | Cyprus Anvil | De Beers Consolidated | Duval Corp (Pennzoil) | * *Some involvement in mining* ** *Major involvement in mining* † *Major involvement in refining* |
|---|---|---|---|---|---|---|---|---|---|---|---|---|---|---|---|
| | *† | | | | | | | | | | | | | | Aluminium |
| | | | | | | | | | | | | | | | Antimony |
| | * | | | | | | **† | | | | | | | | Beryllium |
| | | | | | | | | | *† | | | | | | Bismuth |
| | | | | | | | | | **† | | | | | | Cadmium |
| | | | | | | | | | | | | | | | Chromium |
| | | | * | | | | | | | | | | | | Cobalt |
| | | | | | | | | | | | | | | | Columbium |
| * | * | *† | *† | ** | | * | | | * | * | **† | | | **† | Copper |
| | | | | | | | | | | | | | ** | | Diamond |
| | | | | | | | | | | | | | | | Germanium |
| * | | *† | | * | *† | | | | * | **† | | | | * | Gold |
| | | | | | | | | | † | | | | | | Gallium |
| | | | | | | | | | *† | | | | | | Indium |
| | | | | | *† | | | | | | | | | | Iron |
| | * | *† | | | | *† | | | **† | * | | * | | | Lead |
| | | | | | | | | | | | | | | | Lithium |
| | | | | | | | | | | | | | | | Magnesium |
| | | | | | * | | | | | | | | | | Manganese |
| | | | | | | | | | | | | | | | Mercury |
| * | | | | | | | | | | | **† | | | *† | Molybdenum |
| | *† | | *† | | | | | | | | | | | | Nickel |
| | | | | | | | | | | | | | | | PGMs |
| | | | | | | | | | | | | | | | Rare Earths |
| | | | | | | | | | | | **† | | | * | Rhenium |
| | | *† | | | | | | | | | | | | | Selenium |
| | | | | | | | | | | | | | | | Silicon |
| | | *† | | * | | *† | | | **† | * | | * | | * | Silver |
| | | | | | | | | | | | | | | | Tantalum |
| | | *† | | | | | | | | | | | | | Tellurium |
| | *† | | | | * | | | | | * | | | | | Tin |
| | † | | | | | | | | | * | | | | | Titanium |
| | *† | | | | | | | ** | | | | | | | Tungsten |
| | | | | | | | | | | * | | | | | Uranium |
| | | | | | | | | | | | | | | | Vanadium |
| | *† | *† | | | | * | | | **† | * | | * | | | Zinc |
| | | | | | | | | | | * | | | | | Zirconium |

** Some involvement in mining*
*** Major involvement in mining*
*† Major involvement in refining*

| | EZ Industries | Empresa Minera del Centro del Peru (Centromin-Peru) | Falconbridge | Foote Mineral Co | General Mining Union Corp | Gulf Resources & Chemical | Hanna Mining | Homestake Mining | Hudson Bay Mining & Smelting | Impala Platinum | Inco | Industrias Penoles SA | Johannesburg Consolidated Investments | Johns Manville |
|---|---|---|---|---|---|---|---|---|---|---|---|---|---|---|
| Aluminium | | | | | | | * | | | | | | | |
| Antimony | | | | | | | | | | | | | **† | |
| Beryllium | | | | | | | | | | | | | | |
| Bismuth | | * | | | | | | | | | | *† | | |
| Cadmium | *† | | | | | | | | *† | | | *† | | |
| Chromium | | | | | **† | | | | | | | | *† | |
| Cobalt | | | *† | | | | | | | *† | *† | | | |
| Columbium | | | | | | | | | | | | | | |
| Copper | * | **† | **† | | * | | | | *† | *† | *† | *† | * | *Major asbestos producer* |
| Diamond | | | | | | | | | | | | | ** | |
| Germanium | | | | | | | | | | | | | | |
| Gold | | | * | | **† | | | * | * | | * | *† | **† | |
| Gallium | | | | | | | | | | | | | | |
| Indium | | | | | | | | | | | | | | |
| Iron | | | * | | | | * | | | | | | | |
| Lead | * | | | | | | | * | *† | | | *† | | |
| Lithium | | | | **† | | **† | | | | | | | | |
| Magnesium | | | | | | | | | | | | * | | |
| Manganese | | | | * | **† | | | | | | | | | |
| Mercury | | | | | | | | | | | | | | |
| Molybdenum | | | | | | | | | | | | | | |
| Nickel | | | **† | | | | *† | | | *† | **† | | * | |
| PGMs | | | *† | | * | | | | | **† | *† | | **† | |
| Rare Earths | | | | | | | | | | | | | | |
| Rhenium | | | | | | | | | | | | | | |
| Selenium | | | | | | | | | * | | *† | | | |
| Silicon | | | | *† | | | | | | | | | | |
| Silver | * | *† | *† | | * | | | * | *† | | * | **† | | |
| Tantalum | | | | | | | | | *† | | | | | |
| Tellurium | | | | | | | | | | | | | | |
| Tin | | | | | * | | | | | | | | | |
| Titanium | | | | | * | | | | | | | | | |
| Tungsten | | | | | | | | | | | | | | |
| Uranium | | | | * | * | * | | * | | | *† | | * | |
| Vanadium | | | | * | | | | | | | | | | |
| Zinc | **† | * | * | | | | | * | *† | | * | *† | | |
| Zirconium | | | | | | | | | | | | | | |

| Kaiser Aluminium & Chemical | Kennecott | Lonrho | Lornex (RTZ subsidiary) | MIM Holdings | Malaysia Mining | Marinduque Mining | Metallgesellschaft | Mitsui Mining & Smelting | Molycorp (Union Oil) | Newmont Mining | Nippon Mining | Noranda Mines | Norsk Hydro | * Some involvement in mining<br>** Major involvement in mining<br>† Major involvement in refining |
|---|---|---|---|---|---|---|---|---|---|---|---|---|---|---|
| **† | | | | | | | | | | | | | † | Aluminium |
| | | | | | | | | | | | | | | Antimony |
| | | | | | | | | | | | | | | Beryllium |
| | | | | | | | | *† | | | † | | | Bismuth |
| | | | | | | | *† | *† | | | | **† | | Cadmium |
| | | | | | | | | | | | | | | Chromium |
| | | | | | | * | | | | | † | | | Cobalt |
| | | | | | * | | | | **† | | | | | Columbium |
| | **† | * | * | *† | | * | *† | *† | * | **† | *† | **† | | Copper |
| | | | | | * | | | | | | | | | Diamond |
| | | | | | | | | | | | | | | Germanium |
| | *† | * | | | | * | | *† | | * | | * | | Gold |
| | | | | | | | | | | | | | | Gallium |
| | | | | | | | | *† | | | * | | | Indium |
| | * | | | * | | | | | | | | | | Iron |
| | * | | | **† | | | *† | *† | | * | *† | * | | Lead |
| | | | | | | | | | | **† | | | | Lithium |
| | | | | | | | | | | | | | **† | Magnesium |
| | | | | | | | | | | | | | | Manganese |
| | | | | | | | | | | | | * | | Mercury |
| | *† | | * | | | | | | *† | * | | * | | Molybdenum |
| | | * | | * | | *† | | | | * | *† | | | Nickel |
| | | * | | | | | | | | | | | | PGMs |
| | | | | | | | | † | **† | | | | | Rare Earths |
| | *† | | | | | | | | *† | | | | | Rhenium |
| | *† | | | | | | | *† | | | † | **† | | Selenium |
| | | | | | | | | | | | | | | Silicon |
| | *† | | * | ** | | * | | *† | | * | *† | * | | Silver |
| | | | | | **† | | | | | | | | | Tantalum |
| | | | | | | | | *† | | | † | **† | | Tellurium |
| | | | | | **† | | *† | | | | | | | Tin |
| | ** | | | | | | | | | | | | | Titanium |
| | | | | | * | | * | | | | | | | Tungsten |
| | | | | | | | | | | * | | | | Uranium |
| | | | | | | | | | | | | | | Vanadium |
| | * | | | ** | | | *† | *† | | * | *† | * | | Zinc |
| | * | | | | | | | | | | † | | | Zirconium |

| * *Some involvement in mining*<br>** *Major involvement in mining*<br>† *Major involvement in refining* | Outokumpu Oy | Palabora (RTZ subsidiary) | Pechiney Ugine Kuhlmann | Peko Wallsend | Penarroyd | Phelps Dodge | Placer Development | Preussag | Reynolds Metals | RTZ Corp | Samancor | St Joe Minerals (Flour Corp) | Schweizerische Aluminium (Alusuisse) | Selection Trust (British Petroleum) |
|---|---|---|---|---|---|---|---|---|---|---|---|---|---|---|
| Aluminium | | | **† | | | | | | **† | *† | | | **† | |
| Antimony | | | | | | | | | | | | | | |
| Beryllium | | | † | | | | | | | | | | | * |
| Bismuth | | | | *† | | | | | | | | | | |
| Cadmium | *† | | | | *† | | | * | | *† | | | | |
| Chromium | *† | | *† | | | | | | | * | **† | | | |
| Cobalt | *† | | | | | | | | | | | | | |
| Columbium | | | | | | | | | | | | | | |
| Copper | *† | *† | | *† | | **† | * | | | **† | | * | | * |
| Diamond | | | | | | | | | | | | | | * |
| Germanium | | | | | **† | | | * | | | | | | |
| Gold | | * | | *† | | *† | * | | | **† | | * | | * |
| Gallium | | | | | | | | | | | | | **† | |
| Indium | | | | | | | | *† | | | | | | |
| Iron | | | | | | | | | | *† | | * | | * |
| Lead | * | | | | **† | | * | *† | | **† | | **† | | |
| Lithium | | | | | | | | | | | | | | * |
| Magnesium | | | | | | | | | | | | | | |
| Manganese | | | | | | | | | | * | **† | | | |
| Mercury | | | | | | | **† | * | | | | | | |
| Molybdenum | | | | | | * | *† | | | | | | | |
| Nickel | *† | | | | | | | | | | | | | * |
| PGMs | | * | | | | * | | | | | | | | |
| Rare Earths | | | | | | | | | | | | | | |
| Rhenium | | | | | | | | | | | | | | |
| Selenium | | | | | | *† | | | | | | | | |
| Silicon | | | | | | | | | | | *† | | | |
| Silver | † | * | | | *† | *† | * | *† | | * | | * | | * |
| Tantalum | | | | | | | | | | | | | | |
| Tellurium | | | | | | *† | | | | | | | | |
| Tin | | | | | | | | | | *† | | | | * |
| Titanium | | | | | | | | | | | | | | |
| Tungsten | | | | * | | | | | | | | | | |
| Uranium | | * | *† | | | *† | | | | *† | | | | |
| Vanadium | | | | | | | | | | | * | | | |
| Zinc | *† | | | | **† | | * | **† | | *† | | * | | * |
| Zirconium | | * | † | | | | | | | | | | | |

| | Sherritt Gordon | Societe Metallurgique Le Nickel | Southern Peru Copper Corp | Sumitomo Metal Mining | Teck Corp | Union Carbide | Union Miniere (Societe Generale Metallurgieppe) | Westralian Sands | Zambia Consolidated |
|---|---|---|---|---|---|---|---|---|---|
| Aluminium | | | | | | | | | |
| Antimony | | | | | | | † | | |
| Beryllium | | | | | | | | | |
| Bismuth | | | | | | | | | |
| Cadmium | | | | | | | *† | | |
| Chromium | | | | | | | | | |
| Cobalt | *† | * | | *† | | | | | **† |
| Columbium | | | | † | * | | | | |
| Copper | *† | | **† | | * | | *† | | **† |
| Diamond | | | | | | | | | |
| Germanium | | | | † | | | *† | | |
| Gold | | | | *† | * | | | | |
| Gallium | | | | | | | | | |
| Indium | | | | † | | | **† | | |
| Iron | | | | | | | | | |
| Lead | | | | *† | | | * | | * |
| Lithium | | | | | | | | | |
| Magnesium | | | | | | | | | |
| Manganese | | | | | | | | | |
| Mercury | | | | | | | | | |
| Molybdenum | | | **† | | * | | | | |
| Nickel | *† | **† | | *† | | | | | |
| PGMs | | | | | | | | | |
| Rare Earths | | | | | | | | * | |
| Rhenium | | | | | | | † | | |
| Selenium | | | | | | | † | | * |
| Silicon | | | | | | **† | | | |
| Silver | | | * | *† | * | | | | * |
| Tantalum | | | | | | | | | |
| Tellurium | | | | | | | | | |
| Tin | | | | | | | | | |
| Titanium | | | | | | | | ** | |
| Tungsten | | | | | | **† | | | |
| Uranium | | | | | | * | | | |
| Vanadium | | | | | | * | | | |
| Zinc | * | | | *† | * | | **† | | * |
| Zirconium | | | | | | | | ** | |

** Some involvement in mining*

*** Major involvement in mining*

*† Major involvement in refining*

# DIRECTORY OF PRINCIPAL COMPANIES

*This directory contains details on 62 major international mining companies and groups. The information given on each organisation covers: country of incorporation, the turnover, assets and number of employees for the latest trading year as available, countries of operation and principal subsidiary and associated natural resource companies. A table highlights the organisation's output of main minerals, its share of western and total world output (as applicable) and the organisation's rank numbers by mineral.*

# Alcan Aluminium

*Country of Incorporation:* Canada

*Turnover:* US$5.05 billion (total revenue 1981)
*Total Funds Employed:* US$6,339 million (1981) (total assets)
*Number of Employees:* 66,000

**Main countries of Operation:** Canada, USA, Spain, UK, West Germany, Brazil, Mexico, India, Australia, Japan, Malaysia, Jamaica, Guinea and others.

**Major Products:** Primary aluminium and fabricated products, bauxite, alumina

**Principal Subsidiary and Associated Natural Resource Companies:** Aluminium Co of Canada Ltd (Canada, USA, Caribbean and Ireland), Empresa Nacional del Aluminio (Spain), Alcan Aluminium (UK) Ltd (UK), Indian Aluminium Co Ltd (India), Alcan Australia Ltd (Australia), Alcan Aluminium Werke GmbH (West Germany), Nippon Light Metal Co, and Toyo Aluminium KK (Japan), Alcan Aluminio do Brasil SA (Brazil) and others. Fabricating centres in over 30 countries

**Production of Principal Minerals***

| | *Company Output* | *Percentage of Western World Output* | *Percentage of World Output* | *Company Rank Number* |
|---|---|---|---|---|
| Bauxite *tonnes* | 5,000,000 | 6.5 | 5.7 | 3 |
| Aluminium *tonnes* | 1,395,000 | 11.2 | 8.9 | 2 |

*Alcan's share of total and wholly owned subsidiary output.

# Aluminium Company of America (Alcoa)

*Country of Incorporation:* USA

*Turnover:* US$4,998 million (1981)
*Total Assets:* US$5,654 million (1981)
*Number of Employees:* 44,400

**Countries of Operation:** USA, Brazil, UK, Europe, Australia, Jamaica, Dominican Republic, Surinam, Guinea, Japan, Norway, Mexico

**Major Products:** Aluminium, bauxite, alumina; manufactured products

**Principal Subsidiary and Associated Natural Resource Companies:** Alcoa of Australia Ltd, plus subsidiaries and associated companies in all the above named locations

**Production of Principal Minerals**

| | *Company Output* | *Percentage of Western World Output* | *Percentage of World Output* | *Company Rank Number* |
|---|---|---|---|---|
| Aluminium *tonnes* | 1,792,000 | 14.4 | 11.4 | 1 |
| Bauxite *tonnes* | 7,349,000 | 9.7 | 8.4 | 2 |

# Amax Incorporated

*Country of Incorporation:* USA

*Turnover:* US$2,799 million (1981)
*Total Assets:* US$5,449 million (1981)
*Number of Employees:* 19,927

**Countries of Operation:** USA, Canada, UK, Southern Africa, Europe, Australia.

**Major Products:** Coal, copper, molybdenum, tungsten, oil, iron ore, lead, zinc, silver, cadmium, magnesium, selenium, tellurium

**Principal Subsidiary and Associated Natural Resource Companies:** Canada Tungsten Ltd, Mount Newman Ltd, Anamax Inc, Alumax Inc, BCL Ltd, (subsidiary of Botswana RST Ltd).

**Production of Principal Minerals**

| | *Company Output* | *Percentage of Western World Output* | *Percentage of World Output* | *Company Rank Number* |
|---|---|---|---|---|
| Coal *tonnes* | 32,200,000 | 2.5 | 1.1 | – |
| Copper *tonnes* | 61,000 | 1.0 | <1.0 | – |
| Aluminium *tonnes* | 195,000 | 1.6 | 1.2 | 8 |
| Lead *tonnes* | 64,885 | 2.6 | 1.9 | 5 |
| Zinc *tonnes* | 40,617 | 0.9 | 0.8 | – |
| Silver *tonnes* | 193 | 2.3 | 1.8 | 7 |
| Oil *tonnes* | small | 1.0 | 1.0 | – |
| Molybdenum *tonnes* | 40,978 | 42.4 | 32.9 | 1 |
| Tungsten *tonnes* | 1,633 | 5.9 | 3.1 | 4/5 |
| Iron Ore *tonnes* | 1,800,000 | 1.0 | 1.0 | – |
| Cadmium *tonnes* | 210 | 1.6 | 1.2 | 5/6 |

All the above represent AMAX's share of production. It also operates the only nickel-cobalt refinery in the USA. Using mostly imported materials it produced:

| | | | | |
|---|---|---|---|---|
| Nickel *tonnes* | 35,000 | 7.0 | 5.0 | – |
| Cobalt *tonnes* | 405 | 1.9 | 1.5 | – |
| Copper *tonnes* | 20,400 | 1.0 | 1.0 | – |

Note: The company rank numbers apply to production at Amax mines only and not the equity shares of production shown above.

# Anglo American Corporation of South Africa

*Country of Incorporation:* South Africa

*Turnover:* R770 million (1981 group profit)
*Total Funds Employed:* R6,600 million (1981 net value of companies administered by the Corporation)
*Total Number of Employees:* 231,000 (including South African companies which obtain administrative and technical services from the Corporation)

**Countries of Operation:** South Africa, Zimbabwe, Brazil, Australia, Botswana, Bermuda, etc

**Major Products:** Gold, coal, uranium, plus investments in companies mining diamonds, platinum, copper, nickel, tin, iron ore, vanadium, lead, zinc, manganese and wolfram, chromium, uranium

**Principal Subsidiary and Associated Natural Resource Companies:** Gold: Anglo American Gold Investment Co (South Africa). Diamonds: De Beers (South Africa), Anglo American Investment Trust (South Africa). Mining, finance and development: Minorco (Bermuda), Johannesburg Consolidated Investment (South Africa), Consolidated Gold Fields (South Africa), Charter Consolidated (UK). Gold and nickel: Anglo American Corp do Brasil Ltd (Brazil). Nickel, copper, chromium: Anglo American Corp Zimbabwe Ltd (Zimbabwe)

**Production of Principal Minerals***

| | *Company Output* | *Percentage of Western World Output* | *Percentage of World Output* | *Company Rank Number* |
|---|---|---|---|---|
| Gold *tonnes* ) administered mines | 238.1 | 24.8 | 19.5 | 1 |
| Uranium *tons* ) administered mines | 3,243 | 7.6 | N/A | probably 1/2 |

*Difficult to establish due to vast tangled web of subsidiary and associated companies
Year ending March 1982.

# Asarco Incorporated

*Country of Incorporation:* USA

*Turnover:* US$1,532 million (1981)
*Total Funds Employed:* US$2,093 million (1981)
*Number of Employees:* 12,500 (1981)

**Countries of Operation:** USA, Canada, Australia, Peru, Bolivia, UK, Philippines, Chile

**Major Products:** Asbestos, copper, lead, zinc, silver, molybdenum, coal, nickel, antimony, bismuth, cadmium, indium, selenium, tellurium

**Principal Subsidiary and Associated Natural Resource Companies:** MIM Holdings Ltd, Southern Peru copper Corporation, MEDIMSA SA, Revere Inc.

**Production of Principal Minerals**

| | *Company Output* | *Percentage of Western World Output* | *Percentage of World Output* | *Company Rank Number* |
|---|---|---|---|---|
| Copper *tonnes* | 302,493 | 4.7 | 3.6 | 6 |
| Lead *tonnes* | 118,700 | 4.8 | 3.4 | 6 |
| Zinc *tonnes* | 214,900 | 4.9 | 3.5 | 5 |
| Silver *tonnes* | 733 | 8.3 | 6.5 | 1 |
| Molybdenum *tonnes* | 1,791 | 1.9 | 1.7 | 10 |
| Gold *tonnes* | 1 | <1.0 | <1.0 | – |
| Coal *tonnes* | 870,000 | <1.0 | <1.0 | – |
| Asbestos *tonnes* | 181,000 | 7.0 | 3.6 | – |

The above volumes represent Asarco's 100 per cent share of its own output plus its equity share of subsidiaries and associates. In addition the company's refineries toll-treat outside ores and have a 1981 rated capacity of 376,000 tonnes copper, 1,860 tonnes silver, 240,000 tonnes lead, 100,000 tonnes zinc and 19 tonnes gold.

Note: The company rank numbers apply to production at ASARCO mines only and not to the equity shares of production shown above.

# Associated Manganese Mines of South Africa

*Country of Incorporation:* South Africa

*Turnover:* R123.5 million (1981)
*Total Assets:* R31.1 million (net current assets)
*Number of Employees:* Not available

**Country of Operation:** South Africa

**Major Products:** Manganese, iron, chromium alloys

**Principal Subsidiary and Associated Natural Resource Companies:** Ferro-alloy production: Ferroalloys Ltd (South Africa)

**Sales of Principal Minerals**

| | *Company Output* | *Percentage of Western World Output* | *Percentage of World Output* | *Company Rank Number* |
|---|---|---|---|---|
| Manganese ore (est) *tonnes* | 1,400,000 | 8.8 | 4.8 | 4 |
| Iron ore (est) *tonnes* | 1,300,000 | 0.3 | 0.2 | – |

# Atlantic Richfield Company

*Country of Incorporation:* USA

*Turnover:* US$28,208 million (1981)
*Total Funds Employed:* US$19,733 million
*Number of Employees:* Not stated

**Countries of Operation:** USA, UK, most EEC members, Canada, Indonesia, Argentina, Dubai, Sudan

**Major Products:** Oil, coal, aluminium, copper, molybdenum, cadmium

**Principal Subsidiary and Associated Natural Resource Companies:** Arco International Inc, Arco Exploration Company, Anaconda Minerals Company

**Production of Principal Minerals**

| | *Company Output* | *Percentage of Western World Output* | *Percentage of World Output* | *Company Rank Number* |
|---|---|---|---|---|
| Oil *tonnes* | 33,000,000 | 1.5 | 1.1 | N/A |
| Copper *tonnes* | 135,380 | 2.1 | 1.7 | joint 13 |
| Aluminium *tonnes* | 328,000 | 2.6 | 2.1 | 7 |
| Coal *tonnes* | 14,362,000 | 1.1 | 1.0 | N/A |
| Molybdenum *tonnes* | 1,637 | 1.7 | 1.6 | 11 |
| Silver *tonnes* | 86 | 1.0 | <1.0 | – |
| Uranium *tonnes* | 1,280 | 3.0 | N/A | 5/6 |

ARCO is typical of a major oil company which has diversified into coal and metals. In 1981, 16 per cent of capital was employed in the metals sector.

# Atlas Consolidated Mining & Development Corp

*Country of Incorporation:* Philippines (currency: Pesos)

*Revenue:* P2,558 million (US$312 million)
*Total Funds Employed:* P3,478 million (US$424 million) (1981)
*Number of Employees:*

**Country of Operation:** Philippines

**Major Products:** Copper, gold, silver, pyrites

**Principal Subsidiary and Associated Natural Resource Companies:** 40 per cent Phelps Dodge Philippines Inc, 40 per cent Atlas Fertilizer Corporation

**Production of Principal Minerals**

| | *Company Output* | *Percentage of Western World Output* | *Percentage of World Output* | *Company Rank Number* |
|---|---|---|---|---|
| Copper *tonnes* in concentrate | 138,950 | 2.1 | 1.7 | joint 13 |
| Silver *tonnes* | 17 | <1.0 | <1.0 | – |
| Gold *tonnes* | 6 | 0.6 | 0.5 | – |
| Molybdenum *tonnes* | 2 | <1.0 | <1.0 | – |

# Barlow Rand

*Country of Incorporation:* South Africa

*Turnover:* US$4,790 million (1981*) (16 per cent from mining and exploration)
*Total Funds Employed:* US$716 million (1981 net current assets)
*Number of Employees:* 196,000

**Countries of Operation:** South Africa, UK, USA, most EEC members

**Major Products (mining sector):** Gold, uranium, coal, chromium, fluorspar

**Principal Subsidiary and Associated Natural Resource Companies:** Gold: Durban Roodepoort Deep, Rand Mines, Transvaal Consolidated Land & Exploration, East Rand Proprietary (South Africa). Gold and uranium: Blyvooruitzicht Gold Mining, Harmony Gold Mining (South Africa). Chromium: Millsell Chrome Mines, Winterveld Chrome Mines (South Africa)

**Production of Principal Minerals**

| | *Company Output* | *Percentage of Western World Output* | *Percentage of World Output* | *Company Rank Number* |
|---|---|---|---|---|
| Gold | | | | |
| Uranium | Figures not provided in annual Financial statement of Barlow Rand Ltd | | | |
| Chromium | (probably ranking 3rd in gold output and 2nd/3rd in uranium output) | | | |

*Given for year to 30 September 1981.

# Benguet Corporation

*Country of Incorporation:* Philippines

*Turnover:* P1,992.7 million (Philippine Pesos) (1981 operating revenue)
*Total Funds Employed:* P2,566.3 million (1981 total assets)
*Number of Employees:* 22,781

**Country of Operation:** Philippines

**Major Products:** Gold, copper, silver, chromium ore

**Principal Subsidiary and Associated Natural Resource Companies:** None

**Production of Principal Minerals**

| | *Company Output* | *Percentage of Western World Output* | *Percentage of World Output* | *Company Rank Number* |
|---|---|---|---|---|
| Copper *tonnes* | 27,900 | 0.4 | 0.3 | – |
| Gold *tonnes* | 6.7 | 0.7 | 0.5 | – |
| Silver *tonnes* | 7.2 | 0.1 | <0.1 | – |
| Chromite *tonnes* | 271,000 | 4.8 | 3.0 | – |

# Beralt Tin & Wolfram

*Country of Incorporation:* UK

*Turnover:* £11.02 million (1980)
*Total Funds Employed:* £13.37 million (1980 net current assets)
*Number of Employees:* Not given, with 1,978 employees on the mines

**Country of Operation:** Portugal

**Major Products:** Tungsten, copper, tin, silver

**Principal Subsidiary and Associated Natural Resource Companies:** None

**Production of Principal Minerals**

| | *Company Output* | *Percentage of Western World Output* | *Percentage of World Output* | *Company Rank Number* |
|---|---|---|---|---|
| Tungsten ore concentrates *tonnes* | 2,461 | 8.9 | 4.7 | 2/3 |
| Copper in concentrate *tonnes* | 2,527 | <0.1 | <0.1 | – |
| Tin in concentrate *tonnes* | 133 | <0.1 | <0.1 | – |

# Billiton (Royal Dutch Shell)

*Country of Incorporation:* Netherlands

*Turnover:* £617 million (1981) (revenue from metals sales within the Royal Dutch Shell Group)
*Total Assets:* £13,184 million (1981) (net assets of Royal Dutch Shell Group)
*Number of Employees:* 6,200 (in 1980) (166,000 employees in Royal Dutch Shell)

**Countries of Operation:** Indonesia, Brazil, Ireland, UK, Canada, Colombia, Australia, Netherlands, Peru, Thailand, Surinam, marketing and manufacturing companies worldwide

**Major Products:** Tin, bauxite, alumina, tungsten, zinc, lead, cadmium (projects involving aluminium, nickel, titanium sponge under development in 1981)

**Principal Subsidiary and Associated Natural Resource Companies:** Bauxite: NV Billiton Maatschappij Suriname (Surinam), Shell Brasil (Brazil). Alumina: Aughinish Island (Ireland). Gallium, tantalum, indium, beryllium refining: Kawecki-Billiton Metaalindustrie (Holland). Nickel: Cerro Matoso SA (Colombia), Shell Co of Australia Ltd (Australia). Tin: Thailand Smelting & Refining Co (Thailand), PT Riau Tin Mining (Indonesia). Tungsten/molybdenum: Billiton Canada Ltd (Canada). Copper: Southern Peru Copper Corp (Peru). Zinc/lead ore: Nanisivik Mines Ltd (Canada). Refined zinc/cadmium: Budelco (Netherlands). LME trading: Billiton Enthoven Metals Ltd (UK)

**Sales of Principal Minerals**

| | *Company Output* | *Percentage of Western World Output* | *Percentage of World Output* | *Company Rank Number* |
|---|---|---|---|---|
| Refined tin *tonnes* | 36,000 | 17.6 | 15.1 | 2 (refined) |
| Bauxite *tonnes* | 1,084,000 | 1.4 | 1.2 | 4/5 |
| Alumina *tonnes* | 532,000 | 2.0 | N/A | – |
| Refined zinc *tonnes* | 145,000 | 3.2 | 2.3 | 8 |
| Refined lead *tonnes* | 96,000 | 2.4 | 1.8 | – |
| Refined copper *tonnes* | 75,000 | 1.0 | 0.8 | – |

# Boliden

*Country of Incorporation:* Sweden

*Turnover:* US$1,039 million (1981) (Swedish crowns 5,773 million)
*Total Funds Employed:* US$824 million (1981)
*Number of Employees:* 9,609

**Countries of Operation:** Sweden, Norway, Canada, France, USA

**Major Products:** Copper, lead, zinc, precious metals, selenium, cadmium, tellurium

**Principal Subsidiary and Associated Natural Resource Companies:** Boliden Mineral, Boliden Metall, Boliden Bergsoe, Boliden Canada, Boliden France, Boliden USA, Norzink A/S, PBB

**Production of Principal Minerals**

| | *Company Output* | *Percentage of Western World Output* | *Percentage of World Output* | *Company Rank Number* |
|---|---|---|---|---|
| Copper *tonnes* | 50,100 | 1.0 | 1.0 | – |
| Cadmium *tonnes* | 59 | 0.4 | <0.4 | 10/11 |
| Gold *tonnes* | 2 | 0.2 | 0.1 | – |
| Silver *tonnes* | 161 | 1.8 | 1.6 | 8 |
| Lead *tonnes* | 71,700 | 2.9 | 2.1 | 8 |
| Zinc *tonnes* | 122,400 | 2.7 | 2.0 | 10 |

The above is the output of Boliden Mineral Division. Boliden Metall refines feed from this and other sources.

# Botswana RST

*Country of Incorporation:* Botswana (currency: Pula)

*Turnover:* P80.3 million (US$90.6 million) (1981)
*Total Funds Employed:* $364.1 million (1981 total assets)
*Number of Employees:* 4,066

**Country of Operation:** Botswana

**Major Products:** Nickel, cobalt, copper

**Principal Subsidiary and Associated Natural Resource Companies:** Principally owned by Anglo-American Corp of South Africa and Amax Inc of the USA

**Production of Principal Minerals**

| | *Company Output* | *Percentage of Western World Output* | *Percentage of World Output* | *Company Rank Number* |
|---|---|---|---|---|
| Nickel *tonnes* | 18,273 | 3.7 | 2.6 | 4 |
| Copper *tonnes* | 17,819 | 0.3 | 0.2 | – |
| Cobalt *tonnes* | 254 | 1.2 | 1.0 | 11 |

# Bougainville Copper

*Country of Incorporation:* Papua New Guinea (currency: Kina)

*Turnover:* K335 million (1981)
*Total Funds Employed:* K584.5 million (1981)
*Number of Employees:* 4,293 (1981)

**Country of Operation:** Papua New Guinea

**Major Products:** Copper, gold, silver

**Principal Subsidiary and Associated Natural Resource Companies:** None. The company is owned 53.6 per cent by CRA (a subsidiary of the RTZ Group) and 20.2 per cent by the PNG Government

**Production of Principal Minerals**

| | *Company Output* | *Percentage of Western World Output* | *Percentage of World Output* | *Company Rank Number* |
|---|---|---|---|---|
| Copper *tonnes* in concentrates | 165,000 | 2.6 | 2.0 | 11 |
| Gold *tonnes* in concentrates | 16.8 | 1.7 | 1.4 | 7 |
| Silver *tonnes* in concentrates | 42.4 | 0.5 | 0.3 | – |

# Brunswick Mining & Smelting Corporation

*Country of Incorporation:* Canada

*Turnover:* C$273.3 million (1981 revenue from production)
*Total Funds Employed:* C$299.0 million (1981 net assets)
*Number of Employees:* 1,733 (Mining Division)

**Country of Operation:** Canada

**Major Products:** Zinc, lead, silver, copper

**Principal Subsidiary and Associated Natural Resource Companies:** Base metals: Noranda Mines Ltd (Canada)

**Production of Principal Minerals**

| | *Company Output* | *Percentage of Western World Output* | *Percentage of World Output* | *Company Rank Number* |
|---|---|---|---|---|
| Zinc *tonnes* | 236,061 | 5.3 | 3.9 | 3 |
| Lead *tonnes* | 79,883 | 3.2 | 2.3 | 7 |
| Copper *tonnes* | 4,706 | <0.1 | <0.1 | – |
| Silver *tonnes* | 195.7 | 2.2 | 1.8 | 6 |

# Canada Tungsten Corporation

*Country of Incorporation:* Canada

*Turnover:* C$2.8 million, net income in 1981 (C$23.0 million, net income in 1980)
*Total Assets:* C$91 million
*Number of Employees:* 267

**Country of Operation:** Canada

**Major Products:** Tungsten metal contained in wolframite

**Principal Subsidiary and Associated Natural Resource Companies:** The company has a small interest in Dome Petroleum's oil and gas activities. Amax Inc is the major shareholder in Canada Tungsten Ltd

**Production of Principal Minerals**

| | *Company Output* | *Percentage of Western World Output* | *Percentage of World Output* | *Company Rank Number* |
|---|---|---|---|---|
| Tungsten content *tonnes* | 2,512 (1981) | 9.1 | 4.8 | 2/3 |
| Tungsten content *tonnes* | 4,009 (1980) | 14.5 | 7.4 | 1 |

In 1981 a six months strike curtailed production. 1980 figures are more representative.

# Cominco

*Country of Incorporation:* Canada

*Turnover:* C$500 million (1981)
*Total Funds Employed:* C$367 million (1981)
*Number of Employees:* 12,600 (1981)

**Countries of Operation:** Canada, Greenland, Spain, Brazil, Australia, Ireland

**Major Products:** Copper, lead, zinc, gold, silver, coal. There are also undisclosed amounts of bismuth, cadmium, indium, mercury, gallium and tellurium produced

**Principal Subsidiary and Associated Natural Resource Companies:** Bethlehem Copper Corporation, Valley Copper Mines Ltd, Fording Coal Co, Aberfoyle Ltd, Exminesa SA, Tara Ltd

**Production of Principal Minerals**

| | *Company Output* | *Percentage of Western World Output* | *Percentage of World Output* | *Company Rank Number* |
|---|---|---|---|---|
| Copper *tonnes* | 221,000 | 3.4 | 2.7 | 10 |
| Lead *tonnes* | 177,000 | 7.2 | 5.1 | 4 |
| Zinc *tonnes* | 385,000 | 8.7 | 6.3 | 1 |
| Gold *tonnes* | 3 | < 1.0 | <1.0 | – |
| Silver *tonnes* | 262 | 3.0 | 2.4 | 5 |
| Coal *tonnes* | 1,628,000 | < 1.0 | <1.0 | – |

Tonnages represent group and subsidiary totals. Coal represents Cominco's 40 per cent equity in Fording Coal Co.

# Consolidated Gold Fields

*Country of Incorporation:* UK

*Turnover:* £945 million (1981)
*Total Funds Employed:* £794 million (1981)
*Number of Employees:* 112,779

**Countries of Operation:** USA, South Africa, Australia, UK

**Major Products:** Gold, uranium, coal, tin, iron, copper, titanium and zirconium ores, oil and gas, lead, zinc, silver plus interests in construction materials, manufacturing and commerce

**Principal Subsidiary and Associated Natural Resource Companies:** Gold and uranium: Gold Fields of South Africa (South Africa). Tin: Rooiberg Tin (South Africa), Renison Goldfields Consolidated (Australia). Coal: Apex Mines (South Africa). Titanium and zirconium: Associated Minerals Consolidated (South Africa). Copper and gold: Mount Lyell (Australia). Lead, zinc, silver, copper: Black Mountain Mineral Development Co Pty Ltd (South Africa). Copper and by-products: Newmont Mining Co (USA)

**Production of Principal Minerals**

| | *Company Output* | *Percentage of Western World Output* | *Percentage of World Output* | *Company Rank Number* |
|---|---|---|---|---|
| Gold *tonnes* | 142 | 14.8 | 11.6 | 2 |
| Tin *tonnes* | 7,782 | 3.9 | 3.3 | probably 5 |
| Coal *tonnes* | 2,700,000 | – | – | – |
| Copper *tonnes* | 20,339 | 0.3 | 0.2 | – |
| Refined zinc *tonnes* | 81,000 | 1.8 | 1.3 | – |

*Not including holdings in Newmont

# Corporation Nacional del Cobre de Chile (Codelco)

*Country of Incorporation:* Chile

*Turnover:* US$1,741 million (1981)
*Total Assets:* US$2,517 million (1981)
*Number of Employees:* 29,000 (1981)

**Countries of Operation:** Chile, plus representative offices around the world

**Major Products:** Copper, molybdenum, gold, silver, rhenium

**Principal Subsidiary and Associated Natural Resource Companies:** Codelco has representative offices in USA, UK, Germany, France, Italy, Spain, Japan, Brazil, Argentina, Sweden, South Africa

**Production of Principal Minerals**

| | *Company Output* | *Percentage of Western World Output* | *Percentage of World Output* | *Company Rank Number* |
|---|---|---|---|---|
| Copper *tonnes* | 893,611 | 13.8 | 10.7 | 1 |
| Molybdenum *tonnes* | 15,360 | 16.1 | 14.8 | 2 |

The nationalisation of the copper mining industry in Chile in the early 1970s was followed by the formation of the state-owned CODELCO, to constitute the largest copper mining entity in the world. Plans in hand ensure that this position will be maintained and expanded although private enterprise local and foreign participation in the Chilean mining industry is now encouraged.

# De Beers Consolidated Mines

*Country of Incorporation:* South Africa

*Turnover* (Sales by the Central Selling Organisation): R1,249 million (1981)
*Total Funds Employed:* R1,694 million in investments outside the diamond industry, however there are net current liabilities of R86 million (1981)
*Number of Employees:* 5,050

**Countries of Operation:** South Africa, Botswana, Australia, Lesotho (marketing operations throughout Japan, Europe, Middle East, North and South America)

**Major Products:** Diamonds, plus investment holdings in companies involved in gold, copper, platinum, oil and gas, coal, also synthetic diamonds

**Principal Subsidiary and Associated Natural Resource Companies:** Diamond mining: CDM (Pty) Ltd (South Africa), De Beers Lesotho Mining Co (Pty) Ltd (Lesotho), De Beers Botswana Mining Co (Pty) Ltd (Botswana). Industrial synthetic diamonds: De Beers Industrial Diamond Division (Pty) Ltd (Ireland, Sweden). Marketing diamonds: Central Selling Organisation (worldwide), The Diamond Corp (Lesotho, Botswana). Mining investment holdings: Anglo American Corp of South Africa Ltd (South Africa), Johannesburg Consolidated Investment Co Ltd (South Africa), Minorco (offshore), Barlow Rand Ltd (South Africa), Highveld Steel & Vanadium Corp Ltd (South Africa) plus investments in property, banking, general finance, etc.

**Production of Principal Minerals**

| | *Company Output* | *Percentage of Western World Output* | *Percentage of World Output* | *Company Rank Number* |
|---|---|---|---|---|
| Diamonds *carats* | 15,438,000 | – | – | 1 |

# Duval Group

*Country of Incorporation:* USA

*Pennzoil Turnover:* US$2,681.8 million (1981 net sales) (Duval: net sales $700 million)
*Pennzoil Total Funds Employed:* US$481.7 million (1981 total operating funds)
*Pennzoil Number of Employees:* 10,000

**Countries of Operation:** Pennzoil operates worldwide

**Major Products:** (Duval) copper, sulphur, potash, molybdenum, gold, silver, rhenium

**Principal Subsidiary and Associated Natural Resource Companies:** 100 per cent owned by Pennzoil

**Production of Principal Minerals**

| | *Company Output* | *Percentage of Western World Output* | *Percentage of World Output* | *Company Rank Number* |
|---|---|---|---|---|
| Copper *tonnes* | 112,000 | 1.7 | 1.3 | 11 |
| Molybdenum *tonnes* | 9,980 | 10.3 | 9.5 | 3 |
| Silver *tonnes* | 40.4 | 0.5 | 0.4 | – |
| Gold *tonnes* | 2.1 | 0.2 | <0.2 | – |
| Rhenium *lbs* | 2,000 | 11.4 | 7.3 | within top 5 |

# EZ Industries

*Country of Incorporation:* Australia

*Turnover:* A$218.5 million (1981 Group sales)
*Total Funds Employed:* A$546.3 million (1981)
*Number of Employees:* 3,423

**Countries of Operation:** Australia, USA, most EEC members

**Major Products:** Zinc, lead, copper, gold, silver, cadmium, cobalt oxide

**Principal Subsidiary and Associated Natural Resource Companies:** Marketing: E.Z. America Ltd (USA), E.Z. Europe Ltd (UK)

**Production of Principal Minerals**

| | *Company Output* | *Percentage of Western World Output* | *Percentage of World Output* | *Company Rank Number* |
|---|---|---|---|---|
| Zinc *tonnes* | 186,849 | 4.2 | 3.1 | 7 |
| Cadmium *tonnes* | 584 | 4.4 | 3.3 | 7 |
| Cobalt oxide *tonnes* | 34 | 0.1 | <0.1 | – |
| Lead residue *tonnes* | 42,087 | 1.7 | 1.2 | – |

# Falconbridge Nickel Mines

*Country of Incorporation:* Canada

*Turnover:* C$713 million (1981 consolidated revenues)
*Total Assets:* C$1,253 million (1981)
*Number of Employees:* Not available

**Countries of Operation:** Canada, Dominican Republic, South Africa, Norway, Zimbabwe

**Major Products:** Nickel, copper, cobalt, platinum, silver, gold, zinc, iron, cadmium, selenium, ferro-nickel, other PGMs

**Principal Subsidiary and Associated Natural Resource Companies:** Copper, zinc, gold: Corporation Falconbridge Copper (Canada). Ferro-nickel, nickel: Falconbridge Dominicana C por A (Dominican Republic). Gold: Kiena Gold Mines Ltd (Canada), Giant Yellowknife Mines Ltd (Canada). PGMs: Western Platinum Ltd (South Africa). Silver: United Keno Hill Mines Ltd (Canada). Iron, copper: Westrob Mining Division (Canada). Cobalt refining, nickel refining and alloying: Falconbridge Nikkelverk Aktienselskap (Norway)

**Production of Principal Minerals**

| | *Company Output* | *Percentage of Western World Output* | *Percentage of World Output* | *Company Rank Number* |
|---|---|---|---|---|
| Mined nickel *tonnes* | 47,500 | 9.7 | 6.8 | 2 |
| Copper *tonnes* | 58,500 | 0.9 | 0.7 | – |
| Mined cobalt *tonnes* | 623 | 5.4 | 4.7 | 6 |
| Zinc *tonnes* | 3,444 | <0.1 | <0.1 | – |
| Iron concentrates *tonnes* | 606,000 | 0.1 | 0.1 | – |
| Gold *tonnes* | 1.7 | 0.2 | 0.1 | – |
| Silver *tonnes* | 23.1 | 0.3 | 0.2 | – |
| PGMs *tonnes* | 1.71 | 1.6 | 0.8 | 6 |
| Refined cobalt *tonnes* (est) | 1,445 | – | – | – |

# Foote Mineral Company

*Country of Incorporation:* USA

*Turnover:* US$185.3 million (1981)
*Total Assets:* US$167.0 million (1981)
*Number of Employees:* 1,513

**Countries of Operation:** USA, Chile

**Major Products:** Lithium carbonate, manganese (owns and leases uranium-vanadium properties)

**Principal Subsidiary and Associated Natural Resource Companies:** Lithium: Sociedad Chileno de Litio Ltda. (Chile)

**Production of Principal Minerals**

| | *Company Output* | *Percentage of Western World Output* | *Percentage of World Output* | *Company Rank Number* |
|---|---|---|---|---|
| Lithium carbonate *tonnes** | 13,000 | N/A | N/A | 2 |
| Manganese *tonnes approx* | 5,000 | 0.1 | 0.1 | – |

*translates to approximately 2,250 tonnes lithium.

*(89 per cent common stock is owned by Newmont Mining Corp)*

# General Mining Union Corporation

*Country of Incorporation:* South Africa

*Turnover:* R3,201 million (1981)
*Total Assets:* R2,922 million (1981)
*Number of Employees:* Not given, but estimated at 75,000

**Countries of Operation:** South Africa, Australia, UK

**Major Products:** Gold, uranium, coal, asbestos, chromium, manganese, oil and gas, titanium ores, zirconium ores, copper, silver, nickel, tin

**Principal Subsidiary and Associated Natural Resource Companies:** (including investments) Platinum: Impala Platinum Holdings (South Africa). Chrome: Tubatse Ferrochrome (Pty) Ltd (South Africa), Transvaal Mining & Finance Co (South Africa). Gold and uranium: 35 gold and/or uranium mines in South Africa. Titanium and zirconium ores: Consolidated Rutile Ltd (via Union Corp (Australia) Pty Ltd). Manganese: Electrolytic Metal Corp (Pty) Ltd (South Africa). Tin: Geevor Tin Mines (UK)

**Production of Principal Minerals**

| | *Company Output* | *Percentage of Western World Output* | *Percentage of World Output* | *Company Rank Number* |
|---|---|---|---|---|
| Gold *tonnes* | 108.8 | 11.4 | 8.9 | 4 |
| Uranium *tonnes* | 1,501 | 3.5 | N/A | 3/4 |
| PGMs *tonnes* | 23.3 | 21.2 | 11.1 | 3 |
| Coal *tonnes* | 31,750,000 | – | – | – |
| Chromium *tonnes* (est) | 1,000,000 | – | – | 2 |

# Gulf Resources & Chemical Corp

*Country of Incorporation:* USA

*Turnover:* US$375 million (1981 revenues)
*Total Funds Employed:* US$437 million (1981, total assets)
*Number of Employees:* 2,800

**Country of Operation:** USA

**Major Products:** Lithium, coal, oil and gas, (used to produce lead, zinc, silver) fertilisers, salt

**Principal Subsidiary and Associated Natural Resource Companies:** Lithium: Lithium Corp of America (USA). Oil and gas: Pend Oreille (USA). Lead, zinc, silver: The Bunker Hill Co (USA)–now closed

**Production of Principal Minerals**

| | *Company Output* | *Percentage of Western World Output* | *Percentage of World Output* | *Company Rank Number* |
|---|---|---|---|---|
| Lithium (carbonate) *lbs* | 36,000,000 | – | – | 1 |
| Coal *tonnes* | 5,500,000 | – | – | – |

# Hanna Mining Company

*Country of Incorporation:* USA

*Turnover:* US$290 million (1981)
*Total Funds Employed:* US$627 million (1981)
*Number of Employees:* Not stated

**Countries of Operation:** USA, Guatemala, Colombia, Brazil, Canada

**Major Products:** Iron ore, coal, oil and gas, nickel, aluminium

**Principal Subsidiary and Associated Natural Resource Companies:** Welltech Inc, St John D'el Rey Ltd, Alcoa Alumino SA, Cerro Matoso SA, Iron Ore Company of Canada

***Production of Principal Minerals**

| | *Company Output* | *Percentage of Western World Output* | *Percentage of World Output* | *Company Rank Number* |
|---|---|---|---|---|
| Iron ore *tonnes* | 10,000,000 | 1.0 | 1.0 | – |
| Coal *tonnes* | 3,600,000 | 1.0 | 1.0 | – |
| Aluminium *tonnes* | 25,000 | <1.0 | <1.0 | – |
| Nickel | not stated | | | – |

The above represent the group's equity share of production and do not qualify for a company rank number.

*Hanna share of production.

# Homestake Mining Company

*Country of Incorporation:* USA

*Turnover:* US$262 million (1981)
*Total Funds Employed:* US$352 million (1981)
*Number of Employees:* 2,784

**Countries of Operation:** USA, Australia, Peru

**Major Products:** Gold, silver, lead, uranium, zinc

**Principal Subsidiary and Associated Natural Resource Companies:** Kalgoorlie Mining Associates, C M del M (Peru), United Nuclear-Homestake

**Production of Principal Minerals**

| | *Company Output* | *Percentage of Western World Output* | *Percentage of World Output* | *Company Rank Number* |
|---|---|---|---|---|
| Gold *tonnes* | 10 | 1.0 | <1.0 | – |
| Silver *tonnes* | 44 | 0.5 | 0.3 | – |
| Lead *tonnes* | 21,300 | 0.9 | 0.6 | – |
| Zinc *tonnes* | 7,000 | <1.0 | <1.0 | – |
| Uranium *tonnes* | 700 | 1.6 | N/A | 6 |

*Homestake share in equity of producing companies

# Hudson Bay Mining & Smelting Company

*Country of Incorporation:* Canada

*Turnover:* C$448.05 million (consolidated total revenue 1981)
*Total Funds Employed:* C$950.8 million (total identifiable assets 1981)
*Number of Employees:* Over 6,000 in North America (exact figure not given)

**Countries of Operation:** Canada, USA and South East Asian countries

**Major Products:** Copper, zinc, gold, silver, lead, cadmium, selenium, tantalum, chemical fertilisers, oil and gas, coal

**Principal Subsidiary and Associated Natural Resource Companies:** Copper and by-products: Inspiration Consolidated Copper (USA), Whitehorse Copper Mines (Canada), Minorco (USA, Canada). Tantalum: Tantalum Mining Corp (Canada). Exploration: Inspiration Mines Inc (USA, Mexico), Hudson Bay Exploration & Development Co Ltd (Canada). Non-metals: Francana Oil and Gas Ltd (Canada), Terra Chemicals International Inc (USA)

**Production of Principal Minerals**

| | *Company Output* | *Percentage of Western World Output* | *Percentage of World Output* | *Company Rank Number* |
|---|---|---|---|---|
| Refined copper *tonnes* | 133,750 | 2.0 | 1.6 | 14/15 |
| Slab zinc *tonnes* | 67,700 | 1.5 | 1.1 | – |
| Gold *tonnes* | 2.03 | 0.2 | < 0.2 | – |
| Silver *tonnes* | 39.5 | 0.5 | 0.3 | – |
| Lead in concentrates *tonnes* | 535 | 0.1 | 0.1 | – |
| Cadmium *tonnes* | 67.5 | 0.6 | 0.5 | 7/8 |
| Selenium *tonnes* | 38.1 | 2.7 | 2.7 | – |
| Tantalum $Ta_2O_5$ *(tonnes)* | 134.9 | 30 (est) | N/A | – |

Including subsidiaries, wholly and partly owned

# Impala Platinum Holdings

Year ended 30 June 1982

*Country of Incorporation:* South Africa

*Turnover:* Not given*
*Total Funds Employed:* R385 million (fixed assets)
*Number of Employees:* Not stated

**Countries of Operation:** South Africa, United Kingdom

**Major Products:** Platinum group metals, nickel, copper

**Principal Subsidiary and Associated Natural Resource Companies:** Ayrton Metals Ltd, Metallurgical Processes Ltd. Impala is itself approximately 55 per cent owned by Gencor via an intermediate holding company.

**Production of Principal Minerals**

| | *Company Output* | *Percentage of Western World Output* | *Percentage of World Output* | *Company Rank Number* |
|---|---|---|---|---|
| Platinum group metals *tonnes* | 29 | 27.0 | 13.7 | 2 |

Outputs are not stated. Production of by-product copper, cobalt, nickel is significant in terms of company revenue but not in world rating.

*Turnover estimated at $400 million, given sales of one million oz platinum at $350/oz plus $50 million fabricating and other revenues.

# Inco

*Country of Incorporation:* Canada

*Turnover:* C$1,886 million (1981)
*Total Funds Employed:* C$2,797 million (1981, does not include Exmibal assets written down in 1981)
*Number of Employees:* 24,623

**Countries of Operation:** Canada, USA, UK, Indonesia

**Major Products:** Nickel, copper, precious metals, cobalt, oil and gas

**Principal Subsidiary and Associated Natural Resource Companies:** EXMIBAL (Guatemala), PT INCO (Indonesia), Huntingdon Alloys Inc, Wiggins Alloys Ltd, Daniel Doncaster Ltd, INCO Energy Resources Ltd

**Production of Principal Minerals**

| | *Company Output* | *Percentage of Western World Output* | *Percentage of World Output* | *Company Rank Number* |
|---|---|---|---|---|
| Nickel *tonnes* | 149,660 | 30.4 | 21.2 | 1 |
| Copper *tonnes* | 110,658 | 1.7 | 1.3 | – |
| Cobalt *tonnes* | 1,633 | 7.5 | 5.1 | 3 |
| Platinum Group Metals *tonnes* | 10 | 9.2 | 4.7 | 4 |

Platinum Group includes some gold.

# Johannesburg Consolidated Investment Company

*Country of Incorporation:* South Africa

*Turnover:* Not applicable. A large element is investment income
*Total Funds Employed:* R383 million (1981)
*Number of Employees:* Not stated

**Countries of Operation:** South Africa, Zimbabwe, Namibia, investment interests worldwide

**Major Products:** Platinum, gold, diamonds, coal, nickel, copper, chrome, antimony, uranium

**Principal Subsidiary and Associated Natural Resource Companies:** Rustenburg Platinum Holdings Ltd, Randfontein Estates Ltd, Western Areas Ltd, Elsburg Gold Mining Co Ltd, De Beers Ltd, Tavistock Colleries Ltd, Shangani Mining Corporation Ltd, Consolidated Murchison Ltd, Otjihase Copper Co

**Production of Principal Minerals**

| | *Company Output* | *Percentage of Western World Output* | *Percentage of World Output* | *Company Rank Number* |
|---|---|---|---|---|
| Gold *tonnes** | 40 | 4.2 | 3.2 | 5 |
| Uranium *tonnes* | 646 | 1.5 | N/A | 7/8 |
| Platinum Group Metal *tonnes** | 12 | 11.2 | 5.7 | 3 |
| Coal *tonnes* | 2,000,000 | <1.0 | <1.0 | – |
| Antimony concentrates *tonnes* | 7,885 | 18.8 | 13.0 | 1/2 |
| Copper *tonnes** | 5,000 | <1.0 | <1.0 | – |

Diamond output is not relevant since although large it is represented only by investment income from holdings in the De Beers Group.

Nickel and Chrome outputs are not stated.

* Group production total of mines managed by JCI
** JCI 32.9 per cent equity share in Rustenburg Platinum
***Group equity share of Otjihase Copper Co

A company rank number does not apply as the above are group shares of production.

# Johns–Manville Corporation

*Country of Incorporation:* USA

*Turnover:* US$2,186 million (1981)
*Total Assets:* US$2,298 million (1981)
*Number of Employees:* 27,000

**Countries of Operation:** USA, Canada, Mexico, France, Spain, Iceland

**Major Products:** Asbestos, diatomite, perlite

**Principal Subsidiary and Associated Natural Resource Companies:**

**Production of Principal Minerals**

| | *Company Output* | *Percentage of Western World Output* | *Percentage of World Output* | *Company Rank Number* |
|---|---|---|---|---|
| Asbestos fibre *tonnes* | 411,000 | 20 | 9 | 1 |

# Kaiser Aluminium & Chemical Corporation

*Country of Incorporation:* USA

*Turnover:* US$3,342 million (1981 total revenues)
*Total Funds Employed:* US$3,836 million (1981 total assets)
*Number of Employees:* 26,250

**Countries of Operation:** USA, Australia, Bahrain, Belgium, Canada, Germany, Ghana, India, Jamaica, Swizerland, UK

**Major Products:** Bauxite, alumina, aluminium, fabricated products

**Principal Subsidiary and Associated Natural Resource Companies:** Comalco (Australia), Aluminium Bahrain (Bahrain), Kaiser Aluminium and Chemical of Canada Ltd, Kaiser Aluminium Europe Inc (W Germany), Volta Aluminium (Ghana), Hindustan Aluminium Co (India), Kaiser Jamaica Bauxite Co, Alumina Partners of Jamaica, Anglesey Aluminium Ltd (UK)

**Production of Principal Minerals**

| | *Company Output* | *Percentage of Western World Output* | *Percentage of World Output* | *Company Rank Number* |
|---|---|---|---|---|
| Aluminium *tonnes* | 1,036,300 | 8.3 | 6.6 | 3 |

# Lonrho

*Country of Incorporation:* UK

*Turnover:* £2,456.6 million (1981)
*Total Funds Employed:* £1,044.9 million
*Number of Employees:* 150,000

**Countries of Operation:** Ghana, Zimbabwe, South Africa, investment interests worldwide

**Major Products:** Gold, coal, copper, platinum group metals, cobalt, nickel

**Principal Subsidiary and Associated Natural Resource Companies:** Gold: Ashanti Gold Fields Corp (Ghana), Duiker Exploration Ltd, (South Africa). Gold and copper: Attica Mining (Pvt) Ltd, Corsyn Consolidated Mines Ltd (Zimbabwe). Platinum: Western Platinum (South Africa)

**Production of Principal Minerals**

| | *Company Output* | *Percentage of Western World Output* | *Percentage of World Output* | *Company Rank Number* |
|---|---|---|---|---|
| Platinum *kilos* | 2,863 | 3.5 | 1.8 | 5 |
| Palladium and other PGMs *kilos* | 1,965 | | | |
| Nickel *tonnes* | 729 | 0.1 | < 0.1 | – |
| Copper *tonnes* | Not given | – | – | – |
| Cobalt *tonnes* | 20 | <0.1 | < 0.1 | – |
| Gold *tonnes* | 11.9 | 1.2 | < 0.9 | – |

# Lornex

*Country of Incorporation:* Canada

*Turnover:* C$154 million (1981)
*Total Funds Employed:* C$339 million (1981)
*Number of Employees:* 1,036

**Country of Operation:** Canada

**Major Products:** Copper, molybdenum, silver

**Principal Subsidiary and Associated Natural Resource Companies:** The company is 51 per cent owned by Rio Algom Ltd, a subsidiary of RTZ

**Production of Principal Minerals**

| | *Company Output* | *Percentage of Western World Output* | *Percentage of World Output* | *Company Rank Number* |
|---|---|---|---|---|
| Molybdenum *tonnes* | 1,733 | 1.8 | 1.7 | 9 |
| Silver *tonnes* | 18 | 0.2 | 0.1 | – |
| Copper *tonnes* | 67,759 | 1.0 | <1.0 | – |

# Malaysia Mining Corporation

*Country of Incorporation:* Malaysia

*Turnover:* M$215 million (sales registered by MMC and subsidiaries)
*Total Net Assets:* M$137.7 million
*Number of Employees:* Not given

**Countries of Operation:** Malaysia, Thailand, Nigeria, project in Australia

**Major Products:** Tin, tantalum, tungsten, columbium, diamond project

**Principal Subsidiary and Associated Natural Resource Companies:** Tin and by-products: Kramat Tin Dredging, Southern Kinta Consolidated, Southern Malayan Tin (Malaysia), Tongkah Harbour Tin Dredging, Aokam Tin (Thailand), Amalgamated Tin Mines of Nigeria (Nigeria). Diamonds: Ashton Mining Ltd (Australia).

**Production of Principal Minerals**

| | *Company Output* | *Percentage of Western World Output* | *Percentage of World Output* | *Company Rank Number* |
|---|---|---|---|---|
| Tin (in concentrates) *tonnes (est*)* | 11,400 | 5.7 | 4.9 | 1 |
| Tantalite, Columbite, Tungsten ore } not declared | | | | |
| Refined tin *tonnes (est)* | 50,000 | 24.0 | 20.9 | 1 |

*MMC output in Malaysia and shares of output in Nigeria, Thailand

Note: Figures for 1981 include the results of the Corporation for the year and of the subsidiaries for periods ranging from 12-18 months to 30 June 1981.

# SA Manganese Amcor (Samancor)

*Country of Incorporation:* South Africa

*Turnover:* R391.9m (1981)
*Total Assets:* R115.2m
*Number of Employees:* 8,675

**Countries of Operation:** South Africa, USA

**Major Products:** Manganese, chromium, silicon, ferro-alloys, phosphate products and other chemicals

**Principal Subsidiary and Associated Natural Resource Companies:** Ferro-manganese and other ferro-alloys: Metalloys Ltd (S Africa), Roane Ltd (USA),Ferrometals Ltd and Crometals (Pty) Ltd (S Africa). Silicon: Silicon Smelters (Pty) Ltd (S Africa). Manganese: Middelplaats Manganese Ltd (S Africa).

**Production of Principal Minerals**

| | *Company Output* | *Percentage of Western World Output* | *Percentage of World Output* | *Company Rank Number* |
|---|---|---|---|---|
| Manganese *tonnes* | 3,200,000 | 20.2 | 11.0 | 1 |
| Chromite *tonnes* | 500,000 | 8.9 | 5.6 | 3 |

The company does not publish tonnage figures. The above are taken from Roskill's Metals Databook, third edition 1982.

# Marinduque Mining & Industrial Corporation

*Country of Incorporation:* Philippines

*Turnover:* US$237 million (1981)
*Total Assets:* US$1,494 million (1981)
*Number of Employees:* 6,732

**Country of Operation:** Philippines

**Major Products:** Copper, molybdenum, nickel, gold, silver, cobalt

**Principal Subsidiary and Associated Natural Resource Companies:** None.

**Production of Principal Minerals**

| | *Company Output* | *Percentage of Western World Output* | *Percentage of World Output* | *Company Rank Number* |
|---|---|---|---|---|
| Copper *tonnes* | 23,144 | 0.4 | <0.4 | – |
| Cobalt *tonnes* | 1,171 | 4.7 | 4.4 | 8 |
| Gold *tonnes* | 0.2 | <0.1 | <0.1 | – |
| Molybdenum *tonnes* | 52 | <0.1 | <0.1 | – |
| Nickel *tonnes* | 13,443 | 2.7 | 1.9 | 8 |
| Silver *tonnes* | 5 | <0.1 | 0.1 | – |

# MIM Holdings

*Country of Incorporation:* Australia

*Turnover:* A$721.4 million (1981 sales)
*Total Funds Employed:* A$1,158 million (1981 total assets)
*Number of Employees:* 7,781

**Countries of Operation:** Australia, UK

**Major Products:** Copper, lead, zinc, silver, nickel, coal, iron ore

**Principal Subsidiary and Associated Natural Resource Companies:** Base metals: Asarco Inc (USA), Mount Isa Mines Ltd, Copper Refineries Pty Ltd (Australia). Scrap lead: Britannia Refined Metals Ltd (UK).

**Production of Principal Minerals**

| | *Company Output* | *Percentage of Western World Output* | *Percentage of World Output* | *Company Rank Number* |
|---|---|---|---|---|
| Copper *tonnes* | 143,000 | 2.2 | 1.7 | 11 |
| Lead *tonnes* | 144,000 | 5.8 | 4.2 | 5 |
| Silver *kilos* | 373,000 | 4.2 | 3.3 | 4 |
| Zinc *tonnes* | 99,000 | 2.2 | 1.6 | – |
| Nickel *tonnes* | 4,000 | 0.8 | 0.6 | – |
| Iron ore *tonnes* | 1,260,000 | 0.2 | 0.1 | – |

Year ending 30 June 1981.

# Newmont Mining Corporation

*Country of Incorporation:* USA

*Turnover:* US$375.6 million (1981 net income)
*Total Funds Employed:* US$319.2 million (1981 net current assets)
*Number of Employees:* Not given

**Countries of Operation:** USA, Canada, Peru, South Africa, Australia

**Major Products:** Copper, zinc, lead, molybdenum, gold, silver, lithium, ferro-vanadium, oil, gas, arsenic, selenium, coal, uranium, ferro-silicon, manganese, cadmium, nickel, cobalt, mercury, other non-metallic minerals

**Principal Subsidiary and Associated Natural Resource Companies:** Copper and by-products: Magma Copper Co (USA), Sherritt Gordon Mines Ltd, (Canada), Southern Peru Copper Corp (Peru), O'okiep Copper Co Ltd (South Africa), Palabora Mining Co Ltd (South Africa). Gold: Carlin Gold Mining Co (USA), Newmont Proprietary Ltd (Australia). Lithium and Manganese: Foote Mineral Co (USA). Vanadium: Highveld Steel & Vanadium Co Ltd (South Africa). Lead and Zinc: Tsumeb Corp Ltd (Namibia), Resurrection Mining Co (USA). Uranium: Dawn Mining Co

**Production of Principal Minerals**

| | *Company Output* | *Percentage of Western World Output* | *Percentage of World Output* | *Company Rank Number* |
|---|---|---|---|---|
| Copper *tonnes* | 369,000 | 5.7 | 4.4 | 5 |
| Zinc *tonnes* | 34,580 | 0.8 | 0.6 | – |
| Lead *tonnes* | 7,850 | 0.3 | 0.2 | – |
| Molybdenum *tonnes* | 1,900 | 2.0 | 1.8 | 8 |
| Gold *tonnes* | 10.3 | 1.1 | 0.8 | – |
| Silver *tonnes* | 139 | 1.6 | 1.3 | 11 |
| Lithium carbonate *tonnes* | 13,000 | N/A | N/A | 2 |
| Nickel *tonnes* | 17,350 | 3.5 | 2.5 | 5 |
| Cobalt *tonnes* | 915 | 3.7 | 3.4 | 9 |

Includes total output of wholly and partly owned subsidiaries.

# Nippon Mining Company

*Country of Incorporation:* Japan

*Net Sales:* Y1,306,954 million ($6,224 million)
*Total Funds Employed (total assets):* Y923,926 million ($4,400 million)
*Number of Employees:* Not given

**Countries of Operation:** Japan, Zaire

**Major Products:** Copper, lead, zinc, nickel, cobalt, gold, silver, selenium, tellurium, indium, cadmium, bismuth, titanium metal, oil, ferro-nickel, zirconium

**Principal Subsidiary and Associated Natural Resource Companies:** Gold, silver, lead, zinc: Toyoha Mines Co (Japan), Shakanai Mines Co (Japan). Copper: Compagnie de Developpement Minier du Zaire (Zaire). Titanium metal: Toho Titanium Co (Japan)

**Production of Principal Minerals**

| | *Company Output* | *Percentage of Western World Output* | *Percentage of World Output* | *Company Rank Number* |
|---|---|---|---|---|
| Refined copper *tonnes* | 255,484 | 3.9 | 3.1 | 8 |
| Refined lead *tonnes* | 15,255 | 0.4 | 0.3 | – |
| Refined zinc *tonnes* | 108,168 | 2.4 | 1.7 | – |
| Refined gold *tonnes* | 8.36 | 0.8 | 0.6 | – |
| Silver *tonnes* | 151 | 1.7 | 1.3 | 9 |
| Refined nickel *tonnes* | 3,055 | 0.6 | 0.4 | – |
| Refined cobalt *tonnes* | 1,442 | 5.7 | 4.9 | 4 |
| Refined zirconium *tonnes* | 120 | 1.9 | N/A | 5 |
| Titanium metal* | 12,000 | 23.0 (est) | 12.0 (est) | 3 |

*Annual production capacity, not actual output.

# Noranda Mines

*Country of Incorporation:* Canada

*Turnover:* C$3,030 million (1981 revenue)
*Total Assets:* C$5,249 million (1981)
*Number of Employees:* 79,500

**Countries of Operation:** Canada, USA, Ireland, Guinea, UK, Brazil

**Major Products:** Copper, lead, zinc, alumina, aluminium, gold, silver, molybdenum, mercury, uranium, oil and gas, selenium

**Principal Subsidiary and Associated Natural Resource Companies:** Zinc and cadmium: Canadian Electrolytic Zinc (Canada). Alumina and aluminium: Noranda Aluminium. Molybdenum and copper: Brenda Mines (Canada). Lead and silver: Brunswick Mining & Smelting (Canada). Commodity trading: Rudolf Wolff & Co.(UK). Uranium: Kerr Adison. Gold: Pamour Porcupine Mines. Mercury, molybdenum, copper and gold: Placer Development. Lead, zinc: Tara Exploration & Development (Ireland)

**Production of Principal Minerals**

| | *Company Output* | *Percentage of Western World Output* | *Percentage of World Output* | *Company Rank Number* |
|---|---|---|---|---|
| Refined copper *tonnes* | 681,000 | 10.4 | 8.2 | 2 |
| Refined zinc *tonnes* | 224,700 | 5.0 | 3.6 | 4 |
| Refined lead *tonnes* | 52,700 | 1.3 | 1.0 | – |
| Refined silver *tonnes* | 21,987,000 | 7.8 | 6.1 | 2 |
| Gold *tonnes* | 684 | 1.5 | 1.2 | 8 |
| Cadmium *tonnes* | 14.5 | 2.4 | 1.8 | probably 4 |
| Mined molybdenum *tonnes* | 322 | 5.1 | 4.5 | 5 |

# Norsk Hydro

*Country of Incorporation:* Norway

*Turnover:* N.Kr. 17,482 million (1981)
*Total Funds Employed:* N.Kr. 9,372 million (1981–fixed assets)
*Number of Employees:* 15,174 (1981)

**Countries of Operation:** Norway, Sweden, UK, West Germany

**Major Products:** Oil, gas, petrochemicals, agrochemicals, aluminium, magnesium

**Principal Subsidiary and Associated Natural Resource Companies:** Magnesium: Norsk Hydromagnesiumgesellschaft GmbH (West Germany). Aluminium fabrication: Aluminium Precision Extruders Ltd (UK), Alustock Ltd (UK), Alustock SA (France)

**Production of Principal Minerals**

| | *Company Output* | *Percentage of Western World Output* | *Percentage of World Output* | *Company Rank Number* |
|---|---|---|---|---|
| Magnesium *tonnes* | 48,000 | 20.4 | 15.5 | 3 |
| Aluminium *tonnes* | 110,000 | 0.9 | 0.7 | 9 |

# Outokumpu Oy

*Country of Incorporation:* Finland

*Turnover:* 3,110 million marks (1981)
*Total Funds Employed:* 323.9 million marks (total investments 1981)
*Number of Employees:* 9,368

**Countries of Operation:** Finland, Norway, Canada, Ecuador

**Major Products:** Cobalt, zinc, nickel, copper, gold, silver, ferro-chrome, stainless steel

**Principal Subsidiary and Associated Natural Resource Companies:** Metals marketing: Reynolds European (London) Ltd (UK), Friedrich K Lurk, KG GmbH & Co (West Germany). Copper and zinc: Minera Toachi S.A. (Ecuador), Trout Lake Joint Venture (Canada), Lokken Gruber A/S & Co (Norway)

**Production of Principal Minerals**

| | *Company Output* | *Percentage of Western World Output* | *Percentage of World Output* | *Company Rank Number* |
|---|---|---|---|---|
| Cobalt *tonnes* | 1,229 | 4.9 | 4.6 | 7 |
| Zinc *tonnes* | 139,835 | 3.2 | 2.3 | 9 |
| Nickel *tonnes* | 13,310 | 2.7 | 1.9 | 8 |
| Copper cathodes *tonnes* | 33,796 | 0.5 | 0.4 | – |
| Gold *kilos* | 992 | <0.1 | <0.1 | – |
| Silver *kilos* | 37,805 | 0.4 | 0.3 | – |
| Ferro-chrome *tonnes* | 51,623 | 2.5 (est) | 2 (est) | 18 |

# Palabora Mining Company

*Country of Incorporation:* South Africa

*Turnover:* R214 million (1981)
*Total Funds Employed:* R243 million (1981)
*Number of Employees:* 3,976

**Countries of Operation:** South Africa, plus sales outlets worldwide

**Major Products:** Copper plus by-product magnetite, vermiculite, uranium, precious metals, zircon

**Principal Subsidiary and Associated Natural Resource Companies:** Transvaal Copper Rod Company, American Vermiculite Corporation, Zirconia Sales (Pty) Ltd. Palabora is owned approximately 29 per cent by the Newmont Group and 39 per cent by the RTZ Group.

**Production of Principal Minerals**

| | *Company Output* | *Percentage of Western World Output* | *Percentage of World Output* | *Company Rank Number* |
|---|---|---|---|---|
| Copper *tonnes* | 120,924 | 1.9 | 1.4 | – |
| Magnetite (iron) *tonnes* | 90,544 | 1.0 | 1.0 | – |
| Vermiculite | 192,373 | 35.6 | N/A | 2 |
| Silver *tonnes* | 16 | 0.2 | <0.2 | – |
| Gold/platinum/palladium *tonnes* | 1 | <1.0 | <1.0 | 7/8 |

# Peko-Wallsend

*Country of Incorporation:* Australia

*Turnover:* A$140.5 million (gross sales 1981)
*Total Funds Employed:* A$675.4 million (1981)
*Number of Employees:* 7,982

**Country of Operation:** Australia

**Major Products:** Gold, bismuth, copper, tungsten, silver, titanium and zirconium ores, coal, molybdenum

**Principal Subsidiary and Associated Natural Resource Companies:** Tungsten: King Island Scheelite Pty Ltd (Australia). Uranium: Ranger Uranium Mines Pty Ltd

**Production of Principal Minerals**

| | *Company Output* | *Percentage of Western World Output* | *Percentage of World Output* | *Company Rank Number* |
|---|---|---|---|---|
| Tungsten oxide *tonnes* | 2,519 | 9.0 | 4.8 | 3 |
| Molybdenum *tonnes* | 20 | <0.1 | <0.1 | – |
| Copper concentrate *tonnes* | 66,869 | 1.0 | 0.8 | – |
| Gold *kilos* | 1,317 | 0.1 | <0.1 | – |
| Bismuth concentrate *tonnes* | 9,275 | 39 (est) | 33 (est) | 1 |

# Penarroya

*Country of Incorporation:* France

*Turnover:* FFr 3,990 million (1981)
*Total Funds Employed:* FFr 298,611 million (1981)
*Number of Employees:* 9,219

**Countries of Operation:** France, Spain, Italy, Greece, Brazil, Peru, Morocco

**Major Products:** Zinc, lead, silver, cadmium, germanium, indium, arsenic

**Principal Subsidiary and Associated Natural Resource Companies:** Societe Asturienne, Compagnie Francaise de Mokta, Minemet (France), Preussag Weser Zink (West Germany), Societa Mineraria & Metallurgica di Pertusola, Merali & Metalli (Italy), Sociedade Paulista de Metais (Brazil), Compania Minera Huaron (Peru), Societe Africaine de Metaux et Alliages (Morocco)

**Production of Principal Minerals**

| | *Company Output* | *Percentage of Western World Output* | *Percentage of World Output* | *Company Rank Number* |
|---|---|---|---|---|
| Zinc in concentrates *tonnes* | 76,789 | 1.7 | 1.3 | – |
| Refined zinc *tonnes* | 193,287 | 4.3 | 3.1 | 6 |
| Refined lead *tonnes* | 319,696 | 8.1 | 6.1 | 1 |
| Silver *kilos* | 392,025 | 4.4 | 3.6 | 3 |
| Cadmium *tonnes* | 466 | 3.5 | 2.6 | 3 |
| Germanium *kilos* | 29,254 | 26.6 | 23.4 | 1 |

Including production of partly or wholly owned subsidiaries.

# Placer Development

*Country of Incorporation:* Canada

*Turnover:* C$283.2 million (1981)
*Total Funds Employed:* C$818.1 million (1981)
*Number of Employees:* 2,651

**Countries of Operation:** Canada, USA, Philippines

**Major Products:** Molybdenum, copper, silver, mercury, gold, coal

**Principal Subsidiary and Associated Natural Resource Companies:** Molybdenum: Endako Mines (Canada), Gibraltar Mines (USA). Copper and by-products: Craigmont Mines (USA), Marcopper (Philippines). Mercury: McDermitt Mine (USA). Silver: Equity Silver Mines Ltd (USA), Minera Real de Angeles S.A. (Mexico). Gold: Cortez Gold Mines (USA). Base metals: Noranda (Canada)

**Production of Principal Minerals**

| | *Company Output* | *Percentage of Western World Output* | *Percentage of World Output* | *Company Rank Number* |
|---|---|---|---|---|
| Copper *tonnes* | 51,739 | 0.8 | 0.6 | – |
| Molybdenum *tonnes* | 5,986 | 6.2 | 5.7 | 4 |
| Silver *tonnes* | 159 | 1.9 | 1.5 | Joint 9 |
| Mercury *tonnes* | 489 | 13.7 | 7.5 | 5 |

Placer's equity share of metal produced.

# Preussag

*Country of Incorporation:* West Germany

*Turnover:* DM4,028 million (1981)
*Total Assets:* DM2,961.4 million (1981)
*Number of Employees:* 21,280

**Countries of Operation:** West Germany, Canada, Nigeria, UK, Malaysia, Australia, USA

**Major Products:** Zinc, lead, silver, tin, tantalum, coal, germanium, oil and gas, mercury

**Principal Subsidiary and Associated Natural Resource Companies:** Tin, tantalum and marketing services: AMC Group (UK, Australia, Nigeria and Malaysia). Lead: Preussag Metal Inc, St Catherine (Canada), Preussag-Boliden-Blei GmbH, West Germany. Zinc: Preussag-Weser-Zink GmbH, West Germany

**Production of Principal Minerals**

| | *Company Output* | *Percentage of Western World Output* | *Percentage of World Output* | *Company Rank Number* |
|---|---|---|---|---|
| Lead *tonnes* | 170,906 | 6.9 | 4.9 | – |
| Zinc *tonnes* | 194,078 | 4.4 | 3.2 | – |
| Silver *tonnes* | 114 | 1.4 | 1.0 | – |
| Cadmium *tonnes* | 655 | 4.9 | 3.7 | 1 |
| Mercury *tonnes* | 149 | 4.2 | 2.3 | – |

The above represents largely refinery production and generally do not qualify for a company rank number.

# Reynolds Metals

*Country of Incorporation:* USA

*Turnover:* US$3,481.1 million (1981)
*Total Assets:* US$3,299.5 million (1981 total assets)
*Number of Employees:* 34,200

**Countries of Operation:** USA, Brazil, Australia, Canada, Colombia, Ghana, Jamaica, Venezuela, West Germany

**Major Products:** Bauxite, alumina, aluminium, coal, oil, gas, limestone

**Principal Subsidiary and Associated Natural Resource Companies:** Not detailed since Reynolds has aluminium fabricating subsidiaries in 12 countries and operates worldwide

**Production of Principal Minerals**

| | *Company Output* | *Percentage of Western World Output* | *Percentage of World Output* | *Company Rank Number* |
|---|---|---|---|---|
| Primary Aluminium *tonnes* | 870,750 | 7.7 | 6.1 | 4 |
| Secondary Aluminium *tonnes* | 209,000 | 6.2 | N/A | N/A |

# Rio Tinto Zinc Corporation

*Country of Incorporation:* UK

*Turnover:* £3,020.7 million (1981)
*Total Funds Employed:* £3,709.6 million (1981)
*Number of Employees:* 65,760

**Countries of Operation:** Australia, Canada, Netherlands, Papua New Guinea, South Africa, Spain, UK, USA, West Germany, Zimbabwe

**Major Products:** Lead, zinc, silver, coal, aluminium, iron, copper, gold, uranium, molybdenum, oil, gas, nickel, tin

**Principal Subsidiary and Associated Natural Resource Companies:** Aluminium: Comalco Ltd (Australia), Queensland Alumina Ltd (Australia), Anglesey Aluminium Ltd (UK). Coal, oil and gas: Kembla Coal & Coke Pty Ltd (Australia), Brinco Ltd (Canada), RTZ Oil and Gas Ltd (UK). Copper and gold: Bougainville Copper Ltd (Papua New Guinea), Palabora Mining Co Ltd (South Africa), Rio Tinto Minera SA (Spain). Copper and Molybdenum: Lornex Mining Corp Ltd (USA), US Borax and Chemical Corp. Gold, Nickel and Copper: Rio Tinto Mining (Zimbabwe) Ltd. Lead, silver and zinc: Australian Mining & Smelting Ltd, New Broken Hill Consolidated Ltd (Australia), The Broken Hill Associated Smelters Pty Ltd (Australia), Budel Zinc Plant (Netherlands). Tin: Capper Pass Ltd (UK). Uranium: Mary Kathleen Uranium Ltd (Australia), Rio Algom Ltd (Canada), Rossing Uranium Ltd (South Africa)

**Production of Principal Minerals**

| | *Company Output* | *Percentage of Western World Output* | *Percentage of World Output* | *Company Rank Number* |
|---|---|---|---|---|
| Bauxite *tonnes* | 9.44m | 11.8 | 10.5 | 1 |
| Copper—concentrates *tonnes* | 350,000 | 5.4 | 4.2 | 6 |
| Gold *tonnes* | 16.9 | 1.8 | 1.5 | 6 |
| Lead—refined *tonnes* | 275,000 | 6.8 | 5.1 | 2 |
| Molybdenum *tonnes* | 2,200 | 2.3 | 2.0 | 7 |
| Nickel *tonnes* | 3,500 | 0.6 | 0.5 | — |
| Silver *tonnes* | 83 | 0.9 | 0.8 | — |
| Tin *tonnes* | 2,000 | 1.0 | 0.8 | probably 8 |
| Zinc—slab *tonnes* | 275,000 | 6.3 | 4.5 | 2 |
| Uranium *(tonnes)** | 3,000 (est) | 7.0 | N/A | 2/3 |

*Estimated share of Rossing and Rio Algom

# St Joe Minerals Corporation

*Country of Incorporation:* USA

*Turnover:* US$1.3 billion (1980)
*Total Assets:* US$2.2 billion (1981) (refers to purchase price by Fluor)
*Number of Employees:* Not available

**Countries of Operation:** USA, Argentina, Australia, Chile, Peru

**Major Products:** Lead, zinc, silver, copper, oil and gas, coal, iron ore, gold

**Principal Subsidiary and Associated Natural Resource Companies:** Exploration: St Joe International Corp (South America and Australia). Lead, zinc, silver: Compania Minera Aguilar SA (Argentina). Lead: St Joe Lead Co (USA). Copper, lead, zinc, silver: Woodlawn (Australia). Zinc (not consolidated): Compania Minerales Santander (Peru). Gold: Placer Services Corp (USA). Gold, silver, copper: Compania Minera El Indio (Chile). Oil and gas: St Joe Petroleum Corp (USA), Coquina Oil Corp (USA). Coal: A T Massey Coal Co (USA). Iron: Pea Ridge Iron Ore Co Inc (USA)

**Production of Principal Minerals**

| | *Company Output* | *Percentage of Western World Output* | *Percentage of World Output* | *Company Rank Number* |
|---|---|---|---|---|
| Coal *tonnes* | 6,397,000 | – | – | – |
| Mined lead *tonnes* | 182,550 | 7.4 | 5.3 | 3 |
| Refined lead *tonnes* | 163,250 | 4.5 | 3.4 | 3 |
| Mined zinc *tonnes* | 101,650 | 2.3 | 1.8 | 10 |
| Mined copper *tonnes* | 1,874 | <0.1 | <0.1 | – |
| Mined silver *tonnes* | 52.9 | 0.6 | 0.5 | – |
| Mined gold *tonnes* | 5.3 | 0.6 | 0.4 | – |
| Iron ore pellets *tonnes* | 998,000 | 0.1 | 0.1 | – |

Figures given refer to St Joe Corp for year ended 31 October 1981.
*(acquired by Fluor Corporation, August 1981)*

# Selection Trust (BP Minerals)

*Country of Incorporation:* UK

*Turnover:* £31.4 billion (1981) (gross revenues of BP)
*Total Assets:* £23.9 billion (1981) (capital employed by BP on current cost basis) (£579 million employed by BP Minerals International)
*Number of Employees:* 41,700 in UK, 153,250 worldwide

**Countries of Operation:** (BP Minerals) Australia, Zimbabwe, Norway, Brazil, USA, South Africa. Exploration also in Spain, UK, Botswana, Canada, Germany, Mexico, New Zealand, Portugal, Spain. Investment interests worldwide

**Major Products:** Gold, tin, zinc, iron, nickel, copper, silver, lithium, beryllium ore

**Principal Subsidiary and Associated Natural Resource Companies:** Nickel: Agnew Mining Co Pty Ltd (Australia). Copper, silver, gold: Les Mines Selbaie (Canada). Gold: Unisel Gold Mines Ltd (South Africa). Tin: Brascan Recursos Naturais SA (Brazil). Refined zinc: Norzink (Norway). Copper, zinc, silver: Teutonic Bore (Australia). Copper, gold, uranium project: Olympic Dam, Roxby Downs (Australia). (Sohio, a principal subsidiary of BP, is treated separately, also there are investment interests in Tsumeb Corp, Amax)

**Production of Principal Minerals**

| | *Company Output* | *Percentage of Western World Output* | *Percentage of World Output* | *Company Rank Number* |
|---|---|---|---|---|
| Refined zinc *tonnes* | 40,000 | 0.9 | 0.6 | – |
| Tin *tonnes (est)* | 1,500 (est) | 0.7 | 0.6 | probably 10 |
| Gold *tonnes* | 3.4 | 0.4 | 0.3 | – |
| Nickel *tonnes* | 7,050 | 1.4 | 1.0 | 10 |
| Iron ore *tonnes* | 1.23m | 0.3 | 0.2 | – |
| Copper *tonnes* | 10,000 | 0.2 | 0.1 | – |

BP Mineral's share of output at operating mines estimated from production figures for its main subsidiaries since few details given in the annual report.

# Sherritt Gordon Mines

*Country of Incorporation:* Canada

*Revenue:* C$326.7 million (1981)
*Total Fixed Assets:* C$141.0 million
*Number of Employees:* 2,301

**Countries of Operation:** Canada, Indonesia (ex)

**Major Products:** Copper, zinc, gold, silver, nickel, cobalt, fertilisers and chemicals

**Principal Subsidiary and Associated Natural Resource Companies:** Nickel, cobalt and copper: Marinduque Mining & Industrial Corp (Philippines). Nickel: PT Pacific Nikkel Indonesia (Indonesia, company written off in 1981). Oil and Gas Exploration: Bow Rio Resources Ltd (Canada)

**Production of Principal Minerals**

| | *Company Output* | *Percentage of Western World Output* | *Percentage of World Output* | *Company Rank Number* |
|---|---|---|---|---|
| Refined nickel *tonnes* | 17,330 | 3.5 | 2.5 | 5 |
| Refined cobalt *tonnes* | 914 | 3.7 | 3.4 | 9 |
| Copper concentrates *tonnes* | 9,300 | 0.1 | <0.1 | – |
| Zinc concentrates *tonnes* | 8,775 | <1.0 | <1.0 | – |
| Silver *tonnes* | 11 | 0.1 | <0.1 | – |
| Gold *tonnes* | 0.5 | <0.1 | <0.1 | – |

# SLN (Soc Metallurgique Le Nickel)

*Country of Incorporation:* France

*Turnover:* FFr 276 million (1981)
*Total Capital Investment:* FFr 2,194 million (1981)
*Number of Employees:* 3,603

**Countries of Operation:** France, New Caledonia

**Major Products:** Nickel, cobalt

**Principal Subsidiary and Associated Natural Resource Companies:** Nickel marketing outlets: Societe Selnic (France), Le Nickel Inc (USA), Nickel Penamax GIE (New Caledonia), Nickel/Ferro-nickel Fabrication, Imphy SA (France), Nippon Nickel (Japan)

**Production of Principal Minerals**

| | *Company Output* | *Percentage of Western World Output* | *Percentage of World Output* | *Company Rank Number* |
|---|---|---|---|---|
| Nickel *tonnes* | 43,400 | 8.7 | 6.3 | 3 |
| Cobalt *tonnes (est)* | 350 | 1.5 | 1.3 | 10 |

# Southern Peru Copper Corporation

*Country of Incorporation:* Peru

*Turnover:* US$332 million (1981)
*Total Funds Employed:* US$922 million (1981)
*Number of Employees:* 6,651 (1981)

**Country of Operation:** Peru

**Major Products:** Copper, molybdenum, silver

**Principal Subsidiary and Associated Natural Resource Companies:** The major partners in SPCC are Asarco Inc, Phelps Dodge, Newmont Mining Corporation, Cerro Inc, Asarco controls 52.3 per cent of the equity

**Production of Principal Minerals**

| | *Company Output* | *Percentage of Western World Output* | *Percentage of World Output* | *Company Rank Number* |
|---|---|---|---|---|
| Copper *tonnes* | 225,109 | 3.5 | 2.7 | 9 |
| Molybdenum *tonnes* | 4,153 | 4.3 | 4.0 | 6 |
| Silver *tonnes* | 64 | 0.7 | 0.5 | – |

# Sumitomo Metal Mining

*Country of Incorporation:* Japan

*Turnover:* Y276.5 billion (1981)
*Total Assets:* Y174.4 billion (1981)
*Number of Employees:* 3,216

**Countries of Operation:** Japan, Philippines, Australia, Canada

**Major Products:** Copper, nickel, cobalt, lead, zinc, gold, silver

**Principal Subsidiary and Associated Natural Resource Companies:** Associated Natural Resource companies: Copper Smelting, Toyo Smelter Ltd, Niihama Copper Refinery (Japan), Base Metal Mining, Sumitomo Metal Mining Canada Ltd, Nickel Refining, Niihana Nickel Refinery, Nyuga Smelter Co Ltd (Japan)

**Production of Principal Minerals**

| | *Company Output* | *Percentage of Western World Output* | *Percentage of World Output* | *Company Rank Number* |
|---|---|---|---|---|
| Copper *tonnes* | 145,146 | 2.2 | 1.7 | – |
| Lead *tonnes* | 23,578 | 1.0 | <1.0 | – |
| Zinc *tonnes* | 79,336 | 1.7 | 1.3 | – |
| Cobalt *tonnes* | 1,401 | 5.6 | 5.3 | 5 |
| Ferro-nickel *tonnes* | 14,023 | N/A | N/A | – |
| Nickel *tonnes* | 13,858 | 2.8 | 2.0 | 7 |

The above mostly represents refinery production, largely from imported ores and concentrates, and few qualify for a company rank number

# Teck Corporation

*Country of Incorporation:* Canada

*Turnover:* C$154.4 million (1981 revenue)
*Total Funds Employed:* C$495.6 million (1981 total assets)
*Number of Employees:* 2,083

**Country of Operation:** Canada

**Major Products:** Copper, gold, silver, zinc, niobium, molybdenum

**Principal Subsidiary and Associated Natural Resource Companies:** Copper-molybdenum: Lornex Mines (Canada). Copper-gold: Afton Mines Ltd (Canada)

**Production of Principal Minerals**

| | *Company Output* | *Percentage of Western World Output* | *Percentage of World Output* | *Company Rank Number* |
|---|---|---|---|---|
| Copper *tonnes* | 33,250 | 0.5 | 0.4 | – |
| Gold *tonnes* | 2.0 | 0.2 | 0.2 | – |
| Silver *tonnes* | 35.9 | 0.4 | 0.3 | – |
| Zinc *tonnes* | 26,050 | 0.6 | 0.4 | – |
| Columbium oxide *tonnes* | 1,552 | N/A | N/A | within top 5 |
| Molybdenum *tonnes* | 1,074 | 1.1 | 1.0 | – |

Teck's share of mineral production.

# Westralian Sands

*Country of Incorporation:* Australia

*Turnover:* A$16.4 million (1981 total income)
*Total Funds Employed:* A$29.6 million (1981 total assets)
*Number of Employees:* Not available

**Country of Operation:** Australia

**Major Products:** Ilmenite (titanium ore), zircon (zirconium ore), monazite (rare-earth oxide ore)

**Principal Subsidiary and Associated Natural Resource Companies:** Beach sands of all types: Western Mineral Sands Pty Ltd, Yoganup Pty Ltd (Australia)

**Production of Principal Minerals**

| | *Company Output* | *Percentage of Western World Output* | *Percentage of World Output* | *Company Rank Number* |
|---|---|---|---|---|
| Ilmenite *tonnes* | 437,000 | 10.0 | 9.2 | 5 |
| Zircon *tonnes* | 35,000 | 6.0(e) | N/A | 7 |
| Monazite *tonnes* | 1,900 | 5.6* | 5.2* | 6 |

*estimated as percentage of total rare-earth oxide ore production.

# Zambia Consolidated Copper Mines

*Country of Incorporation:* Zambia

*Turnover:* Kwacha 977.1 million (1981) (total sales)
*Total Assets:* Kwacha 2,081.8 million (1981)
*Number of Employees:* 58,665

**Country of Operation:** Zambia

**Major Products:** Copper, cobalt, lead, zinc, silver, selenium

**Principal Subsidiary and Associated Natural Resource Companies:** Formed in 1981 by merger between Roan Consolidated Mines Ltd and Nchanga Consolidated Copper Mines Ltd. State owned, via Zambia Industrial and Mining Corp (a State corporation). Subsidiaries: Copperbelt Power Co Ltd, other industrial product companies.

**Production of Principal Minerals**

| | *Company Output* | *Percentage of Western World Output* | *Percentage of World Output* | *Company Rank Number* |
|---|---|---|---|---|
| Refined copper *tonnes* | 591,853 | 8.1 | 6.1 | 2 |
| Refined cobalt *tonnes* | 2,686 | 10.7 | 9.8 | 2 |
| Refined lead *tonnes* | 11,407 | 0.3 | 0.2 | – |
| Refined zinc *tonnes* | 36,106 | 0.8 | 0.6 | – |
| Selenium *tonnes* | 22.45 | 1.6 | 1.5 | – |
| Gold *tonnes* | 0.3 | <0.1 | <0.1 | – |
| Silver *tonnes* | 21.9 | 0.2 | 0.2 | – |

For year ended 31 March 1982.

# APPENDIX: SOURCES FOR FURTHER INFORMATION

## General Information and Statistics

ABMS, American Bureau of Metal Statistics Inc, Non-Ferrous Metal Data 1981, 420 Lexington Avenue, New York NY 10170

Availability of Strategic Minerals, The Institution of Mining & Metallurgy, 44 Portland Place, London W1 (also library)

CRU, Commodities Research Unit, series of monthly publications covering individual metals, 31 Mount Pleasant, London WC1

Financial Times Mining International Year Book 1982, Longman Group Ltd (1981) ISBN 0141 3244

International Financial Statistics (monthly) prepared by the Bureau of Statistics of the IMF, Washington DC 20431, USA

International Minerals/Metals Review 1982, published by McGraw-Hill

Metallgesellschaft, AG Metal Statistics, annual statistical compilation, published by Metallgesellschaft, Reuterweg 14, 6000 Frankfurt Am Main, Federal Republic of Germany

Mining Annual Review, published by Mining Journal Ltd, 15 Wilson Street, London EC2

Roskill's Metals Databook, Third Edition 1982, Roskill Information Services Ltd, 2 Clapham Road, London SW9

US Bureau of Mines Mineral Commodity Summaries 1982, obtained from US Bureau of Mines, Division of Publications, 4900 La Salle Road, Avondale, Maryland, 20782

World Metal Statistics, monthly publication by World Bureau of Metal Statistics, 41 Doughty Street, London WC1

## General Interest Periodicals

Commodity Year Book 1982, published annually by Commodity Research Bureau Inc, One Liberty Plaza, New York NY 10006

Elements of Materials Science & Engineering, Fourth Edition, Van Vlack, published by Addision-Wesley Publishing Co, Reading, Massachusetts

Engineering & Mining Journal (monthly magazine), published by McGraw Hill Inc

Eurocharts Metals Report (weekly charts and technical information), published by Eurocharts, Rudolf Wolff Ltd, Plantation House, 10-15 Mincing Lane, London EC3

Guide to Non-Ferrous Metals and their Markets, Peter Robins and John Edwards, published by Kogan Page, London 1980

Metal Bulletin (twice weekly magazine plus monthly supplement), published by Metal Bulletin plc, 45/46 Lower Marsh, London SE1

Metal Markets Weekly Review, Shearson American Express Ltd, St Alphage House, 2 Fore Street, London EC2

Metals Week (weekly publication) published by McGraw Hill Inc, 1221 Avenue of the Americas, New York NY 10020

Mining Journal (weekly publication), published by Mining Journal Ltd, 15 Wilson Street, London EC2. Also publishes Mining Magazine (monthly)

New Scientist (weekly magazine), published by IPC Magazines Ltd, Kings Reach Tower, Stamford Street, London SE1

Non-Ferrous Metals in Japan 1982, published by Sumitomo Corp, 11-1 Kanda Nishikicho Sanchome, Chiyoda-ku, Tokyo, Japan

Northern Miner (weekly newspaper), published by Northern Miner Press Ltd, 7 Labatt Aavenue, Toronto, Ontario MSA 3P2 (also publishes Canadian Mines Handbook, on annual basis)

Resources Policy (quarterly magazine) published by Butterworth Scientific Ltd, 88 Kingsway, London WC2

Reuter Metal Newsletter (daily), published by Reuters Ltd, 85 Fleet Street, London EC4

## Aluminium, Alumina & Bauxite

Aluminium (periodical), published by Aluminium-Verlag GmbH, POB 1207, Dusseldorf 1, West Germany

Aluminium and its Future Pattern of Use in Great Britain, 1981, The Materials Forum, published by the Metals Society, 1 Carlton House Terrace, London SW1

Aluminium Federation Library, Broadway House, Calthorpe Road, Five Ways, Birmingham B15 1TN

International Primary Aluminium Institute (monthly and annual statistics on production and trade), New Zealand House, Ninth Floor, Haymarket, London SW1

The Aluminium Situation (monthly publication), The Aluminium Association Inc, 818 Connecticut Avenue, NW, Washington DC 20006

US Bureau of Mines Aluminium Industry Survey (monthly publication)

## Antimony

Antimony Newsletter, Battelle Memorial Institute, 505 King Avenue, Columbus, Ohio 43201, USA

Consolidated Murchison, company annual reports

CRU Tin and Antimony Monitor (monthly report)

US Bureau of Mines Antimony Industry Survey (quarterly publication)

## Beryllium

Brush Wellman Company data

US Bureau of Mines Beryllium Industry Survey (annual publication)

## Bismuth

Bismuth Institute Information Centre, periodic publications, 47 Rue de Ligne, B1000 Brussels, Belgium

US Bureau of Mines Bismuth Industry Survey (quarterly)

## Cadmium

Cadmium Association, periodic publications, 34 Berkeley Square, London W1

US Bureau of Mines Cadmium Industry Survey

## Chromium

Chromium in the Steel Industry, IISI, Avenue Hamour 12/14, B-1180 Brussels, Belgium

The Ferro-alloys Association (periodic publications and statistics), 1612 K Street N.W., Washington DC 20006, USA

US Bureau of Mines Chromium Industry Survey (monthly)

## Cobalt

Cobalt Development Institute (periodic publications), Rue Ravenstein 3, B1000 Brussels, Belgium

International Conference on Cobalt—Metallurgy & Uses, papers from Centre d'Information des Meteaux non Ferreux, Brussels, Belgium

US Bureau of Mines Cobalt Industry Survey (monthly)

## Columbium

Niobium/Columbium—versatile, cost effective, plentiful: booklet from Niobium Products Co GmbH, Wagnerstrasse 4, 4000 Dusseldorf 1, West Germany

US Bureau of Mines Columbium Industry Survey

## Copper

CIPEC (occasional reports and publications), 177 Avenue du Roule, F92200 Neuilly/Seine, France

Copper Development Association, periodic publications, Orchard House, Mutton Lane, Potters Bar, Herts

Copper: the Anatomy of an Industry, Sir R Prain, published by Mining Journal Books Ltd, London 1975

International Copper Research Association (INCRA), 708 Third Avenue, New York NY 10017

The Impact of Fibre Optics on Copper Wire Markets, Hugh Douglas & Co, California

US Bureau of Mines Copper Industry Surveys (monthly)

## Diamond

De Beers and other, company annual reports

The World Diamond Industry, October 1982, James Capel & Co, Winchester House, 100 Old Broad Street, London EC2

The World of Diamonds: Timothy Green, published by Weidenfeld and Nicolson, London 1981

## Gallium

Dataquest UK Ltd, (periodic publications on the electronics industry, also useful for germanium, indium and selenium), 144 New Bond Street, London W1

Impact of gallium arsenide, Strategic Inc, 134 Holland Park Avenue, London

US Bureau of Mines Gallium Industry Surveys

## Germanium

The Economics of Germanium: Roskill Information Services

US Bureau of Mines Germanium Industry Survey

## Gold

Chamber of Mines of South Africa, quarterly leaflet, 5 Holland Street, Johannesburg, South Africa

Consolidated Gold Fields Ltd, Gold 1982, Moorgate, London

International Gold Corporation, associate company of Chamber of Mines of South Africa (periodic publications), 30 St George Street, London W1

Sharps, Pixley Monthly Market Report, Sharps, Pixley Ltd, 10 Rood Lane, London EC3

The Gold Institute, periodic publications, 1001 Connecticut Avenue, NW, Washington DC 2001

US Bureau of Mines Gold Industry Survey

## Indium

US Bureau of Mines Indium Industry Survey

## Iron & Steel

Institute of Scrap Iron & Steel (periodic publications), 1627 K Street N.W., Washington DC 20006, USA

International Iron & Steel Institute (periodic publications), Avenue Hamoir 12/12, B-1180 Brussels, Belgium

International Steel Trade Association (periodic information), 69 Cannon Street, London EC4

Metal Bulletin Handbook 1982, annual statistical compilation, published by Metal Bulletin P/C

US Bureau of Mines Iron & Steel Industry Survey

## Lead

Lead & Zinc Statistics (monthly bulletin), International Lead & Zinc Study Group, Metro House, 58 St James's Street, London SW1

Solders and Soldering: A Primer (2nd edition), Lead Industries Association, 292 Madison Avenue, New York, NY 10017, USA

US Bureau of Mines Lead Industry Survey (monthly)

'World Car Forecasts Report', DRI Europe Ltd, 30 Old Queen Street, London SW1

## Lithium

Lithium Corporation, Foote Minerals, company annual reports

US Bureau of Mines Lithium Industry Survey

## Magnesium

International Magnesium Association (periodic information and statistics), 1406 Third National Building, Dayton, Ohio 45402, USA

US Bureau of Mines Magnesium Industry Survey

## Manganese

Economist Intelligence Unit (periodic reports), 27 St James's Place, London SW1

The Manganese Centre (periodic information), 17 Avenue Hoche, F75008 Paris

US Bureau of Mines Manganese Industry Surveys

## Mercury

Policy Responses to Resource Depletion: The Case of Mercury: Nigel Roxborough, JAI Press, Greenwich, CT, 1980

US Bureau of Mines Mercury Industry Survey

## Molybdenum

Intermet Molybdenum Yearbook 1982 (and other publications), Casilla 1978, Santiago, Chile

US Bureau of Mines, Molybdenum Industry Survey

## Nickel

Alloy Metals & Steel Market Research GmbH (periodic publications), Wesel, West Germany

Inco, Falconbridge and other, company annual reports

Petroleum and Hard Minerals from the Sea: F.C.F. Earney, published by Edward Arnold, London 1980

The Outlook for Nickel and Accompanying Metals, Metall, December 1982 (monthly publication), Berlin 33, West Germany

US Bureau of Mines Nickel Industry Survey

## Platinum Group Metals

Annual Platinum/Palladium Reveiw, J Aron Commodities Corporation, New York

Ayrton Metals Ltd (monthly market report), 30 Ely place, London EC1

Guide to Precious Metals and their Markets, Peter Robbins & Douglass Lee, published by Kogan Page, London 1979

Impala Platinum, Rustenburg and other, company annual reports

Platinum Metals Review (monthly publication), Johnson Matthey plc, Hatton Garden, London EC1

US Bureau of Mines Platinum Group Metals Industry Survey (quarterly)

## Rare Earth Metals

Molycorp Inc, USA, company annual reports

US Bureau of Mines Rare-Earth Elements, Yttrium & Thorium Industry Surveys

## Rhenium

Intermet Publications, Rhenium Market Study 1981-85, Casilla 1976, Santiago, Chile

US Bureau of Mines Rhenium Industry Survey

## Selenium

Selenium-Tellurium Development Association (periodic publications), PO Box 3096, Darien, Connecticut 06820, USA

US Bureau of Mines Selenium Industry Survey

World Minor Metals Survey (mainly trading detail and price histories, also covers other minor metals), Metal Bulletin Plc, 1982

## Silicon

The Economics of Silicon Ceramics, Roskill Information Services, 1982

US Bureau of Mines Silicon Industry Surveys (monthly)

## Silver

The Great Silver Bubble, Stephen Fay, published by Hodder & Stoughton

The Silver Institute (monthly newsletter and periodic publications), 1001 Connecticut Avenue, NW, Washington DC 20036

The Silver Market (annual magazine), Handy & Harman, 850 Third Avenue, NY 10022

Sharps, Pixley Ltd (monthly market reports), 10 Rood Lane, London EC2

US Bureau of Mines Silver Industry Survey

## Tantalum

Markets for Cutting Tools in the 1980s: Gorham International Inc, Gorham, ME 04038, USA (also for tungsten)

Tantalum Mining Co of Canada, Greenbushes Tin Ltd and others, company annual reports

Tantalum Producers International Study Centre (periodic publications and quarterly bulletin), 1 Rue Aux Laines, 1000 Brussels, Belgium

US Bureau of Mines Tantalum Industry Survey

## Tellurium

Selenium-Tellurium Development Association (periodic publications)

The Economics of Tellurium: Roskill Information Services

US Bureau of Mines Tellurium Industry Survey

## Tin

Annual Review of the World Tin Industry 1982-83, Shearson/American Express Ltd, St Alphage House, 2 Fore Street, London EC2

International Tin Council Monthly Statistical Bulletin (and other periodic publications), Haymarket House, 1 Oxendon Street, London SW1

Packaging Review (monthly magazine), published by IPC Industrial Press Ltd, Quadrant House, the Quadrant, Sutton, Surrey

Paper & Packaging Bulletin 1982, Economist Intelligence Unit, 27 St James's Place, London SW1

Tin and its Uses (quarterly publication), International Tin Research Institute, Fraser Road, Greenford, Middx

Tin International (monthly magazine), Tin Publications Ltd, 7 High Road, London W4

## Titanium

First Titanium & Superalloys Conference Papers, held by Metal Bulletin

Japan Titanium Society (periodic publications and statistics), Konwa Building, 1-21-22, Tsukiji Chuo-ku, Tokyo

Titanium Statistics: Heinz Pariser via Metal Bulletin Plc

US Bureau of Mines Titanium Industry Surveys

## Tungsten

Tungsten Statistics (quarterly bulletin), UNCTAD Committee on Tungsten, United Nations, Geneva

Tungsten Today (periodic newsletter), Amax Tungsten Division, POB 261, 1000 AG Amsterdam, Netherlands

US Bureau of Mines Tungsten Industry Survey

## Uranium

International Energy Agency (periodic publications), c/o OECD, 2 Rue Andre Pascal, F75775 Paris, France

Metallurgical Technology of Uranium and Uranium Alloys; American Society of Metals, Metals Park, OH 44073, USA

The Uranium Institute (library and occasional publications), New Zealand House, Haymarket, London SW1 4TE

UK Atomic Energy Authority (general information on atomic energy), 11 Charles 2nd Street, London SW1

## Vanadium

Ferro-Alloy Statistics 1975-1980 (see also for chromium, manganese, silicon, iron), Heinz Pariser via Metal Bulletin Plc

The Economics of Vanadium, 4th edition: Roskill Information Services

US Bureau of Mines Vanadium Industry Survey

## Zinc

International Lead & Zinc Study Group (monthly publications), Metro House, 58 St James's Street, London SW1

US Bureau of Mines Zinc Industry Survey

World Mineral Markets: Zinc (separate volumes for copper, aluminium, nickel), Ministry of Natural Resources, Whitney Block, Queen's Park, Toronto, Canada, 1981

## Zirconium

Mineral Sands Producers' Association Ltd (periodic information and statistics), London Assurance House, 20 Bridge Street, Sydney, NSW 2000

US Bureau of Mines Zirconium Industry Survey